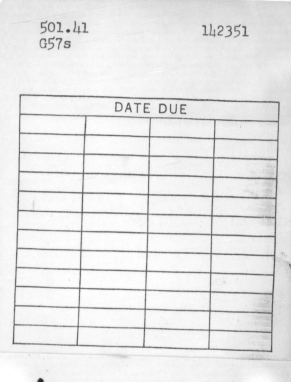

THE SCIENCE CRITIC

CONTENTS

FOREWORD

In our world of complex communication systems, transmitting everything from a heart beat to a space walk, we conduct our daily affairs with an alpha numeric pager on our belt, a radio telephone in our bag, and a faxed copy of the daily newspaper under our arm. We use lap-held computers to pass text at data transmission speed down telephone lines between countries at the cost of a local call.

But it is the message and not the medium of communication and expression that should be our concern. The novelty of the medium is continuous. It is now possible for the artist to express his vision using a digitiser, with nothing in his pen but the means to paint in pixels with a range of colours and a freedom never before available to manipulate an endless expression of his truth. Is what the artist is doing art or science? Is he not showing that the 'two cultures' are fusing into a new oneness? Surely, now is the time to create the science critic, who will perform a role parallel with, but more difficult than that of, the film, TV or theatre critic.

We must beware of the enthronement of new high priests lest they create 'laws' designed to enforce only their perception and prejudice. I have discussed aspects of this book over the years with the author, my father, and insist that now is the time to challenge, as he does, our accepted perceptions and prejudices.

David A. Goldsmith,
Television Director,
BAFTA Award Winner

ACKNOWLEDGMENTS

There are many who helped to shape me, and whom I must thank: Ritchie Calder, Bill Dick, Arthur Haslett, C.P. Snow, Jim Crowther, François Le Lionnais and Jacques Bergier. Others who share a broad 'responsibility' for what appears in these pages are A. Rahman and Dennis Flanagan and Earl Ubell.

I am indebted particularly and immediately to those with whom I discussed, orally and/or in writing, many key matters and who gave me advice and practical help: in the UK – Peter Cooper (information chief at the Royal Society), David Jollands (wise book editor), Geoff Deehan (BBC Science Radio head), John Halas (animation film doyen), Roger Silverstone (insightful TV sociologist), Roger Miles (exhibition 'philosopher' at the Natural History Museum, London), Jerry Timmins (BBC Overseas radio expert), George Hay (science fiction know-much), Brian Wilson (mathematics-in-education adviser), John Brunner (creative science fiction writer), Philip Hills (communications specialist), Jack Meadows (heavenly communicator), David Goldsmith (TV BAFTA winner), Chris Roberts (photo specialist), Richard Gregory (*Exploratory* insightful), Mick Rhodes (limitless *Horizon* chief); in the USA – Jim Cornell (ISWA inspirer), Diane McGurgan (NASW administrator), Gene Garfield (citations King); and Jacques Richardson (recently UNESCO, Paris), and Mike Baker (ICSU, Paris).

I am grateful to the following people and organizations for permission to quote copyright material from the works cited: Blackie & Son for *The Scientist and You*; Routledge & Kegan Paul and Noonday for *The Revolution in Physics* and Routledge & Kegan Paul and Humanities Press for *British Scientists of the 19th Century*; *The Observer* for 'The Origin of the Universe'; Chatto & Windus and Harper & Row for *A Banned Broadcast*; Shell International Petroleum for *Biotechnology*; Viking Penguin for *Explorations in Science*.

I wish, also, to acknowledge my debt to the many publications from which I have quoted brief extracts, the sources of which are detailed in my text.

My special thanks are due to the encouraging insights and practical advice of David Stonestreet (RKP).

CHAPTER 1

THE PROPER PUBLIC FOR SCIENCE

It is not a question of understanding for other people, but of putting other people in a situation in which they are able to understand. To understand is as important for each one of us as it is to love; it is an activity that cannot be delegated; we cannot put Casanova in charge of our love affairs – let's not saddle scientists with the responsibility for our understanding.

Albert Jacquart,
geneticist

How to achieve such inaccuracies, such anomalies, such alterations and refashionings of reality, that what comes out of it is lies if you like, but lies that are more true than literal truth.

Van Gogh

Science is not popular in Britain: never has been, not in the sense that its basic principles and logic are part of everyday understanding. There is, always has been, and will continue to be, a gap of understanding between the scientist (the professional performer in the laboratory) and the non-scientist. Compare, however, the organic social link between, say, the footballer (the professional performer on the playing field) and the non-footballer. Such a link – essentially necessary – I believe to be impossible to secure between the scientist and the general public, unless . . . .

This is what this book is about: thoughts that belong to me after four decades of effort in science writing, thoughts gained on the peaks and in the valleys of experiences which cause me to abjure the terms 'popularization of science', and to insist, through the mix each into the other of the paradigms of history and the clouds of philosophy, on 'the popular presentation of science'.

I define first where I stand.

When, in December 1965, Lord Florey became a past president

of the Royal Society, he referred in his valedictory address to the ancient custom of having meetings of the Royal frequently on Thursdays. He wondered whether they still served a useful purpose. This was a question he put to himself as he sat in his presidential chair 'in the privileged position of not understanding the paper being given and so of being able to survey the audience'. He believed his doubts seemed to be shared by Fellows, for often there were few around and he had long 'learnt on such occasions not to ask if there is a quorum for conducting the business of the society.'

Lord Florey may not have become aware that his successor, Lord Blackett, often would find himself in a similar position. I recall discussing with Blackett the periodicals he read. He surprised me by confessing that he had given up reading *Nature*; he only glanced through it, for most of it was incomprehensible to him.

These stories are not new. They reflect the condition of the times. A condition with which we are all familiar, which we describe as 'the age of specialization', 'the age of the information explosion', or 'the age of the fragmentation of science', etc. All true, but being generalizations, unlikely to help us solve the basic problem they reflect: how to keep up with what's new that is not in one's own narrow field.

We are familiar, also, with the ignorance about science and technology among the general population. That reflects the condition of the times, too. What is this condition? So far as the mass of the people are concerned, whether in a first or third world country, it is that we have in the scientific method an approach to a coherent philosophy, but we cannot make this a living reality, an accepted – therefore experienced – focus of our lives.

Why is it that the existing conditions are so unfavourable? Are we failing in first world countries because we cannot succeed in embracing science intimately? And it is important for third world countries to understand the reasons so that they do not merely imitate what is useless to them.

Too few people have an understanding of what science is. They see it through spectacles which have lenses grounded in clichés, so they cannot see its many faces. Nor can they hear its many voices as they listen through mono-tuned hearing aids.

The word 'science' began to enter the English language during the Middle Ages (the beginning of the seventeenth century). It came from France, and was synonymous with knowledge. But the early Latin translators of Aristotle gave the adjective 'scientificus' a technical meaning, and this was transferred to science to mean accurate and systematized knowledge. The Aristotelian theory of knowledge was the guide: you had 'scientific knowledge' when you arrived at it demonstratively, not by experiment.

But with the beginning of modern science – first with the Copernican revolution in the mid-sixteenth century, and then with the great age of science inspired by Galileo and Newton in the seventeenth century – the expression 'scientific knowledge' came into use to distinguish this form of knowledge from common knowledge: that is, science and knowledge were coming to be regarded as no longer synonymous in England. Science came to stand for a particular kind of knowledge, whether derived by straight deductive logic (as in Euclid), or whether using observation and experiment (as in Bacon and Harvey.) But the full realization of this did not come until almost the mid-nineteenth century, as expressed, for example, in Herschel's *Discourse on the Study of Natural Philosophy* in 1830.

In the seventeenth century, with the emergence of modern science, the new systems of the world began to be part of the education of the well-read person, and the new ideas were made available to the aristocracy and the middle class. It was at this period that what has been described as 'the first systematic written attempts at scientific popularization' appeared. The first of the great popular expositors was Bernard le Bovier de Fontenelle (1657–1757). He must have been a delightful person. Voltaire considered him the most universal genius he had met. He was a rare person in his combination of scientific knowledge and love of literature. He was a nephew of the distinguished dramatist Pierre Corneille, and a contributor to the *Mercure Galant*, a paper to which another uncle, Thomas Corneille, was an active contributor. He was appointed secretary of the Académie des Sciences in 1699, and like Oldenburg, the secretary of the Royal Society in London, he had wide contacts bringing him information on the new developments in science throughout Europe.

Fontenelle wrote particularly for the public of the salons. His *Entretiens sur la pluralité des mondes* (1686) is in the form of a dialogue and is addressed to the needs of an imaginary marquise, 'jeune, aimable et ignorante'. His intention, he declared, 'is to deal with philosophy (*he meant principally astronomy and physics*) in the least philosophical manner possible. I have tried to develop it to a point where it shall be neither too dry for the gentry nor too superficial for the scientists . . .'

For Fontenelle, popularization was a class matter. The plebs had no place in his dissemination. His works and those of the writers who followed him, such as the *Spectacle de la Nature* (1732) by the Abbé Pluche, were primarily for the aristocracy, the wealthy bourgeoisie and the ladies of the Court. The story is told of Fontenelle that one day at the Café Procope, the famous intellectual centre of Paris in the seventeenth century, he declared: 'If my hand were crammed full of knowledge, I wouldn't open it for the people.' And on the sixth day of his *Entretiens* he advises the marquise:

'Let us content ourselves with being a select little band and not disclose our mysteries to the people.'

In fact, the popular presentation of science to a general public could not come about until public forms of education had made literacy more general, until the media of dissemination were more widespread, and until there was a demand. People needed to realize that there was something to know which they were missing. The *Entretiens* may have been written for a need already expressed: for example, Molière's *Femmes Savantes* appeared 15 years before the *Entretiens*, and his middle-class women spent evenings in their attics trying with a telescope to see 'the people on the Moon'. However, I do not regard Fontenelle as a popularizer of science, not unless we broaden the function to describe 'a popularizer' as a person concerned with dissemination not purely to a mass, popular audience, but also to a specialist, intellectual audience. I distinguish, therefore, between the 'mass popularizer' and the 'specialist popularizer'. Historically, this distinction reflects the status and development of science in the last decades of this century. It could not have arisen before.

During the early part of the nineteenth century there were efforts – following the impetus of the Industrial Revolution – to make science available to mechanics, etc., but the movement led by Henry Brougham, founder in 1826 of the Society for the Dissemination of Useful Knowledge, to give working men a scientific education was doomed to failure. It could not be secured by imposing instruction in the sciences upon a highly inadequate system of state education.

Similarly, the Royal Institution founded in 1799 to train mechanics within a few years became a centre of scientific 'entertainment' for a select few. Thomas Webster, who was Clerk of the Works in the early days of that Institution, wrote to Count Rumford in 1799 with a proposal to found a school for mechanics in the house of the Royal Institution. Rumford was delighted with the scheme and it was accepted by him – and by the Managers after some hesitation. But by 1802, Webster had been granted sick leave and the project was abandoned. Why? I quote Webster's words, but we must bear in mind that these were written in 1837, 35 years after the events, and Webster was by then an old and rather frustrated man. He wrote:

But this project for improving mechanics, well intended as it was, which promised to be so useful, and which had already gained for the Institution 'golden opinions' was doomed to be crushed by the timidity (for I shall forbear to speak more harshly) of a few. I was asked rudely (by an individual I shall not name) what I meant by instructing the lower classes in science. I was told likewise it was resolved upon that the plan must be dropped as quietly as possible. It was thought to have a dangerous political tendency. I was told that if I persisted I would become a marked man.

The popularization of science was still designed for a few and not for the masses.[1] The widely popular lectures of men such as Huxley and Tyndall were of significance only to those who could read and write, and thus had little effect on the working masses.

From my shelves I have selected at random a number of definitions of science, all given during recent decades. For the biologist J. B. S. Haldane, himself a distinguished popularizer, science is first,

> the free activity of man's divine faculties of reason and imagination. Second, it is the answer of the few to the demands of the many for wealth, comfort and victory. Third, it is man's gradual conquest, first of space and time, then of matter as such, then of his own body and those of other living beings, and finally the subjugation of the dark and evil elements in his own soul.
>
> (Haldane, 1939)

For another distinguished biologist, C. H. Waddington, 'Science is the organized attempt of mankind to discover how things work out as causal systems.' For the writer Aldous Huxley, 'Science may be defined as a device for investigating, ordering and communicating the more public of human experience.'

For the American, Warren Weaver,

> Science is not technology, it is not gadgetry, it is not some mysterious cult, it is not a great mechanical monster. Science is an adventure of the human spirit. It is essentially an artistic enterprise, stimulated largely by curiosity, served largely by disciplined imagination, and based largely on faith in the reasonableness, order and beauty of the universe of which man is part.

These definitions are acceptable. They have the virtue of not being imbued with a late nineteenth-century emotion which perpetuates the myth that there is really only one form of science, called variously 'basic,' or 'pure,' or 'fundamental,' or 'science for its own sake.'

There are other definitions: for example, that of Sir Karl Popper,

> I think we shall have to get accustomed to the idea that we

must not look upon science as a 'body of knowledge', but rather as a system of hypotheses: that is to say, as a system of guesses or anticipations which in principle cannot be justified, but with which we work as long as they stand up to tests; and of which we are never justified in saying that we know they are 'true' or 'more or less certain' or even 'probable'.

What is important is an understanding that the idea of definition does not apply strictly to a human activity which has undergone many changes in its history: an activity which, as the crystallographer J. D. Bernal expressed it, 'is so linked at every point with other social activities, that any attempted definition can only express more or less inadequately one of the aspects, often a minor one, that it has had at some period of its growth.'

If we ask: 'What is the purpose and meaning of science?' we find that today's answer is different from that given yesterday; and different sorts of people would give different answers at different times. Thus, as Bernal put it, no anatomical definition of science is worthwhile. To be understood, science requires to be seen as a social phenomenon, linked organically with all other forms of human behaviour. All human activity is charged with change: today's Hamlet is a being different from the Hamlet of a generation ago. Similarly, today's science is different from yesterday's, and will be different again tomorrow.

Science is part of the process of social and cultural evolution, therefore is as subjective and psychologically conditioned as any other branch of human endeavour. Thus, the proper public for science will vary from period to period, dependent upon the answer given to the question, 'What is the purpose and meaning of science?'

In the Western world the attitude of 'science for its own sake' can be traced to the influence of the classical Greek contempt for manual labour, which was expressed by the 'pure scientist,' who, aloof in his ivory tower, disdained those whom he regarded as using tradesmen's standards to judge the usefulness of knowledge. The 'pure scientist' could avoid having to ask awkward questions on the effects of his researches: would not need to consider, 'What is my science for?' 'What is my responsibility as a scientist not only to society, but also to science itself?' In this context, the 'applied scientist' was a lesser being.

That was the social atmosphere in Cambridge, England, in the years before 1939 (outbreak of the Second World War) when students, such as C. P. Snow, distinguished as a novelist, began their scientific work. Snow was different: in a sense, therefore, more fortunate. He was doing research in the 1920s in a renowned laboratory, the Cavendish, and he wanted to write books. Thus, he said,

> There have been plenty of days when I spent the working hours with some scientists and then gone off at night with some literary colleagues. . . . It was through living amongst these groups and back again that I got occupied with the problem of what, long before I put it on paper, I christened to myself as the 'two cultures.'

That most felicitous phrase – a phrase that to his surprise so caught the imagination that it has become incorporated into our everyday language – he defined in 1959 in the Rede lecture in Cambridge to describe 'the gulf of mutual incomprehension' that separated literary intellectuals at one pole and scientists at the other. He believed the intellectual life of the whole of Western society was being split increasingly into those two polar groups.

Snow ended his lecture with a plea for the closing of the gap as 'a necessity in the most abstract intellectual sense, as well as in the most practical. When these two senses have grown apart, then no society is going to be able to think with wisdom'.[2]

In his 'second look' at 'the two cultures' in 1964, Snow 're-phrased the essence of the lecture' as follows:

> In our society (that is, advanced western society) we have lost even the pretence of a common culture. Persons educated with the greatest intensity we know can no longer communicate with each other on the plane of their major intellectual concern. This is serious for our creative, intellectual and, above all, our normal life. It is leading us to interpret the past wrongly, to misinterpret the present, and to deny our hopes of the future. It is making it difficult or impossible for us to take good action.
>
> I gave the most pointed example of this lack of communication in the shape of two groups of people, representing what I have christened 'the two cultures.' One of these contained the scientists, whose weight, achievement and influence did not need stressing. The other contained the literary intellectuals. I did not mean that literary intellectuals act as the main decision-makers of the western world. I meant that literary intellectuals represent, vocalise, and to some extent shape and predict the mood of the non-scientific culture: they do not make the decisions, but their words seep into the minds of those who do. Between these two groups – the scientists and the literary intellectuals – there is little communication and, instead of fellow-feeling, something like hostility.
>
> (Snow, 1969)

How far during these past decades have we come to bridging the gulf, to ending the dangerous division into the 'two cultures', and to securing a 'one culture' in which, whatever our specialist academic background, we can communicate understandably with each other? The answer is – not at all. The gulf today is wider than it was in 1959. We heard Snow's warning, but we could not heed it.

Historically, the rejection was inevitable. The gulf is too deeply rooted in Western cultural experience to be bridged overnight.

One root has grown from a meeting of the Fourth Lateran Council in 1215, at which it was agreed that the dogma of transubstantiation should be explained, and it could be in terms of the then existing theory of matter.[3] The question to be answered was 'How is it possible for Christ to be present in the consecrated bread and wine of the altar?' The answer lay in the distinction between the substance of matter and its transient qualities. For instance, the substance of water remained unchanged although it could appear, also, either as steam or ice.

It was logical, therefore, to assume that in supernatural circumstances the substance of matter might be changed without affecting its external characteristics. Thus, the bread and wine of the altar continued to look like bread and wine even though in the hands of the priest they had become the body and blood of the incarnate God.

In the years that followed more emphasis was laid on this deifying matter, and on making an understanding of this a prime spiritual concern. Coincidentally, Europe was becoming the leader in the development of quantitative mathematics. There was no better approach to an understanding of matter than through measurement and numbers. This emphasis had an important cultural effect through the different attitude it brought to the traditional seven liberal arts. These were divided into the Trivium, made up of grammar, rhetoric and logic: that is, concerned with verbal methods and analysis, and the Quadrivium – arithmetic, geometry, astronomy, and music (really the study of acoustical proportions): that is, concerned with measurement and calculation.

Until the end of the thirteenth century, theologians and those concerned with science were brought up in a common routine: they shared the same academic background, and could discuss easily with each other using the same cultural language. However, as the effects of deifying matter became clearer, there followed increasing emphasis on the quadrivial approach, and a decline in trivial argument. The scientists tended to narrow their research methods to the mathematical, and their subjects of study to the physical. They tended to become concerned with the tangible. And problems of the intangible, of values, of such non-measurables as beauty and goodness became the concern of others.

Over the centuries, the alleged antithetical character of scientific and moral (or aesthetic) modes of thought was intensified, and expressed in our educational system. The gulf that took shape was based on the late mediaeval emphasis on the importance of matter, and on mathematics as the only basis for rational certainty.

Intertwined with this root is another which has developed within the past two hundred years. It is a new phase of what is known as 'the romantic reactions'. It is an expression of the inability to co-ordinate two modes of knowing, to marry fact and value, matter and mind. For example, the partnership of science, engineering and capitalism in the last century led to Thomas Carlyle's choice of the machine as the key symbol of the emergent industrial system. Novelists and poets have used the metaphor of the machine to convey the sense of human inadequacy in the face of an alleged growing industrial totalitarianism.

Throughout the 1800s, the machine as symbol served as a figure for the complex force shaping the age, and became also an explicit measure of value. Writers in Victorian England consistently used the word 'machine' in a pejorative sense, because it was for them a manifestation of the scientific habit of mind, seeking always to reduce the complexities of society to a few simple measurable laws. And that symbol of the machine as representing the scientific mind is opposed in Victorian writing to what might be termed 'organic' metaphors, expressed in William Morris's belief in an instinctive creative ability degraded by mechanized production.

William Morris's *News from Nowhere* is the most radical critique of mechanized life, in which work as a value in itself is rejected in favour of the spontaneous release of vital, strongly sexual, energy. It was part of the aesthetic distaste for the machine, an expression of basic conflict between scientific and organic modes of thought, between rationalism and intuitionism. The conflict has grown sharper: there is a growing discrepancy between what science provides in the way of certain, verifiable knowledge, and what we understand by the quality of life.[4] There is a widespread view that there is no possibility of a more harmonious relationship between man and nature, and science has come to be identified with acute forms of alienation from nature.

It would be foolish not to recognize that the gulf between the 'two cultures' has grown, so that the scientist is seen as a stranger in the society in which he lives, and not to recognize that a debate is going on which in its significance far overshadows all previous discussions. We have moved far from the simple polarization of knowledge. The new debate is concerned with the awesome problem of the limits of scientific inquiry: with questions such as,

'What is the status of new knowledge?' 'Do we need to encourage or restrain its growth?' Not so many years ago the intellectual climate of the day would not have conceived of such questions being posed. Only a few voiced such thoughts, and then most privately.

The public for science has been regarded always as the bourgeois educated who were contrasted with the supposed illiterate masses. Science for the latter had to be made popular, a condescending approach which has failed miserably to achieve any real understanding of the nature of modern science, or of the basic laws underlying present-day scientific achievement. But nor could science be made popular for the alleged literate middle class. The British educational system is concerned with producing specialists, not with culturally rounded beings.

Thus, the public for science remains undefined and confused. Illiteracy in science and innumeracy are the hallmarks of a social system in which the engineer is still very low in public esteem compared with the financier, property developer, pop star, or pure scientist.

A new debate began in 1945 with the first practical demonstration of the capture of nuclear energy. The Bomb at Hiroshima started off that debate, which demanded a new definition of the public for science. Two waves of thinking began: one, a wave of anti-scientific response which has taken us well outside of the 'two cultures' concept. It was fuelled by the horrors of the Vietnam War in which the new techniques of devastation, both physical and mental, alienated a large part of the young in our industrialized world from involvement in scientific and technological work performed in the service of the government.

Indeed, Snow's 'two cultures' statement was made on the eve of a most powerful formulation of the counterculture which arose during the 1950s. It found vivid expression in the writings of the American, Theodore Roszak. He singled out the scientific world view as the root cause of what is most alarming about modern science. 'I have insisted,' he wrote in 1973,

'that there is something radically and systematically wrong with our culture, a flaw that lies deeper than any class or race analysis probes and which frustrates our best efforts to achieve wholeness. I am convinced it is our ingrained commitment to the scientific picture of nature that hangs us up. The scientific style of mind has become the one form of experience our society is willing to define as knowledge. It is our reality principle, and as such the governing mystique of urban industrial culture.'

(Roszak, 1973)

For Roszak, the scientific world view is the critical variable in what he regards as a suicidal pattern of collective behaviour. He identifies the quasi-religious belief in Reason as the motive force behind urban industrialism. His supporters use the terms 'technology', 'the system', and 'the machine' interchangeably. And they point the finger at the controlling network of large-scale institutions – government, business corporations, universities – as the real sources of power because they control 'the machine'. Roszak's arguments are persuasive in terms of emotions. What has helped the counterculture is the loss of confidence in themselves which sections of the scientific community are displaying in public. This is linked up with an intense debate on the social responsibility of the scientists, and the ethics of their behaviour.

Advances in science and technology seem to have escaped our control, and they are regarded as acting not only independently of us, but even against our will. In philosophical terms, we are witnessing the alienation of science and technology from culture generally. Thus, a UNESCO document – *La Science et la Diversité des Cultures* 1974 – stated,

For more than a century, the sector of scientific activity has grown so rapidly within the available cultural context that it now seems to supplant that culture. Surely, in spite of the rapidity of this growth, the lines of force of this culture will not be slow in rising to conquer it in their service of man. And yet, the recent triumph of science has conferred upon it the right to govern the culture in its entirety. In this perspective the main objective of culture becomes the diffusion of science. Finally, there are those who, frightened by the manipulation to which man and society are subjected under the power of science, see in science the spectre of cultural defeat.

According to this view, science is a foreign body which culture must fight and control; it must be always an outsider because it is the incarnation of the rational and foreign to any cultural tradition. This is a nonsense, but it is part of the second wave of thinking for which the Bomb was responsible. It is in the field of biological research that the problem is posed sharply. Questions posed are, 'Should limits be placed on biological research because of the danger that new knowledge can present to the established, or desired, order of society?' 'Is placing limits on biological research a threat to intellectual freedom and creativity?'

The wider debate goes on. Attention is drawn to other fields of research in which it is claimed new knowledge may have the most considerable implications. For example, in the development of nuclear energy devices, or the widespread introduction of microprocessors (the chips). Are there then areas of research that should be forbidden, or considered taboo, because of their potential 'harmfulness' to our society? I put the word in quotes because desired change is harmful, too, when society is not prepared for it with the necessary social mechanisms. I believe an answer cannot be provided by the scientific community alone, for it concerns all of us, and the answer is laden with political and social values. Now we are beginning to arrive at a contemporary definition of the proper public for science.

Our world is changing dramatically and drastically. Nothing is free anymore. When I was young there was free air, free water, almost free energy and land. The industrial system was optimized on that concept, and Western technology was developed on that basis, so that management, engineers, scientists and economists were trained to think in that way. There was, also, a colonial system which provided raw materials and labour cheaply. All that is gone. Everything has become polluted in some way, and we have come to understand that if we are to survive in this planet then each system generating materials, goods and services has to be a closed one. The inputs and outputs must be essentially identical, or at least inert and non-invasive to the surroundings. The motor car, electrical machinery, electronics, plastics and computers each in its turn made various industries obsolete. Microchips will have an even more revolutionary effect. The emergence of the closed system and the age of conservation will combine to make existing processes obsolete. It is that which has to be understood.

We are at the beginning of the age of public participation in science – whether we like it or not, whether we wish it or not, whether we are aware of it or not. This has occurred as it has become clearer that science is not a private game, but is for everybody: that it has a function in society: and that if used in a planned way could improve our condition immeasurably. Opposition to this viewpoint is mostly on the grounds that any attempt to constrain scientific research, to canalize it into ordained channels, would introduce authoritarianism, a form of Lysenkoism. This, it is alleged, would stifle creativity, and would have consequences harmful to science itself. This charge is highly debatable. In practice, a general problem of social importance provides as good a basis for the unplanned discovery as does complete freedom. Real novelty seems to spring more readily from consideration of some practical problem than from out of the blue.

But the situation is not easy to resolve. We need to ask, 'What forms can such public participation take?' 'What are the implications of these demands not only for science, but also for the dissemination of scientific information?' If through successful research we arrive at such an understanding of genetic inheritance that obstacles to genetic engineering are removed, which is happening, then I suggest the demand for public participation is clear and acceptable.

The once widespread feeling that scientists alone should have domain over the scientific enterprise is being challenged. The question posed, and fought out in many arenas, is, 'Who should be responsible ultimately for decisions about controversial scientific and technical issues?' When scientists seek to maintain as much autonomy as possible, it is quite natural to regard them as a political interest group with powerful career concerns. To speak more plainly, the autonomy of science up to now has been the result of a promiscuous collusion between the scientist and the politician. New knowledge can no longer serve as a justification for introducing risks to the public, unless an informed public is willing to accept those risks.

Because we have to establish a viable relationship between science and society, the controversy is no longer whether there will be public participation in the control of science, but over who will take part in the establishment of controls, how such controls will be organized, and how far they will influence detailed decisions concerning the nature and procedures of research. Thus, the proper public for science is the people.

An important implication is that we have now reached the stage at which we can no longer continue to regard the popular presentation of science as being *a favour* which the scientist is prepared, with condescension, to grant us, but instead to recognize that it has become an element in the scientific research process.

I began first to glimpse this when, each Tuesday night from September 1961 to June 1964, as tutorial lecturer in science at London University, and as a member of the central executive of the WEA (Workers' Educational Association), I ran a course together with H. J. Fyrth, resident staff tutor in the Extra-Mural Department at London University, on science, history and technology.[5] There were a number of science and history teachers from secondary schools and colleges of education and a few research workers who attended regularly, and their presence, interest and comments caused me to speculate that not only was there a place for the popular presenter of science in the tradition-bound, science-excitement-starved educational system, but also that we might be having some effect on attitudes in the research laboratory. Of course, what was possible administratively in adult and continuing education was not where it was needed greatly − in the primary and secondary school system.

I believe that science should be presented in a popular form in the classroom. Science today may be regarded as having two forms, both equally valid. There is the purely academic science performed in the laboratory, and there is the 'popularized' science, the science for the people, what the mass of the population understands by science, for it is the only science they can understand. It may be regarded as 'academic science at secondhand', but it is more than that. It has a national and cultural validity which academic science has not. It is expressed in the ordinary, everyday language which people speak and understand, not in the standardized, cosmopolitan pattern required by the learned journal. And because it is expressed in this way, it is patterned by the local culture and can be taught and seen to be understood in the classroom.

I agree with Dr Richard Whitley, of the Manchester Business School, that the popular presentation of scientific knowledge has become part of the expository phase of scientific practice. He uses the term 'popularization', and as a sociologist defines it widely to include transmission of scientific knowledge to: specialists in the same scientific field; 'general experts' with similar skills and orientations working on related problems; educated amateurs in other sciences, research funding agencies, etc.; students at all levels of the education system; and the general, lay public. Thus, he wants to treat 'popularization as an integral component of the knowledge production process.' (Whitley, 1985)

The specialist fragmentation of science and the speed of change renders invalid the view that scientific ideas can be made clear through popularization. The popularizer of science cannot disseminate the subtle ideas of science. These really cannot be understood without hard, disciplined effort. The trouble with most science writing is that it seeks to present science as a collection of facts. And scientific explanation tends to be incomprehensible to anyone except the expert. Popularization of science as a social undertaking designed for a privileged minority has no place in a democratic and rapidly changing society. We must rid ourselves of this concept of 'popularization', which is derived from Victorian, condescending, do-goodism, from a patronizing aloofness irrelevant to today's needs. We do not need to make science 'popular' in a world in which science is of key importance. A single experience of space technology is more educative in this respect than millions of words. Besides, we cannot make science 'popular' in the sense that Huxley and Tyndall wished to. The Bomb has put an end to that. The catastrophic power of the applications of modern science confronts each one of us with profound moral issues. Popularization of science has tended to alienate people from science. What we need is more public understanding of science, and more public appreciation of the impact of science.

To help to bring this about, we must concern ourselves with the view that there is but 'one culture', not two. That is, that there are different roads of endeavour within the one culture, roads built and under construction, along which go humanists, philosophers, theologians, artists, and so on. That these different roads lead to but one central point where mankind is to be found. The different travellers along these different roads have their own conception of reality. I am concerned with seeking how far the reality made apparent by the scientist is in accordance with the reality revealed, for instance, by the artist, and by the travellers along the other roads. There is a relationship between discovery made along these different roads, but it is not a linear relationship.

Almost 20 years ago, Harvey Brooks in the USA expressed a most delightful idea which has been with me ever since. He said, 'It would be interesting to know whether some psychologist, by studying current tastes in art and poetry, could predict what *kinds* of theories were likely to be acceptable in elementary particle physics, or perhaps vice versa.' It has been suggested to me by crystallographer Alan Mackay that both simply sample the *zeitgeist*. But the question remains, 'Who perceived what first?'

When I read the following verse by Jan Skacel, the Czech poet,

> Poets don't invent poems,
> The poem is somewhere behind
> It's been there for a long time,
> The poem discovers it only.[6]

I wrote,

> Scientists don't invent laws
> The law is somewhere around
> It's been there for a long time,
> The scientist only discovers it.

And then I read that John Campbell, the distinguished American science fiction editor, had written to John Blish, the British SF writer, in a letter dated 27 May 1955: 'It's now generally recognized that "physical reality" – meaning objects, magnetic, electric fields, etc. – is something man discovers, but doesn't invent. Faraday didn't invent electro-magnetic induction; he discovered it'.[7]

Within our 'one culture', I see the relationship between events as rather like the physical events surrounding an earthquake. From a starting point, its focus, an earthquake is responsible for happenings over a very wide area. For example, in 1960 an earthquake in Chile caused old volcanoes to erupt again, and giant waves to race across the Pacific to faraway Australia. There is an earthquake somewhere almost every day. It may be only a slight tremor, or a series of violent upheavals, or a single shock that may kill hundreds of thousands of people.

Let us assume that the Second Law of Thermodynamics or Quantum Theory is equivalent to a cultural earthquake. Shall we be able to pinpoint the focus, which may be along a different cultural road, say that trodden by the artist? Again, the music of Stravinsky or the art of Picasso is sufficiently original to be regarded as an earthquake. What and where was the focus which triggered them off, and where in other roads of endeavour can we trace the effects? Put another way – as by Professor Ilya Prigogine and his colleague Isabelle Stengers – 'Science can no longer claim the right to deny the relevance and interest of other points of view, and least of all can it refuse to listen to the viewpoints of the human sciences, philosophy and art' (Prigogine and Stengers, 1979).

The popularizer of science as he functions today cannot be of help. His function is not to cope with the world *problématique*, not to challenge those who continue to use a scientific style developed for manipulating Euclidean straight lines and Newtonian mass points.

That is why my case in this book leads to an urgent call for the creation of the science critic, a public policy generalist alerting us to the growing-pains of future worlds through the day-to-day discoveries of the present.

CHAPTER 2
TECHNIQUES IN THE MEDIA CULTURE

What the mass media offer is not popular art, but entertainment which is intended to be consumed like food, forgotten, and replaced by a new dish.

W. H. Auden ('The Poet and the City,' *The Dyer's Hand*, 1962)

The public has become used to conflicting opinion on health issues and position reversals on topics ranging from dietary recommendations to the depletion of the ozone layer. Many have come to feel that for every Ph.D. there is an equal and opposite Ph.D.

Tim Hammonds (Food Marketing Institute, USA) (quoted in the *NASW Newsletter*, vol. 32, no. 2, April 1984)

We live and conduct our activities in 'a media culture'. That is, we are influenced constantly in our thinking and behaviour by what is presented through the mass media. Each medium has its own grammar, essential for the language used by its particular communicators, who have their personal perspectives with which they select and interpret the phenomena they seek to communicate. To understand how and why science is presented, to grasp the frustration and annoyance so many scientists express when they describe the presentation of their laboratory work as 'inaccurate', to sense the inadequacy most non-scientists experience as they bypass the routine presentation of science, we need to comprehend that for the major media (newspapers, radio, TV, and film) the translation of science from the language of the laboratory can occur only under conditions laid down by the 'media culture'. The exponents of that 'culture' regulate our desires and actions according to framework constraints imposed by each medium. Thus, Newton's apple (hence, gravity) has different expression on TV and radio. The framework is real: it determines the form in which the subject is communicated. 'Accuracy' will be determined by the medium's particular technical constraints.

We need to accept, also, that today the mass media basically are forms of 'entertainment'. TV has seen to that. There can be no 'people involvement' otherwise. Facts are facts are facts, but they are meaningless in the media context unless presented as drama. Science rates as drama only exceptionally: e.g. in the story of Mme Curie, or the astronauts 'trapped' in a spaceship. That is why most science stories appear in a submerged or disguised form, or in non-peak hours. Given that 'communication, nowadays, is a matter of human rights' (Macbride, Leary et al., 1980), or,

> If we accept that technical issues, including those of medical care, environmental pollution, alternative sources of energy, and nuclear weapons, are an important fraction of all public controversies, then popular scientific communication is closely linked to the governance of our societies, and thus is worthy of intensified investigation by humanistic and social science scholars
>
> (McElheny, 1985)

it is rare to find a person who is not in favour of the public understanding of science.

But this alluringly desirable end cannot be approached without appropriate presentation. To rephrase a cliché: this is a task too important to be left to the popular presenters alone, or to the scientists alone. It requires to be undertaken as a joint enterprise, for apart from a few very rare individuals the task is too demanding given the constraints under which the mass communicators or scientists have to function. And even appropriate presentation is unlikely to secure 'understanding'.

When, in September 1984, I discussed this in New York with Earl Ubell, science correspondent of WCBS-TV, and a former president of the NASW, he said,

> I do not believe that one can ever 'deliver' to the public an understanding of science through the mass media. We serve other functions – we alert, we set agendas, and we provide materials for imaginative play. It is up to the other communication channels – schools, books, clubs – to provide the kind of experience that leads to understanding of any subject, particularly of science. Our functions in the mass media can lead to the better utilization of the other channels that lead to learning and comprehension, but I do not think one can rely on us for the transmission of the gut material. And

by us, I mean television, magazines, and all (including *The Times*) newspapers.

What of the scientists, why do they not more actively present their work to a popular audience? Basically, most cannot write the simple English that makes for compulsive reading. Their training conditions them to present their research findings in a eunuch-like language from which all personal feeling is absent. Further, they could not care less about the *hoi polloi*. Their rewards, such as promotion and grants, can come only from their peers. To 'popularize' is to sin. That remarkable, exceptionally talented presenter to mass audiences, Dr J. Bronowski, suffered much from the jealous criticism of his peers, as did Margaret Mead, the distinguished American cultural anthropologist. And when a leading epidemiologist first appeared as anchor man on a BBC TV series, *Your Life in Their Hands*, he was warned that if he persisted he would not get promotion.

My experience has been largely with the popular presentation of the natural sciences. What of the social sciences? That is dangerous ground on which to tread. It is highly marshy, with no straight and firm paths along which to march. Eugene Garfield, director of the Institute for Scientific Information in Philadelphia, distinguishes between science as being concerned with the impersonal processes of the universe, and the humanists with human achievements that make up the cultural heritage.

Simplification is difficult enough with the 'hard sciences'; with the 'soft' it is rare to avoid over-simplification. That is why I suggest it is more necessary to 'train' the social scientist in the techniques of popular presentation than it is the natural scientist. The social scientist deals with people in society: for instance, social influences on child development (the relation between home and school, the development of social cognitions); or teacher education (the response to changing needs, the importance of understanding the epistemological bases of subjects in the school curriculum).

The routine science correspondent or reporter is not equipped to handle these complex problems. Indeed, many published stories tend to point the finger of ridicule at the social scientist for investigating what are regarded as areas of 'absurdity:' for instance, phonetic aspects of tribal communication in Tonga. This is a favourite tactic by, for example, MPs hungry for some publicity: they single out a research project funded by, say, the Economic & Social Research Council, and in simulated indignation denounce this alleged waste of public funds.

By and large, the mass media do only a fair job in communicating science, whether hard or soft, although their influence is great. We need a large-scale, intensive survey of the impact of popular presentation through the mass media, based on indicators of public awareness. This could be a follow-up to the 'first integrated study' in the UK done by the Primary Communication Research Centre, and published in March 1978 (Jones, Connell, Meadows, 1978). It dealt with newspapers, radio and television. My proposed survey would need to take off with recognition that the institutions of the mass media and of education are not able to adjust themselves or their audiences sufficiently speedily, therefore adequately, to the inflationary expressions of new knowledge in the sciences. What is required is not only an elucidation of what actually happens through the impact of presentation, but also a definition of the 'survival needs' necessary to promote understanding, consciousness, adjustment to changing environments, and involvement in decisions about areas of scientific development.

A PRINTED WORDS

(i) The Press

(a) *Newspapers*

Whereas science is accurate to 10 decimal points, newspapers like to settle for round figures.

Art Snider (science editor, *Chicago Daily News*)

When distant and unfamiliar and complex things are communicated to great masses of people, the truth suffers a considerable and often radical distortion. The complex is made over into the simple, the hypothetical into the dogmatic and the relative into an absolute.

Walter Lippmann (*The Public Philosophy*, 1955)

The distinguished British biometrist, J. B. S. Haldane, delighted in writing popular scientific articles. He did so for a number of years for the Communist daily, the *Daily Worker* (now the *Morning Star*). He was convinced that it was 'the duty of those scientists who have a gift for writing to make their subject intelligible to the ordinary man and woman'. Besides, he added, those who write on science for the general public, can increase their income (Haldane, 1939).

How did he do his writing? I had the good fortune to spend a few hours alone with him when I discussed this. It was in Paris – incredibly many years ago – when I was Science Editor at UNESCO, and he was staying with his friend, Lewis Gielgud, then UNESCO's head of personnel. We had lunched in Gielgud's apartment, after which J. B. S. and I went out for a long walk.

I asked him whether he had a model for his writing style. He replied, 'Of course. I model myself on Daniel Defoe.' 'Any particular work?' 'Yes, Moll Flanders.' He went on to describe his method, which he insisted was personal, was only 'one way,' and possibly worked for him alone.[1]

Haldane was not, and still is not, a typical science writer. In each generation there are a few scientists who are good mass communicators, and enjoy being so. However, as almost all newspaper science writers began as general reporters, most do not have a scientific qualification. They 'learn' their science on the job. I became a science writer because I have always been concerned with words, their organization and impact. At University I had been assistant, then acting, editor of the College magazine. I refused the job of editor because I was in my final degree year, which I was taking in two not three years. After I graduated, because I was interested in writing and communication I went into journalism, not to begin with as a science writer. But when the opportunity arose, I suggested to the editor of the mass circulation Sunday paper on which I worked, that I could undertake this on a part-time basis. He agreed.

Most science writers began in this opportunistic, unsystematic way. One of Britain's leading science writers, P. Ritchie Calder, began as a crime reporter[2] and another, J. G. Crowther, was a teacher.[3] Others began in more astonishing circumstances. One news editor shouted at a reporter, 'Do you know what an atom is?' 'Yes.' 'OK. Go to this press conference, and if you find a good story, you're our science correspondent.' This is not entirely apocryphal. It is in the tradition of the Army Sergeant-Major who asked the new squad, 'Anyone 'ere like music?' A hand went up. 'OK, you go to the canteen and move the piano.'

Science journalists may be regarded as the middlemen in the social operation made up of the scientist on the one side and the news editor (representing the particular newspaper's tradition) on the other. The 'market' is defined for them. They must supply the 'goods' that sell in that market, or they are out of business.

What of the subject matter, the marketable commodity? That is normally left to the journalist to decide. Once he is regarded as 'a specialist' he will be treated as an expert and left to his own devices. Of course, suggestions will continue to be made to him, but he will be free, and expected, to build up his own contacts, to receive the appropriate publications, and generally to function as a responsible staff journalist: able to 'deliver the goods' as urgently as required.

What sells newspapers is 'human interest'. In other words, names make news. Most scientists are not 'names'. They make news in terms of their discoveries and involvements. A successful cure for one of the forms of cancer will turn a nonentity into a star immediately, and quite rightly. Most scientists cannot – and will not – understand this. It is becoming less true as science journalists as a profession begin to come of age, but the itch to overwrite or to sensationalize does exist.

Lord Zuckerman, the doyen of British science advisers, in his 1976 Romanes Lecture at Oxford on 'Advice and Responsibility', made some harsh criticism of the press and media presentation of science and technology. He said that scientific advice to government was becoming increasingly complex in part 'because the daily Press and the broadcasting services are continually subjecting the government, Parliament and the electorate to a barrage of comment about scientific and technological matters, with the result that official advice is almost always tendered against a background of what is usually well-publicized, but not infrequently superficial, fact and opinion.'

He went on to say that a major problem facing governments and their scientific advisers was, 'How to sift responsible from irresponsible advice, how to prevent public knowledge about science from becoming overwhelmingly tinged by emotion or even from assuming some political intent.' Examples of major scientific issues which he regarded as having been media-distorted included atmospheric pollution, drugs of all kinds, genetic engineering, pesticides, nutrition, nuclear energy and plutonium. (Zuckerman, 1976.)

The sociologist, Edward Shils, has made the same criticism:

> Journalists, especially those involved in reporting and commenting on science policy . . . delight in finding the 'establishment' of science in the wrong. . . . Their animosity against government makes them critical of government science and of the scientists who perform it. They like to catch them in contradiction or impugn the veracity of their statements.[4]

No amount of hectoring or lecturing can cause the science journalist to change the cultural environment in which he operates, due to the constraints imposed by his medium's language and particular grammar. A fine example of the different language of the scientist is given in the story of the struggle of the people who lived in the Love Canal area of Niagara Falls, when in August 1978 the New York State health commissioner acknowledged officially the existence 'of a great and imminent peril' to their health due to poisoning by the chemical wastes.

When local residents asked 'fearful questions' about the presence of dioxin, the official response was, 'no evidence'. There was no explanation of what this might mean.

Whereas the health department officials might have meant that it is extremely difficult to test for the minute quantities of dioxin that can be toxic to human beings, and that such tests might *not* have been performed or analyzed at the time, the citizens treated their answers as a bald attempt to deny that dioxin was present at all. Often, the phrase 'no evidence' has been interpreted in the public press, and even by the educated lay and scientific public interested in Love Canal, as meaning, 'All the tests *have* been performed and there is no evidence,' a very different meaning indeed from 'No evidence because no tests have been performed,' or 'The tests have been performed and no *statistically significant* differences have been found.' The DOH (department of health) frequently failed to distinguish these meanings or clarify exactly what they had done and had found.

(Levine, 1982)

The scientist's reasons for wishing to present his research to the public often differ from those of the journalist. Basically, they may both express an idealistic desire to inform the public, the major source of funds of science in these flourishing democratic days. They may believe, also, that science, like music, art and sport, is an integral part of our culture. But the scientist may be more concerned with a Nobel prize or with appearing on financially rewarding television programmes. The journalist may hope for 'a scoop,' or at the least a good story – and editorial commendation.

But even as a law in science is not immutably valid, so there can be no guarantee that a science story once written will be published. Or, even if it does appear, that it has not been cut. The science journalist may not be present when this happens. Given good sub-editing, it is usual, when cutting to fit a story into the space allocated, to start from the end and to discard lines as necessary until the required number of column inches is reached. This often means that the caution required by the scientist tends to be removed.

Journalists do wish to be given favoured treatment for one basic reason only – so that they can perform with accuracy and with moderate ease the task of transmitting to their papers or TV and radio stations, or news or wire services, hence to the public, what is going on at a given meeting. They are not seduced by a good supply of drink: they are interested in news and views. Obviously, they are good drinkers, and they are grateful for food and drink during a hard day when working under pressure they may not have had time even for a hasty sandwich.

All journalists have deadlines, or time limits, to meet: some more immediate than others. Reporters, particularly news agency men, may have to do two jobs at once: to listen to an ongoing debate, and simultaneously to write up for transmission what has gone before. They will provide running stories, that is short sections known as 'takes', which they hand either to a messenger or phone through themselves. Usually, a friendly colleague will fill them in on their return. In any event, a report from a morning conference in London may be read that afternoon or the following morning in papers throughout the world. TV and radio reporters are concerned with finding personalities who can talk lucidly and directly. They, too, have deadlines to meet.

Advance texts are important as aids in the early preparation of stories. Thus, a handbook of the NASW points out, if afternoon newspaper reporters have to wait until a speaker presents a paper at, say 11 a.m. before starting to write and transmit, most of them will miss their early edition times. In the USA, due to the country's wide geographical spread, wire service reporters write and transmit several hours ahead to reach morning and afternoon papers published in several time zones.

Laboratory precision is out, and rightly so. To communicate to a news editor (that is, to the public) terms must be simple, details limited, and the on-the-other-hand syndrome omitted. In an American study in 1974, newspaper clips were sent to the scientists named in them, together with a questionnaire about errors. Most frequently mentioned were omissions such as, relevant information about method of study and results, names of other investigators on the research term, and qualifications made to statements.

The British Association for the Advancement of Science (BA) has, since its beginnings in 1831, been concerned with the communication of scientific knowledge, particularly to the public. There is normally a good attendance of science writers and journalists at the annual meeting, held usually in a university town. In 1963 I decided to 'quantify' the extent of the popular impact of the BA proceedings. I did so by charting the numbers, lengths and subjects of the science items reported in the 'quality' (Q) and 'popular' (P) press. My analysis was published in *Discovery*, a monthly magazine since incorporated in the *New Scientist*.

I summarize this by a re-statement of the Matthew Effect: he who has information shall be given more; he that has not shall be given diversion.

I repeated this simple analysis for the BA's 146th meeting in September 1984 in Norwich. It confirms my previous analysis. The only substantial volume of reporting was in the Q papers – *The*

Times, Guardian, Daily Telegraph and *Financial Times* in that order. The P press as a whole had about 70 column inches and an average age of two citations per newspaper.

Reporting was greater on the middle three days of the five-day meeting. On the first day, the programme consisted of symposia, and on the other days mainly of Section meetings. In subject terms, the Life Sciences, especially the Biomedical Sciences and Zoology, and Engineering evoked most press coverage. Agriculture, Botany and Forestry received only five mentions between them. Chemistry and Physics received little coverage, as did Economics. Surprisingly, in view of the educational orientation of the BA, and current government budgeting constraints, there was little interest in Education (five mentions).

That year's meeting in Aberdeen was attended by 3,647 people to whom about 300 papers were presented in the 15 sections by 307 speakers. The media were well represented, but, I asked, how effective was the outward flow of information from the meeting? In what ways did it bring 'science' more before the public eye? In an effort to answer these questions I looked at the ways in which the Press reported the meeting.

The *Daily Express*, one of the large-circulation popular papers, gave the meeting only 9 column inches throughout the week. A local paper, the *Glasgow Herald*, gave 646 inches, and was closely followed by the *Guardian*, a paper in the 'quality' (Q) class. These papers – *The Times*, the *Financial Times*, the *Guardian*, the *Daily Telegraph*, and the *Yorkshire Morning Post* – gave the most coverage (49.5 per cent of the total), the two main Scottish papers because of their obvious special interest followed with 36.9 per cent and the popular Press gave only 13.6 per cent.

The main source of news stories was provided by the social sciences – psychology, education and anthropology. Undoubtedly, this was because it is easier to project generalizations, especially when they concern human behaviour, than to interpret the esoteric language of the exact sciences. But in 1963 there were no reports from the Chemistry section (except in the local Scottish papers), and almost none from Geology, Geography and General. In fact, the majority of the lectures went unreported.

As far as communication through the Press was concerned, the BA did not talk to the people. So far as the Press was concerned, most lecturers had nothing to say of wide public interest.[5]

Interest in the General Section X was due to its more open and less discipline-oriented nature: The Future Envisaged, Environment in a Turbulent World, Fateful Techniques, Information Technology, and Science and Parascience.

To sum up: two of the P tabloids – the *Sun* and the *Star* – did not treat the meeting as newsworthy. The Q press provided for the academic, the industrialist and the professional more of what they had already or had easy access to.

Two studies done in 1972 and 1978, found that the Q papers published more science, whether measured in items or column inches, than did the P papers.[6] They found, also, that biomedical topics were the most presented, particularly in the P press. R. A. Logan, at the University of Iowa, in an unpublished PhD thesis in 1977, made an interesting suggestion: that a dimension could be added to the simple measurement approach by determining whether a practical item was 'acquaintance-with' (that is, straight factual reporting), or 'knowledge-about' (providing background information). Barry White, of the Technology Policy Unit, University of Aston in Birmingham, believes that 'although both types of items result in a general awareness of what is happening in science and technology, it is only those articles which provide background material and comment which give the non-scientifically trained layman the chance of being involved in informed debate about new technological developments.' For him, only that kind of article performs the function thought so 'vital' by the British Association.[7]

White used the distinction to make a more quantitative analysis of the differences in the science coverage in the press. He examined the treatment of nuclear energy in 26 issues of four newspapers – *The Times*, *Guardian*, *Daily Mirror* and *Daily Express* – published in November 1978, and again from 28 March to 30 April 1979, immediately after the accident in March 1979 at the Three Mile Island (TMI) nuclear plant. He looked through 28 issues of the *Guardian* and *Daily Mirror*, and 27 of the *Daily Express*. *The Times* was not included as it was not published due to industrial action. In the latter period, nuclear energy items were classified, also, into one of four other categories: two acquaintance-with and two knowledge-about. These were: (i) news about TMI, (ii) other news about nuclear energy, (iii) information about nuclear energy in general, (iv) information about particular reactor types, notably the PWR (pressurized water reactor) and the AGR (advanced gas-cooled reactor).

The results support conclusions made in other studies. The Q papers provided greater science coverage than did the P press. The latter emphasized the biomedical sciences, and proportionally published more than double the amount appearing in the *Guardian* or *The Times*. There was more coverage of nuclear energy in March/April 1979 than in November 1978: a five-fold increase in the *Daily Express*, almost the same rise in the *Daily Mirror*, and 2.5 times in the *Guardian*. But in absolute terms, *Guardian* coverage was more than five times that of the other two papers put together.

However, the total amount of space devoted to science after the

TMI incident was not appreciably more than in the first period examined. The P press covered the accident with an acquaintance-with news approach: the Q papers with a mixture of acquaintance-with and knowledge-about. The difference in the supply of scientific information is expressed in the knowledge levels of the different readers. The revised Matthew Effect is operative.

In the USA, science writers were all but ignored in covering TMI, wrote David M. Rubin, associate professor in the Department of Journalism and Mass Communication at New York University, and head of the Task Force on Public Information for the President's Commission on the Accident at TMI (writing in the *NASW Newsletter*, vol. 28, no. 1, January 1980, pp. 1f.) 'Trapped in disorganized, mass press conferences where only the aggressive were heard, and isolated (along with other reporters) from many sources which could have answered technical questions, science writers had little opportunity to ask sophisticated questions of knowledgeable sources.'

And when informed science writers were able to ask relevant questions 'they just couldn't get answers, either because sources were ignorant and unprepared, or they were not being made available.' The director of public affairs of the Nuclear Regulatory Commission (NRC) recognized that the needs of science writers were not being served by any of the institutions responsible for the accident. There was, wrote Rubin, 'a lack of sources, no press center, no support services, no background information, and few official spokespersons.' Without such facilities, science writers could not be expected 'to explain a technologically complex event under deadline pressure'.

The Task Force concluded that 'the major and initial burden in the area of public information falls on those who operate and regulate the nuclear plants. They failed to serve the people's right to know at Three Mile Island'.[8]

By coincidence, at about the same time (November 1984) French science journalists were complaining about the continuing difficulties they were facing when they tried to get information from their atomic energy agency, the CEA (Commissariat à l'énergie atomique). The Association des Journalistes scientifiques de la Presse d'Information (AJSPI) complained to the Minister of Research and Technology that not only was the monthly magazine, *Science et Vie*, which was highly critical of nuclear power developments, being denied CEA facilities, but also that a CEA public relations officer had phoned a New York magazine to state that a particular *Science et Vie* staff writer, on freelance assignment for the American magazine, 'doesn't write accurate stories'! Fortunately, the Minister intervened in favour of the AJSPI.

By contrast, when, during the weekend of December 1/2 1984,

the dreadful release of methyl isocyanate (MIC) occurred at Bhopal, Central India, the British Chemical Industries Association (CIA) moved into insightful action as 'the dimensions of the story changed with the serious press pursuing three main angles – that of UK safety and the alleged off-loading by international companies of dangerous plant and processes on the Third World. Also the issue of exporting domestically prohibited pesticides, pharmaceuticals, and other chemicals to developing countries began to creep in.'

When, within the week, technical errors – such as 'cyanide gas', 'use of MIC in the plastics industry', 'manufacture of MIC in the UK' – were being repeated regularly in the media, the CIA introduced a Journalists' Incident Fact Pack (including an MIC 'Get It Right' card for quick and easy reference) which was distributed to every specialist journal and major provincial paper in the country and to broadcast outlets. In addition, chemical manufacturers were sent a detailed description of the types of questions the CIA press office were being asked and the kinds of replies being given. A press office team was made available to deal with technical aspects, and the whole exercise appeared to have been wise since the media statements, according to the CIA, became largely correct.

The CIA believes that the long-term effects and legends depend largely on how the story is handled in the first 48 hours, and in the 10 days following. The lesson is to secure discussion 'on a factual and informed basis – rather than a response on the level of hysteria based on ignorance.'

The science writer as a specialist journalist is a comparatively recent newcomer to a newspaper. However, his position is secure. What is not is his status: on most papers he ranks low in the hierarchy, usually below the salt, mistrusted by senior executives, tolerated by the news editor, an *ad hoc* filler of space for the Features editor. Of course, personality counts. A science writer with charisma does sit at High Table, well above the salt, and does have a comparatively easy entry into the nobility conferred by regular front page stories. But such writers are few. And they have no illusions that their field cannot be invaded by a first-class general reporter. Local reporters without any specialist science reporting background played a key part in alerting people to, and providing continuing information about, the Love Canal problems. Some played roles beyond news reporting: for example, in tracing the source of the toxic chemicals, and highlighting existing health problems. They were responsible for turning a local toxic waste issue into a nationwide problem.

Specialist papers such as the *Financial Times* and the *Wall Street Journal* have increased steadily the space devoted to science. This was in response to demand from readers. For the *Financial Times*, science coverage began in about 1950 when the late, distinguished physicist, Sir Francis Simon, began to contribute an occasional scientific article to the leader page, and provided short notes from time to time under the title, 'Science and Industry'. Readers began to seek further information, and from that the idea of science as an integral part of the *Financial Times* was born. There followed the formation of a department dealing mainly with technological developments, and the organization of specialists covering every aspect of a particular industry. The development of a fairly complex machinery was inevitable as the demands by industry on the *Financial Times* staff increased: plants and laboratories to be visited, conferences to be present at, new products to be examined, and new processes to be discussed.

The *Financial Times* found that a new definition of the word 'accurate' was needed. It was clear that the harm done by any story the paper published which was wrong, or gave a wrong impression, could far outweigh the good will obtained by publishing many which were right. 'When discussing a new development, for example,' wrote a former editor of the paper,

'it is possible to make an exciting story of it – worthy of page 1 – if one overlooks the fact that it may still be in the laboratory stage and that it will require years of expensive research before it has any practical applications, if it ever does. The temptation to publish this sort of story is always there but it has to be resisted. Publish the story, yes, but only against its proper background. And as the editor is unlikely to be an expert on these matters himself it is vital that scientific correspondents should be among the most responsible men on the staff.'

(Newton, 1962)

This approach is reinforced by the current *FT* science editor, David Fishlock. His experience is that the professional scientists' or engineers' complaint about science reporters is that they 'get it wrong'. Soon after joining the paper in 1967 he was told by a distinguished physicist that he did not 'waste time reading newspaper reporting of science'. Fishlock has found that 'indignation has tended to be the keynote of much science writing in the past few years.' He is critical of those science writers who have 'joined enthusiastically in the assault on scientists' motives.' Their writing on such topics as 'the hazards' of pesticides, herbicides, drugs, toiletries, nuclear reactors and their by-products were 'all-too-often characterized largely by indignation and prejudice'.[9]

I do not agree entirely with Fishlock. Of course, facts are sacred, but what is important is the environment in which the facts are presented. A different outlook causes a different conclusion to be drawn from the same facts. And I believe that science writers have a special function, similar to that performed by the theatre or art critic: fearlessly to report on the activities of the scientists and the social implications of their work, whether purely intellectual or applied. Further, the science writer has tended to ignore the profesional and political influences of the organizations of scientists and of science – bodies such as the Royal Society, the Research Councils in the UK, and their equivalents elsewhere – especially in these days when public funding is of such importance. Science writers can no longer leave science policy to the scientist or the politician: they must comment on the advice given by scientists to the government of the day, and be prepared to 'expose' the personality of the scientific advice-giver, despite Lord Zuckerman.

As the pace of technological change becomes ever more rapid, science writers have a specific task not only to report on what is new, but also to comment on the impact on our ethics and ethos, and to look critically at our social mechanisms which should function to help cope with change. Professor Bryan Jennett, neurosurgeon and Dean of the Faculty of Medicine at the University of Glasgow, has made a call for a new institution to provide a coherent approach to the assessment and application of medical technology (Jennett, 1984). The science writer has a key part to play because of his links with the general public. In the USA and in some European countries, journalists are involved in 'consensus conferences', which consider the effectiveness and application of medical technologies. 'Experts present evidence before a wide audience, and both represent a broad range of backgrounds and disciplines. Twelve panelists, or "jurors," including doctors, but also such others as a nurse, statistician, laywer, economist, philosopher and journalist, listen, question and then draft a consensus statement which is widely disseminated.'[10]

(b) Magazines

I distinguish between four major types of publication in the magazine fields, which I call: Public, Scholarly, Industrial and Technical, House. The audience for each of these types is different, so, therefore, is the form of presentation. The concern of the magazines in this broad field is not primarily with mass understanding, but with the process of specialist communication. They are important sources of information for the popular science communicator.

Public magazines
These are available on public sale by subscription or through a newsagent primarily to a non-specialist readership. They vary in

quality of presentation. The best known internationally is the monthly *Scientific American*, which has a paid circulation of over a million and appears in eight different language editions. Each English-language issue carries eight major articles. The coverage is wide: for example, in December 1984 there were articles on Children and the Elderly in the US; Atomic Memory; How Embryonic Nerve Cells Recognize One Another; the Digital Reproduction of Sound; Prey Detection by the Sand Scorpion; the World's Deepest Well; Turning Something Over in the Mind; and the Spanish Ship of the Line. The authors are academics, either at a university or working in industry as scientists or engineers. There are some eight regular departments, including Computer Recreations, the Amateur Scientist, Science and the Citizen, and Books. Illustrations in colour and black/white are lavish, but considered, and based on original artwork.

There is a standard fee of about $1000 for a 4,000/5,000 word article, although this may vary. The author understands that his article may be, and frequently is, re-written by the staff. Like all Public magazines, the *Scientific American* depends upon advertising for its viability.

The inspirers and guiding lights of the magazine were Gerard Piel and Dennis Flanagan, both originally journalists on *Time* magazine, who in 1948 began the new format and approach to the presentation of science when they took over an existing publication. Both retired in 1984: Piel is now chairman of the publishing company, and Flanagan is Editor Emeritus.

Their success is obvious in making it apparent that there is a large market for the appropriate presentation of science. They charted a route now followed by publications in the US and other industrialized countries.

In Britain, however, there is no equivalent. There is the unique weekly, news magazine type publication, the *New Scientist* which combines science news and feature articles.

I believe the future for magazines of this Public type is good, not necessarily in the printed form but more likely as electronic journals. Whatever the technical form, the necessity for the interpretation and presentation of new science will always be urgent.

Scholarly magazines
These are a key factor in the communication process, developed over the years, whereby the scholar can be provided with the information necessary for his work. Almost half-a-century ago, J. D. Bernal drew attention to the problem arising from information overload and the economics of publishing:

In the old ideal of science, communications were the only link between scientists. Now the very quantity of scientific information has made its diffusion an enormous problem, with which the existing machinery has utterly failed to cope. The present mode of scientific publication is predominantly through the 33,000 odd scientific journals. It is . . . incredibly cumbersome and wasteful and is in danger of breaking down on account of expense.

(Bernal, 1939)

These publications insofar as they are linked with learned societies are not concerned with the popular presentation of science. I mention them only because they are primary sources of information for the science writer, and in this indirect way are of significance in the popular communication chain. And major learned bodies have specialist information/press officers and staff to help the press to deal with scientific developments. Thus, in August 1984, the American Chemical Society at the national meeting arranged for press conferences on 'The problems and promises surrounding genetic engineering', and produced a special issue of *Chemical Engineering News* on the subject.

There are some scholarly magazines which, although not published by learned bodies, have international reputations as publications of primary papers. These are, for instance, the weeklies *Nature*, published by a commercial enterprise, in the UK and *Science* published by a professional body – the AAAS, in the US. Their links are intimate, not only with the scientific community worldwide, but also with science journalists, who use them as major sources of information. Many of their staff may themselves be members of the national science writers' association. They are concerned, also, with providing regular news and features on science policy and the social impact of science. In this regard, they demonstrate their keen awareness of the social function of science, and of the need to link laboratory research with the industrial and political worlds in which it operates and which determine its behaviour.

In absolute numbers, the greatest number of these Scholarly magazines are produced in Europe and North America. Throughout the world, there are between 57,000 and 90,000 specialized journals dealing with some aspect of science.

UNESCO's J. G. Richardson points out that the international distribution of a journal is a function of the accessibility of the readers to the journal's language. World-wide the English language journals have the widest circulation, because of the near-ubiquity of English as the *lingua franca* of science. Journals published in

Chinese, Japanese and Russian circulate relatively little outside the countries where they are published, although the number of readers within the country of circulation may be very high indeed.

Industrial and technical magazines
Essentially, these publications are devoted to advancing the technical, industrial and profit-making interests of their subscribers. They provide information on, for instance, new methods and equipment, through the editorial pages and through the advertisements. They supply news, data and technological information that the reader can apply to his own operations. They are, also, amongst the sources used by science writers when information is sought on a particular topic: for example, *Computer Weekly*, one of the many publications in this rapidly growing field, provides regularly in tabloid form news analyses, micro information, a software file, company news, new products, and job vacancies. In one sense, such a magazine provides a continuing education process for the readers.

It is difficult to estimate the total number of magazines in this field throughout the world, but they run into hundreds of thousands. They constitute a vast intercommunication network, exchanging data, information and experience among many thousands of companies and individuals, each working in some particular field of industry.

House magazines
These are published by large industrial and commercial organizations, particularly the multinational corporations, and by the United Nations and the specialized agencies, such as the FAO, WHO and UNESCO. There are, in addition, publications of regional bodies, such as CERN (the European Laboratory for Particle Physics), whose CERN Courier is published ten times a year in English and French.

These House magazines are invaluable non-primary sources of information, and sources, also, of excellent illustrations.

Finally, I mention the development of the transdisciplinary journals distributed throughout a given region. These are largely a European product, developed in response to the changed cultural background begun in the 1970s. They include *Ambio*, published in Stockholm and dealing with social analysis of ecological change; *Biofutur*, a bilingual Paris monthly, concerned with industrial technologies; the tri-lingual (English, French, German) *Acta Oceanographica*, published under the auspices of a French research institute; and *Die Naturwissenschaften*, a German and English monthly in the basic sciences, published in Heidelberg.

In summary, in recent years there has been a marked growth

in the specialized publication of scientific and technical journals, particularly in Europe and North America. In Africa and the Middle East, the situation is 'stable', and 'rising somewhat' in the Asian-Pacific and Latin American-Caribbean regions.[11]

(ii) Books

> After making herself very agreeable, Lady Constance took up a book which was at hand, and said, 'Do you know this?'
> And Tancred, opening a volume which he had never seen, and then turning to its title-page, found it was 'The Revelations of Chaos,' a startling work just published, and of which a rumour had reached him.
> 'No,' he replied; 'I have not seen it.'
> 'I will lend it to you if you like: it is one of those books one must read. It explains everything, and is written in a very agreeable style.'
> 'It explains everything!' said Tancred; 'it must, indeed, be a very remarkable book!'
>
> Benjamin Disraeli, *Tancred*

I ask myself, 'Why in my consideration of these forms of presentation have I put Books at the last?' Certainly, it was not deliberate. I love books: life without them is unimaginable. Yet – I suppose that deep within me there is the fear that the future for the traditional printed work is limited. Access to, and participation in, forms of communication of knowledge are changing significantly. And clearly due to make considerable impact is interactive video, the technology which enables pictures, whether still or moving, to perform as does the book.

There is a tremendous variety of scientific and technical books: ranging from the colourful, illustrated book for children, through the popular science texts and do-it-yourself handbooks to the academic textbooks. Only a few – very few – and in the main textbooks for students, provide an adequate financial regard both for the author and for the publisher. And as a result of limitless exposure to TV and to graphics in advertising the demand for colourful presentation means that the cost of producing such books has increased.

No publisher can deny that publishing is about making money. He works to financial parameters in a business which, as with any other producer of consumer products, has to declare an acceptable profit at the end of the financial year. In commissioning books, he seeks out the popular, the topical and the exciting: the potential money-makers. In science, such 'products' are in short supply, mainly because there are few competent and imaginative science writers. There is, also, a special complication. By the time a book is commissioned, written and appears in the bookshop at least two years will have passed from the moment of imaginative conception. This is a challenge to the author: he knows the science scene changes rapidly, and he must take the time lag into account in his writing.

An experienced science writer will know that it is possible to devise a basic vocabulary of 200 English words which occur so regularly in our spoken and written language that together they would make up 60 per cent of the words used in a *Times* leading article. This does not mean that a foreigner or an infant familiar with these 200 words would understand 60 per cent of a *Times* leader, since the remaining 40 per cent, drawn from a vocabulary of many thousands of words *regularly* used by leader writers, are the words used to convey meaning, and a point of view. If the reader is in any doubt he can refer to a dictionary, and the chances are that the meaning of the unfamiliar word will be pinned without ambiguity.

The trouble with scientific and technical vocabulary and language is that the dictionary, even to a non-educated scientist, is usually of very little help. The definitions of scientific words which we find in technical dictionaries, and often in general dictionaries, are frequently unintelligible to the non-scientist and might even confuse a scientist specializing in a totally different field.

The difficulty lies in the concepts which are involved, and in the scientific processes which have to be referred to as an indispensable part of a definition. The scientists who write definitions for dictionaries are concerned with exact and logical expression. Unless their definitions are to be mini-textbook length, they have to use defining words which are meaningless to non-scientists. These words are usually italicized, the reader having to refer elsewhere in the dictionary for a definition of the words in question. The process may have to be continued *ad infinitum*, a nightmare kind of Cerberus' head.

There is another problem which has become aggravated in recent years. The advance of science during the last few centuries has been so rapid that no language has been capable of providing, ready-made, all the new words which have been required. Hence, scientists have had to invent new words and have usually turned to the classical languages for their raw material. As a result, a very significant part of scientific language has been put together by a modular process, with the Latin and Greek roots, prefixes and suffixes joined together to form the words needed. Thus, from the Greek, the *pterodactyl* had 'winged fingers', and an *isosceles* triangle stands on 'equal legs'. *Isomers* are chemical compounds with 'equal parts', *polymers* those with 'many parts'.

The decline of Latin and Greek, and even of Classical Studies, as subjects to be taught in schools and universities, has resulted in a generation or two of young people who have less chance of understanding scientific language than their forebears. Sadly, the Classicists, in their entrenched positions, have failed to establish a rearguard action using scientific literacy as part of their *raison d'être*. The inherent difficulties of the concepts and of the language itself lie at the root of the problem of popular scientific presentation, whether in the spoken or the written word. In recent years, we have become involved with a phenomenon of mass communication which is probably unrivalled in history. I refer to the computer – in itself a dull object, visually unattractive and incredibly complicated. I ask, 'How is it that within less than a decade practically every child in the country has learnt "about computers" and what they are able to do?' Although they may not know how a silicon chip becomes an integrated circuit, they know the language associated with computers, and how to make and use a program. How has presentation been so successful with the computer, which is essentially a product of science and technology, and yet has failed in popular science presentation generally?

Understanding computers in the early stages was certainly not helped excessively by the media, nor by any particular media personality. The popular presentation of astronomy has had a personality such as Patrick Moore, nature and the environment David Attenborough, and biology and geology David Bellamy, but computers have had no single comparable TV personality. An early book on computers, that is about 1978, has a stultifying dullness and a visual unattractiveness.

The home and school acceptance of microcomputers came about through the motivation provided by games and space invaders. Michael Faraday's Christmas Lectures at the Royal Institution in the last century provided their own motivation in the form of 'tricks' and 'magic'. His famous lectures in 1849 on *The Chemical History of a Candle*, which are regarded as outstanding examples of popular presentation, were delivered with highly motivating experiments.[12]

By contrast, presenting science to the general public through books has many hazards, not least of which is the limited budget, and the uncertainty of the market. For example, consider an encyclopedia dealing with the whole of science, suitable for 11-year-olds and interested 'non-scientists' beyond that age. Assuming that the science has to be correct and contemporary, where do we start? Dealing first with language, difficult concepts have to be avoided, and those that are included should be explained in context. Short sentences, short paragraphs and short words are essential. But short words are seldom possible. Difficult and new words should be defined in a glossary, which is essential for each popular science book. The level of readability can be calculated and monitored by an editor, using one or other of the readability measures.

The single factor which provides a reader with motivation to continue reading, and to learn and understand what is being read, is visual presentation. With the coming of colour television, this is more important than ever. Because good artwork and photographs are immensely expensive, it becomes more difficult for publishers to compete with television. Many of the most effective popular science books for adults have been closely linked with TV productions, or with media personalities.

There are few popular science programmes specially designed for young children. And we have few examples of juvenile presentation to match the adult books. The problem is exacerbated by an economy which prevents schools from keeping their libraries up to date, and public libraries from replenishing their stocks. Courageous publishers can continue only if they go deliberately for TV-type visual presentation.

Well-written popular science books do have a positive effect on a student population. For example, a study in 1967 in the USA at Tulane University, New Orleans, Louisiana, compared two groups of eighth-grade students. One group used a standard science textbook. The other did not have a textbook, but did have unlimited access to the school library for outside reading in science. On examination, the library group scored as high or higher than the textbook group in tests measuring science achievement, attitudes, writing and critical thinking (Barrileaux, 1967). A 1980 study at Central Michigan University reported the same finding in comparing seventh-grade students who used either a single science textbook or a range of outside reading materials (Fisher, 1980).

About 10 per cent of the approximately 3,000 children's books published each year in the US are popular science. There are a number of publications which help to evaluate the quality of such books. *Science Books & Films*, published five times a year by the American Association for the Advancement of Science, has a section on children's books in which a rating from 'not recommended' to 'highly recommended' is given. The New England Roundtable of Children's Librarians is responsible for a three-times-a-year *Appraisal: Science Books for Young People*; and 'Outstanding Science Trade Books for Children' is published in a Spring issue of the National Science Teachers' Association magazine *Science and Children*; this is compiled in collaboration with the Children's Book Council.

The criteria these organizations use for evaluating science books include: accuracy; correct, complete and up-to-date information presented clearly and understandably; the quality and relevance of the illustrations; appropriateness of the book for particular groups of readers; and the value of the book compared with similar titles. Facts and theories must be clearly distinguished, generalizations supported by facts, and significant facts not omitted. Anthropomorphism is inappropriate, as is an approach to make science appear magical and fantastical rather than a rational process of inquiry and investigation.

Britain, of course, has nothing like this coverage. There are, however, two review publications: *Ways of Knowing: Information Books for 7 to 9 Year Olds*, in which Peggy Hicks, Berkshire assistant county librarian, presents an annotated listing of more than 100 non-fiction titles, including many books on science, published by Thimble Press, Stroud, Gloucestershire; and *The Good Book Guide to Children's Books*, published by Penguin Books in association with Braithwaite & Taylor.[13]

The criteria for evaluation of what is a 'good' popular science book for adults are much the same as those for the young.

B AUDIO-VISUALS

(i) Television

Television is an admirable medium for expression in several ways: powerful and immediate to the eye, able to take the spectator bodily into the places and processes that are described, and conversational enough to make him conscious that what he witnesses are not events but actual people.

Jacob Bronowski, 1973

Television might not be the unique and essentially contemporary form of communication which we think it is. Maybe it shares a good deal in form with other non-literate communications, in other societies at other times. The point has been to identify a persistence in human culture to which television in our time gives expression; the communication of texts which are, singly or together, simple, authoritative, powerful, reassuring, justifying, accessible, unambiguous, coherent, appealing, relevant, literal, magical, true.

Roger Silverstone, 1985

Television, the newest of the media, shapes all. No other medium dare ignore it: each recognizes that survival depends on successful adaptation of television's language and grammar to its needs. Thus, today's tabloid newspaper is modelled on a TV soap opera format. 'Bingo' money is everywhere.

The Promenade concerts are probably the most popular of BBC annual events, successful in artistic and box-office criteria. But even these have to bow to the needs of TV. At the beginning of the 1984 season the first night's programme was changed because the BBC had already arranged a TV recording of the same work (*The Dream of Gerontius*), and on the last night the interval in the Albert Hall concert had to be nearly doubled to wait for the finish of the first episode of the new series of *Dynasty*, the American soap opera. On election day 1964, Sir Hugh Greene, the BBC chairman, agreed to accede to a request from Harold Wilson, the Labour Party leader, to change the timing of the very popular television programme, *Steptoe and Son*, due to go out that night. The following morning, Wilson rang and thanked him, and said that the change had probably given Labour an extra 12 parliamentary seats.

There is little popular discourse through the media: what there is resides mainly in a few broadcast programmes. The presentation of science is limited. There is, however, a significant difference between what appears on American and British screens. In Britain, the BBC, which is financed by the public through payments of annual licence fees for radio and TV sets, ensures a high popular quality of science coverage. And Independent Television is not far behind: 'commercial' TV has always had a more 'common touch', necessary because it is dependent on advertising, which again is dependent on mass viewers.

I was responsible for one of the first popular science programmes on the commercial network. Lew Grade (now Lord Grade), then responsible for programmes at ATV, after months of negotiation, finally gave me and two colleagues the go-ahead to produce a weekly half-hour series called *Meet the Professor*. My approach was simple. I contacted two competent Fleet Street journalists who 'knew nothing about science' − Lewis Greifer and Maurice Wiltshire. I explained my particular approach to the presentation of science in the form of everyday entertainment but in which science was not distorted in 'miracle' terms.[14]

The series, which began on 23 February 1956, had some good press notices and viewing ratings. It ran for a first 13 weeks, and then for another 13. By that time, I was exhausted − acting as producer and person responsible for the science contents, I decided not to continue, and we had also another pioneer programme under way. This was Greifer's inspiration. Called *Fantasies of the Night*, it was an attempt to popularize the psychoanalytic approach to behaviour. It was well enough received, but presented certain basic problems which decided us not to continue after 13 weeks.

I had, also, begun to develop serious doubts about the validity of my approach. Bernard Levin, then TV critic of the *Manchester Guardian*, reinforced my doubts in a characteristically outspoken criticism (17 March 1956), describing the programme as 'unashamedly aimed at the moron belt'. It is true that *Meet the Professor* was made up of clichés and stereotypes. But this was a pioneer experiment. The problems were fascinatingly great. I could only begin to define them as we went along. I recognized that my desire to inform through TV in particular had to be expressed in terms of entertainment: that is, the requirement to inform is more correctly the recognition of the necessity to present the information as entertainment.

I had stumbled on a basic truth, but I could not then formulate it. I knew that we had demonstrated that we could make science interesting for a mass audience,[15] but I was not yet seized of the awful seduction of TV.

I was concerned with relating science more intimately with the everyday world. In fact, the final programme attempted an answer to this. As Maurice Richardson wrote in the *Observer* (3 June 1956), it was 'an excellent, post-fabricated case-history: diagnosis, treatment and cure by radio-active iodine of cancer of the thyroid'. And Philip Purser in the *Daily Mail* (1 June 1956) stated, 'With this final edition in this series of "Meet the Professor" we were back with television as an expression of reality. Indeed, it is difficult to think of anything more realistic . . . than the dramatisation of a cancer cure to which most of the programme was devoted . . .'

Of course, I now recognize that I was cheating with that programme. Anyone can succeed with the presentation of medicine and health. It was 'an expression of reality', but only insofar as 'reality' was presented as entertainment. That is the basic truth apparent today which I had not been able to recognize in the beginning.

Since those pioneer efforts, TV has become the major agenda-setting mechanism in all fields of behaviour. It affects directly not only which topics are discussed publicly, but also gives 'notoriety' to those who are invited to express their views, and defines and manages the way in which the discussion is conducted. This is most obvious in the handling of current affairs. But what of science? Of course, the basic televisual programme criteria apply. What appears finally on the screen must be as near to 'compulsive viewing' as possible. A 1978 British Library report (Jones, Connell, Meadows, 1978) stresses there are other criteria specific to science and not broadcasting to be considered. For instance, the science broadcaster must pay particular attention to his sources. It is they who determine 'the nature of the coverage of science'. The Report claims that this was established early in the development of the field.

The significance of this argument is apparent. Sources are always of importance in the selection of topics to be presented, whatever the medium. As science came increasingly to be recognized as a force within society, it was inevitable that as a reflection of this a science correspondent came to be appointed in 1959 to the BBC News Division (BBC Handbook, 1960). The general coverage of science has grown rapidly. It is presented not only as News, but also in Current Affairs, and a specific topic will be presented as a major programme in itself: for instance, a BBC Spectacular on, say, the Universe or the Brain.

Much of Independent Television's coverage of science is contained within a regular series of news, news magazines and current affairs programmes. For example, *First Tuesday* (Yorkshire) is a documentary series which often deals with scientific matters; *The Real World* (TVS) deals with a variety of scientific, technological, and medical subjects; *Start Here* (Channel 4) introduces children to physics and chemistry; *Well Being* (Ch. 4) deals with medicine and everyday health and *Where There's Life* (Yorkshire) looks at the human and social sides of medicine. Medical information is available, also, in some form in soap operas, such as *Coronation Street*. In *Earth Year 2050* (Ch. 4) an attempt was made to present visions of the future. *Survival* (Anglia), now in its 25th year, is screened in every country which has TV.

The most consistent presentation is that provided by BBC Science and Features Unit, set up in 1963. The flag-bearers are the magazines *Horizon* and *Tomorrow's World*, concerned since July 1965 with the presentation of technology. These provide faithful-to-science and good viewing programmes. *Horizon* began as a monthly in May 1964. It appears now fortnightly. Its objective is 'to explore the scientific attitude', and it is 'more concerned with ideas and philosophies of science than with techniques, or even discoveries'.

Of course, the basic philosophy as laid down by Aubrey Singer, the first head of BBC Science and Features, still holds: 'the televising of science is a *process of television*, subject to principles of programme, structure, and the demands of dramatic form. Therefore, in taking programme decisions, priority must be given to the medium rather than to scientific pedantry.' Programmes have to be 'entertainingly informative at an intellectual level'.

How is this attempted? I am indebted to the three authors of the British Library Report (PCRC, Leicester) for the following summarized statement of their in-depth analysis. Their work is based on a study of *Horizon* (Jones et al., 1978).

The accounts given in the programmes 'always articulate certain propositions about science, about scientific work and about scientific workers.' The meaning given to 'each utterance in the verbal discourse' results from 'a shared agreement, a reciprocity of perspective' between those involved with the public communication of science. To avoid criticism by scientists of the ways in which their work has been characterized, broadcasters use a variety of means to seek to ensure that only the intended meaning is realized.

The presentation of the material begins typically with a medium close-up (mcu) of the person being interviewed; the interviewers are only rarely seen or overheard. The next shot is a close-up (cu), done by either a cut or zoom. The cu is used to concentrate the interest of the viewer.

The technique of presentation has changed over the years. In the early days it was the expert who was the dominant figure. He spoke directly to the viewers by talking to the camera. He was the centrepiece, introduced and lauded by the presenter. He determined which aspects of the subject were to be discussed. Today, the expression of authority is the presenter, or interviewer. He speaks directly to camera, and the expert answers *his* questions, and does so at an angle to camera. This is regarded as minimizing the viewer's potential identification with the speaker. 'Rather than promote a sense of inclusion, this mode of depiction promotes a *sense of witnessing* – being present at, but not being directly involved in – an ongoing communicative situation.' The viewers are *on*lookers.

An overall perspective can be constructed only through combining the visual and verbal discourses. For example:

vcu	verbal
lymphocytes attacking bacteria	During three million years of evolution we've built up a complex system of defence against micro-organisms. White cells keep watch for invading microbes. These black dots are bacteria. It's a direct attack. There's another line of defence – less direct but just as deadly.
Cut to animated graphics	When a patrolling white cell sees an invader it passes the alarm to a whole army of other cells. At the alarm some of these cells turn into chemical factories and move into the attack. (*Blueprints in the Bloodstream*. 9.ix.77)

The commentary makes no specific reference to the visual

material (microscopic film and animated graphics), but it is clear that the 'complex system of defence' mentioned is being shown. The words, 'These black dots are bacteria. It's a direct attack' halts the flow of the exposition to identify what the visual discourse signifies. The graphics are animated only when the statement is made. The full, intended meaning results only from the combination of the verbal/visual discourse, and it is the verbal which is the dominant. The commentary initiates and winds up particular aspects of the topic.

In practice, *Horizon* programmes, which are compilations of discrete sequence of verbal and visual discourse dealing with different aspects of the topic, provide a sense of unity (how one aspect relates to another) in 'the boundary exchanges' located at the beginning and end of the sequences of commentary. This gives viewers an orientation to the main body of a given sequence. These boundary exchanges consist of framing (to indicate a change is to occur) and focusing moves.

For example:

Visual	Verbal
mcu Van Rood talking, right of camera	... she will be making antibodies against these bits and pieces. Now in this case these antibodies will not, generally speaking, reach the baby. (*Blueprints in the Bloodstream*)

˄The framing move here is a marker (Now) and a silent stress ( ). A visual transition may take the place of a verbal frame. For example:

Visual	Verbal
cu woman talking right of camera	... and again, of course, I think I was regarded as being a bit of a lead swinger which is rather unfortunate. And then, eventually I had to give up work
	Visual transition
Cut to ls of team walking to camera; pan to follow	In 1971, a team at London's Westminster hospital made an unexpected discovery. It was to shake the world of rheumatology. (*Blueprints in the Bloodstream*)

A focusing move is the opening statement, 'In 1971 . . . of rheumatology.' It indicates what is to follow, and gives some measure of its significance. *Horizon* often secures transitions from one aspect to another by using an interrogative. For example:

Visual	Verbal
C/over	But how we do acquire these antigens – these fingerprints on our cells? (Here follows an explanation. Then the next question:) If our cell types are so different could that be why we reject transplants? (*Blueprints in the Bloodstream*)

The questions raise problems, and placed at the beginning and end of thematically distinct segments indicate that answers will be given later.

Horizon programmes present scientific inquiry as puzzle solving: accordingly, 'the act of doing scientific work and the act of telling stories about that work in the form of a detective story are considered to be homologous, implicitly, at least.' This has certain consequences for the characterization of how scientific work is done. A programme will begin with a summary of accomplishment, and then go on to a detailed presentation of how this was achieved. However, this detailed and polished reconstruction re-casts the actual work of research, and 'gives to it a *sense of consistency and coherence not present in the original undertaking*.' (My emphasis – MG.)

Because of the televisual need to see science as 'a neatly integrated and collaborative enterprise' the breaks and discontinuities in scientific research and the rivalries between different teams are smoothed out. It is this which tends to upset the professional scientists, who find that the dead and loose ends and contradictory results that are an integral part of research are absent.

How objective is a science programme, such as *Horizon*?[16] This is an impossible question. What is really meant is, 'How balanced is the programme?' Put in this way it is clear that because there is always *a priori* a subjective choice of the topic, objectivity cannot be maintained. So far as balance is concerned, all points of view cannot possibly be presented on topics such as the impact of nuclear power stations on the environment, or the origins and impact of acid rain. It is possible to reconstruct and film a research project as though it were happening now, but there is still a selection of events.

Further, the TV team and equipment will interfere with or even change the phenomenon under investigation. The light of the microscope may influence the biological rhythm of a tissue culture;

and the camera may modify behaviour in certain environments. Virgilio Tosi, Italian film-maker, adds to these the following aspects of the so-called objectivity of images: the changing angle of the camera and the spatial composition of the framing; the continual variation of the field being shot (pans, travelling shots, widening and closing-in zoom shots); successions of pictures created through editing and with special effects (fading, lap dissolve, mixtures of images from real life with graphics); special techniques which make the invisible visible; temporal duration of a visual message which differs from real time.

These are derived from the structure of the language used, and are based on the physical and psychological differences between the visual perception of the human eye and the visual interpretation of reality reproduced by a camera. Thus, TV-video only engages partially the complete angle of vision of the eye, which is about 120° in the horizontal sense (Tosi, 1984).

As with the other media, very little is known in any country about the impact of science programmes on viewers and on the makers of such programmes. Much research is needed.

(ii) Film

If I have been able to fly, it is because I have read *Le vol des oiseaux*.

Wilbur Wright, the first aviator

Scientific cinema predates cinema as entertainment. When, on 28 December 1895, the first public projection of the Lumière cinematograph took place in Paris, giving birth to cinema as entertainment, scientific film had already made great advances.

In 1874, Jules Janssen, a French astronomer, had had built in Japan 'a photographic revolver' to help record the phases of transit of Venus across the sun's disc. This was the first primitive example of what we know as 'time-lapse' filming, in which a series of successive images are recorded at extremely short intervals. Two years later he foresaw that 'the characteristics of the revolver of providing, automatically, a series of images in close succession of a phenomenon with rapid variations, will enable us to deal with the interesting questions of physiological mechanics relating to walking, flight and the various movements of animals.'

He was aware of the experiments carried out in the 1870s by the British photographer, Eadweard Muybridge, at Palo Alto in the USA, in which the movements of a horse during a race were revealed by a series of cameras along the track, each being triggered by the horse as it ran. These encouraged the French physiologist, Etienne-Jules Marey, in 1882 to use his 'photographic gun', inspired by Janssen's revolver, to take the first pictures of the flight of birds, actually wing movements of gulls. He was helped by the new silver bromide dry plates, more sensitive than the commonly used wet collodion type.

For several years after cinema as entertainment began, new research in filming was done almost exclusively by scientists within their laboratories. For instance, in October 1888 Marey presented to the Académie des Sciences his film roll chronophotograph, which has the basic features of the modern ciné camera.

His pioneer work on the physiology of movement is detailed in his publications, including *Le vol des oiseaux* (The flight of birds), published in 1890. As a scientific weapon, we have come to accept cinema as of the greatest value. As Raymond Spottiswoode wrote,

> To watch the opening of a flower in the morning demands the
> utmost patience and attention . . . to watch the growth of a
> tree is impossible, save by comparing many trees at different
> stages of their development. The cinema has altered all this.
> By taking a film at the rate of, say, one frame an hour (instead
> of 24 a second) an increase of speed of nearly 90,000 times
> is obtained, and this can easily be exceeded where necessary.
> If the process is a continuous one, this not only enables it to
> be watched as a single whole . . . it ensures that a complete
> record is obtained.

Further, he went on, there is an aesthetic aspect, 'The beauty of a flower is half realized when it is seen with the eye; the camera reveals fully another half . . . delights which have been imagined by poets, but never before seen by other men' (Spottiswoode, 1935).

However, I am concerned here not with research but with the popular science film designed for a mass public. Thousands of such films have been produced, sponsored by government and industry: made by such bodies as the Shell Film Unit in Britain, the National Film Board in Canada, and the Scientific Film Studios in Moscow. There is a problem of definition. Popular science films are often confused with teaching or educational films. But there is a basic distinction: and this is true for all the media concerned with the popular presentation of science, they cannot ever 'deliver' to the public an understanding of science: their function is to alert, to set agendas, and to provide materials for imaginative play. As G. Sieler, German Democratic Republic popular science TV producer, put it at the 1982 Jena meeting of the International Scientific Film

Association (ISFA): 'the challenge is one of achieving an artistic conversion of scientific reality by uniting logic and poetry'.

This became clear to me when, in the 60s, I was involved intimately with the British Scientific Film Association and the International Scientific Film Association as editor-in-chief of *Scientific Film*, the official journal of ISFA. At each annual congress of ISFA, there is a Popular Science Section devoted to the selection of prizewinning films. At the 14th congress in Prague in 1960, for the first time there were special showings of TV productions. I believe that popular science TV programmes have brought an end, by and large, to the public showing of popular science films.

Standard film has many advantages over videotape. It is true that film and video each have a place. For example, before a large assembled audience, film projection is the technique of choice. Video projection is still unsuitable for audiences of above 100 people. Film projection provides a crisper image and a greater tonal range. In its own technical field, that is broadcast TV or videocassette, the choice lies with videotape – so obviously, that before transmission film prints are usually transferred to videotape.

Where images require a special quality, as in advertising on broadcast TV, film is used. It has this special creative quality at the moment. In addition, contrary to general belief, video is not necessarily cheaper and faster than film. It is true that video editing is more speedy than film editing, but the cost of video editing suites is high and puts great pressures on the editor to rush through his work. Editing techniques and styles differ for film and for TV. But I believe, also, that in the near future these differences will tend to disappear.

Film must be treated as part of the audio-visual presentation of information. In this regard, training is important. In its final report, a UNESCO Working Group on the Popularization of Science through Television, meeting in May 1978 in Luxembourg, stressed that although UNESCO should not undertake direct action itself, it should encourage a number of organizations already active in the field to offer training opportunities. The ISFA has proposed that there should be created Audio-Visual (AV) centres, and the British University Film and Video Council organizes periodic seminars and meetings to update qualified people in the existing AV centres. The British Open University, in training staff for the production of teaching audio-visuals, has found it wiser to select those with a scientific background rather than those who are already specialists and technicians in AV fields. It may be that the training of the scientist as a popular presenter will be one of the major developments in the future in each of the media.

The leading international body concerned with the popular presentation of film and TV is the ISFA (38 Avenue de Ternes, 75017

Paris, France.) It was created in 1947 to provide 'a critical, scholarly apparatus which could categorize and evaluate existing scientific films and establish suitable standards of production'. At the annual seven-day congress, an average of one hundred films are presented in each of the three sections – research, higher education, and popular science. The regular publication of films presented – specified by categories, with technical data and summaries – is an invaluable information source.

A most useful guide to national organizations in many countries throughout the world, and an extensive bibliography, appears in 'How to make scientific audio-visuals for research, teaching and popularization' by the distinguished Italian teacher and film-maker, Virgilio Tosi (UNESCO, 1984), to whom I am indebted for much of what appears above.

(iii) Radio

If you can describe clearly without a diagram the proper way of making this or that knot, then you are a master of the English tongue.

Hilaire Belloc

If a science broadcast is not properly understood and remembered what, it may be asked, is the point of it? Is not understanding the whole aim of science? Without understanding there can be no real science. This would certainly be true if the main aim of scientific broadcasting were to impart a thoroughgoing education in science. Although it may seem surprising, I think that any comprehensive policy must seek justification not in the field of science itself, but in its sociological implications.

Fred Hoyle, *BBC Quarterly*, 1952

On the evening of 30 October 1938, the rising theatre director, Orson Welles, broadcast from CBS in New York a radio script, based on H. G. Wells' *War of the Worlds*, describing in graphic detail how invaders from Mars with their superior technical equipment had landed in specific places in New Jersey. The story of the panic that followed this seemingly authentic happening is well known (Cantril, 1940). An event of this kind is never likely to happen again, not with radio as the medium of communication: that is, in the sophisticated industrialized world.

However, that does not mean that listeners do not still get involved in radio broadcasts. The continued success of soap operas is an example. But those responsible for science broadcasts in the first world, who seek to present in an interesting way the achievements of scientists, their methods and attitudes, and to demonstrate the relation between the advance of science and the social well-being of the community, recognize that they will always find it difficult to convey to a general audience what has really been achieved.

The main problem is that in many instances the broadcaster is talking about what really needs to be seen to be understood. Thus, where a piece of machinery has to be described, the only effective way of helping the listener who wants more detail is to provide a back-up service of written material.

Revolutionary ideas and new concepts are not easy to present. This is true in all the media, but particularly radio. The way broadcasters use language has some bearing on this. To communicate effectively on radio you have to break away from the linear form of languages as we are accustomed to it in books and newspapers. People do not think or speak in such terms. Good broadcasters tend to speak in extended blocks of meaning, using ungrammatical phrases, gradually building up the idea they are communicating. They will rely heavily on analogies to stimulate the imagination. It is not a very precise way of communicating, but it can be very evocative, and it is appropriate for discussing such subjects as atomic physics, since such a subject tends to be rather 'conceptual'. There is nothing the lay listener can see, so he has to use his imagination and think in terms of ideas and analogies.

The nature of science to a considerable extent determines how and to what extent it can be treated, and also in what programmes. However, in popular presentation all the broadcaster can hope for is to provoke enough interest among listeners to encourage them to find out more.

In radio, there is no opportunity for a listener to go back and listen to something again, unless he has taped it. Once the broadcaster has said something it is gone for ever, so listeners have to be led very carefully through a topic, since there will always be distractions for the individual listener. Long magazine or discussion programmes are designed to allow listeners to join late or leave half-way through.

The different situation in the third world is described below (p. 77).

The intelligibility of a subject is related to the skill of the communicator. Few practising scientists have the right skill. The science producer is the key person. The BBC Science Unit, responsible for radio broadcasting talks and documents, has four producers, of whom three are trained scientists, and the other is an English graduate. Their training enables them to understand what the scientist is trying to say, and their understanding of the audience helps the scientist to express what is more likely to be understood. This relationship between producer and speaker is probably unique in the whole field of communication.

There are few ongoing research programmes into the impact of science as presented on sound broadcasting in industrialized countries. Radio is regarded as 'an old friend' and is taken for granted. This is a pity as, despite the sweeping impact of television, there are almost no households without at least one radio set, and in the majority of them the radio is usually switched on for the whole of the non-sleeping hours.

Professor Jack Meadows, of the Leicester University Communications Research Centre, believing that the physical sciences were under-represented on radio, suggested that a series of programmes should be devised for use by BBC local radio stations. He was seeking answers to such questions as whether the announcement of a programme on a physical science subject generated a barrier which caused viewers to switch off, or whether that was a viewpoint peculiar to radio producers.

His proposal was to devise a form of programme which, as Dr Neil Ryder of Chelsea College, University of London, who is responsible for the 'evaluation' of the project, puts it, 'might fly under such defences – low profile, informal kind of programming, not strongly headlined and advertised, based on everyday topical issues.' Thus it was that during the early summer of 1984, eight BBC radio stations across the North and Midlands transmitted six short programmes based on ideas suggested by Meadows. They were produced in a variety of formats by BBC Local Radio, Programme Services Unit. Each programme had a distinct element of scientific information or explanation, and, deliberately, a mix of biological and physical science stories to determine how the listeners distinguished, strongly or otherwise, between the two.

Meadows in devising his outline scripts provided details of experts to be found in different regions on the topics he had chosen. The first four programmes were based on what appeared on the front cover of the weekly BBC *Radio Times* for the week of transmission: in the week of the first cricket Test match, Willow trees and cricket bats; in Ascot week, Breeding horses; in Wimbledon week, Grass courts; and in the week of the finals of 'Come Dancing,' Clothes design for dancers; the final two programmes were based on topics of current interest at any time of the year: the meaning of weather maps; and why churches don't fall down.

Ryder's evaluation on a part-time basis is not an easy task. He was still engaged on it during 1985. As he sees it, 'My task, "evaluation" had placed upon it by different parties aims which pulled in significantly different directions, and which did not emerge until specific methods were suggested.' It is clear that 'the scientist's wish is to talk direct to the listener in his/her kitchen. But to do this he has to meet the criteria of station managers, central service directors and producers as well as maintaining his own reputation amongst his colleagues. What, therefore, do we want to know, and from whom, in order to carry out this function of evaluation?'

He has detailed the variety of questions to which the scientist wants answers, the BBC in their approach to listeners, and his own about listeners, producers and local stations. The key problem was 'the difficulty of locating an interestingly large number of listeners who may hear the programme go out live.' In the end, he had to abandon the attempt to reach the authentic live audience.

Ryder's report is likely to be a notable contribution to the socio-logical methodology involved in seeking to find answers to what at first appear to be simple questions, such as listeners' constructs of science presentation on radio. What is clear is that more resources, full-time staff and money, are needed urgently if those concerned with the presentation of science are not to continue to base behaviour on uninformed guessing.

C MUSEUMS

'In the old days,' a venerable Russian scientist once remarked to me, 'they used to make museums for savants; now they make them for children.' But it is not only children who benefit by the economy and clarity of the new museums. We have come to realize that the same material may have a number of radically different possibilities of arrangement, and that each arrangement will itself bring out some new fact often unsuspected until that arrangement is made. The task of the curator will therefore be just as much to manipulate his collection as to guard it.

Bernal, 1939

Most museums are dull places: uninspiring in their presentation, intimidating in their platitudinous boredom. The word itself bears memories of endless trekking through tiring corridors, with occasional halts to peep hastily into glass containers hiding – not exposing – objects, whose public presence is explained to bored ears. I exaggerate, but not grossly. The connotation of a cultural mausoleum, repository of the dead, continues to be relevant.[17]

However, it is true much has changed, and is changing, in presentation and in the concept of the museum: so that it is becoming a forum where the genius of the times and the spirit of the people find expression. It aspires to be a laboratory, a novel happening, a transitory experience . . . a place where the public holds discourse with artists, where it can express its doubts, seek information, make a start in art or science, and freely question what it is seeing and experiencing.

Solana, 1980

The 'grand museums' – the National Museums in London, Paris, and Washington, and their counterparts in New York, Chicago and Munich – are the oldest and best known and the biggest of their sort. They get the largest number of visitors. But museums everywhere operate under a number of constraints.

(1): in terms of structure, museum buildings seem often to have been designed to intimidate the public. They are neo-classical and gothic palaces or cathedrals intended to instil a sense of awe and reverence, and to make the visitor feel humble in the presence of Authority. Traditionally, this architecture is coupled with displays that make no concession to the difficulties faced by the visitor, whatever his age, to make him feel that the subject is forever beyond his understanding. This is another aspect of Victorian 'popularization,' when it was taken for granted that the discourse of 'high culture' could be grasped only by an elite, but that a gesture of aid should be made to the masses. This creates both a 'putdown' and a 'putoff' effect (Lewis, 1980).

The older museums have little room for originality of presentation. They have had the architecture thrust upon them. The architect designed the building, the décor of the galleries and the individual showcases, which he had positioned on artistic grounds. The exhibit designer is limited, therefore, to placing his objects in cases and to add explanatory labels in the spaces left. It is extremely difficult, therefore costly and not attempted, to design the exhibitions from the concepts-to-be-communicated upwards. It is possible sometimes to remove the showcases, but that presents only a partial solution.

Clearly, the older buildings, regarded as historic monuments, inspire resistance to change. There are many who take this attitude. They insist, for instance, that the British Museum (Natural History) is a Victorian building with an Edwardian display and should stay that way.

(2): most museums are starved of funds. The result is that what is available tends to go into the collections and research, rather than into public involvement. In protest against this, there was 'a revolution' in the 1930s at the Chicago Museum of Science &

Industry. Curators were dismissed, collecting for its own sake was halted, and the Museum began deliberately to serve the public.

But underfunding has been always the great problem. Sir William Flower, director of the BM (Natural History), pointed out almost a century ago, ' . . . the largest museum yet erected, with all its internal fittings, has not cost so much as a single, fully-equipped line-of-battle ship, which in a few years may be either at the bottom of the sea, or so obsolete in construction as to be worth no more than the materials of which it is made.' And Sir Roy Strong, director of the Victoria & Albert Museum in London, wrote in the *Financial Times*, 26 April 1984 of ' . . . leaking roofs, crumbling masonry, deficient drainage, outmoded heating and ventilation, and disintegrating electrical circuits' in his museum. The same conditions are apparent in most big science museums. In 1984, funds were reduced for the popular presentation of science at the Natural History Museum in London, and at other places. One consequence is that museums have had to accept substandard exhibitions from commercial sponsors to fill their galleries. However, from mid-1985 Britain's museums started making admission charges, with government approval, on a voluntary basis: that is, trustees of the individual museum decide whether to charge in the knowledge that all profits will go to the museum itself.

(3): museums have three functions – conservation and ordering of the collections; research, theoretical and practical, based on the collections; and service to the public. These functions do not bed well together. Rivalries develop, and tensions are created – usually at the expense of the public. Museums owe, and justify, their existence to collections of objects, and showing at least some of these to the public is an acknowledged central activity. However, in the mind of many museum workers a mystique has developed around the objects. This attitude sometimes is given a moral justification: 'The essence . . . is that one can see for oneself the actual evidence . . . Visitors have the possibility of deriving their own concepts, drawing their own conclusions' (Halstead, 1978). Whatever else is untenable about this argument, it is self-restricting to limit the communication of science to showing objects, because on the whole science is less and less concerned with the existence of things in nature and more with understanding processes and relationships. Museum staff often see the task of communicating with the public as less demanding and worthy than management of collections or scholarly research. It is true that many museum specialists are unable to empathize with the lay visitor.

Creative, original work in mounting public exhibitions and developing educational programmes is difficult. Generally, staff have to contend with hierarchical management systems and rigid procedures. This reinforces a bureaucratic fear of error, the need to play safe, and frequent referral upwards for approval during the development of the project. Contributory factors are time-scale and cost, as it takes many months to produce a big exhibition, and it is not easy to change its structure once it is opened.

The cost of a good science exhibition, including payment for one director, is from £350 to £500 per sq. metre. A sizeable, prestigious exhibition would cover some 500 sq. metres. It would on the average take about two years to have the exhibition open to the public: one year for design and one year to produce, although these activities would overlap. The inertia of large organizations is based upon unwritten rules and precedents that have to be observed and that deter innovation. The traditional conveyor-belt system for producing exhibitions is:

Curator	**Designer**	**Education officer**	Open to public
Selects object Writes caption	Dresses case	Organizes educational activities	

As R. S. Miles, head of the Department of Public Services at London's Natural History Museum, wrote,

> This is not a system that encourages self-examination or self-renewal, or one that encourages any sort of innovation. The absence of feedback of information rules out the possibility of improvement by trial and error, and the rigid separation of the various activities engenders mutual suspicion rather than team work, and thus rules out synergic advance.[18]

During recent years, increasing attention has been paid to the importance of exhibit evaluation. When carried out, particularly during the process of developing exhibits, evaluation can do much to improve the quality of the communication. There is no doubt that formative evaluation (or developmental testing) is worthwhile. But, it has not been adopted widely, primarily because it needs a different management system. In practice it means submitting work to visitors for approval, rather than to some hierarchical chief, before going on to the next stage.

Evaluation is one technique for helping staff to assess the quality of their performance. It would be helpful if linked with this there were independently expressed statements. In the museum world, these hardly exist. There is no accepted 'language' for reviewing exhibitions. Art and history museums are regarded as worthy of serious attention for adults, but science museums are regarded as for children. Art shows have their regular reviewers in newspapers and weeklies by devoted, organized critics. Science exhibitions may be noticed in specialist journals, but the reviewers limit themselves usually to descriptions of the exhibits and perhaps a few superficial comments often on whether children would like them or not.

(4): museum staff are not aware that they are behaving in amateurish ways when seeking to educate the public. 'To the professional educator, the most conspicuous feature of conventional museum exhibitions is their communicative incompetence' (Lewis, 1980). Until 1982, the literature on museum exhibits contained no single authoritative handbook for the practical guidance of museum staff (Miles et al., 1982). And museums have been slow to come to terms with television, which has accustomed the public to having subjects presented in a popular way, with little regard to the traditional academic disciplines, and to lively and professional presentation.

In fact, most museums know little – almost nothing – about their audience. Only a few have the basic management information provided by visitor surveys. Exceptionally, the Natural History Museum has carried out surveys over an extended period (eight years) to monitor the reception of new exhibitions and changes in audience. Preliminary 'market research' is almost unheard of elsewhere. It is no surprise that much of the communication fails to find its audience.

Traditionally, museums communicate by the methods of demonstration and definition: that is, objects are put out as evidence that such things exist, and labels are added to say what they are. One trouble with definitions is that they introduce new terms which they do not define, so the visitor is left baffled, unless there is a special attempt to make connections with his familiar world. This is not easy in many areas of science, especially if dealing with abstract or counter-intuitive ideas.

In recent years, 'a new way' has begun with the development of Science Centres, which are of importance as a reaction to the limitations of the traditional museums. They are visitor-oriented, and acknowledge openly the advantages of not having collections to curate. They have put much effort into developing interactive (also termed participatory or hands-on) exhibits. This is the rationale of the Exploratorium,[19] Frank Oppenheimer's Los Angeles Science Center. Unfortunately, elsewhere, it has sometimes led to the introduction of rather trivial button-pushing machines.

In the physical sciences and engineering, machines, devices, laboratory experiments and demonstrations exist often in a form that can be adapted readily for public use in a science centre. This is not true, on the whole, for the biological and earth sciences: that is why the natural history displays in science centres are weaker than are the other exhibits. It is usually necessary to invent analogues for natural processes in natural history exhibits, so inventiveness in developing the 'software' and the 'hardware' is of critical importance. In this regard, microcomputers are of help. They have made the development of interactive exhibits a great deal easier. The use of simulation has great potential in dealing with both the physical and biological sciences, where, for instance, the original experiment is too slow, too quick, too dangerous, too expensive or too unreliable to show live, 'in the flesh'. The problem is to secure good software.

Museums and Science Centres do make use of the dynamic media: film, videotape and videodisc, slide/tape systems and computers. However, such displays are costly to set up and maintain, and require special skills, knowledge and creativity to obtain a good professional product. These qualities are usually beyond the means of most museums. Maintenance may suffer, and the exhibition get a bad press, although sometimes this may be due to poor organization and management.

The communication of a scientific argument may require the development of a long chain of ideas. For example, Darwin's hypothesis about the origin of species needs, first, an exposition of the four premises of his theory of natural selection, then the articulation of the theory, and a further exposition to show how the theory could account for the formation of new species. A museum is not a good place for this sort of communication. People cannot move easily back and forth to relish the argument as they can in a book. Also, walking to and fro around a large exhibition is tiring. Some Science Centres have introduced novel communication techniques. Thus, at the Exploratorium, 'Explainers' who are red-jacketed, high-school students, recruited and trained every four months, stationed among the displays to help visitors, and at the Science Museum of Minnesota extensive use of theatre as an interpretative technique. Demonstrations are used extensively at the Ontario Science Centre; a question-and-answer session takes place each day at the Boston Aquarium with a member of staff sitting casually on the edge of the big tank (this is unusual because aquaria as a whole make little effort to communicate); most of the exhibits in the Exploratorium and the Ontario Science Centre are designed and constructed in their workshops, and because they are not enclosed, the public can view the exhibits being designed; and the Philadelphia Academy of Sciences turned its temporary exhibition into a workshop for the preparation of its new dinosaur exhibit so that the public could see the specimens being released from the rock and mounted in life-like postures.

(5): museums grossly overestimate the public's knowledge of science, their ability to take up new information under museum conditions, and their interest in detail. This contributes to failure in communication. Museums have strongly heterogeneous audiences:

visitors vary in age, knowledge, reasoning ability and academic achievement. They have a wide variety of reasons for being in the museum. At the Natural History Museum, at most 5 per cent come with education in mind: the remainder are 'window shoppers,' those seeking the equivalent of escaping from the rain. What can museums do about this? In practice, they do nothing. They do not understand that it is a waste of time to talk about the public's understanding of science, but that they would be wiser to concentrate on the misunderstandings that prevent communication, and on finding some basic concepts on which to build. Only one museum in Britain (Natural History) carries out research on its visitors' knowledge, expectations, interests and so on, during the planning of new exhibits.

Museums are tiring mentally. 'Museum fatigue' (object satiation) can set in within four minutes (Melton, 1972). This strengthens the case for using a variety of media, especially an audio-visual theatre in which the visitor can sit down and rest. And some museums have begun to make available relevant books at their major new exhibitions to encourage study at home. However, this practice is still rare in science museums.

A major source of inspiration in the search for a new approach has been Otto Neurath. He is known best for the introduction of Isotype (the International System of Typographic Picture Education). His first attempt at popular presentation was centred on the Museum of Housing and Town Planning in Vienna in 1923, which became the Social and Economic Museum. He sought to humanize knowledge by making technical matters accessible through the quality and structure of the communication. He saw humanization as the opposite of popularization, which he regarded as aiming at a semblance of understanding by translating technical terms into popular language.

The other major inspiration is Frank Oppenheimer's Exploratorium in San Francisco. Since 1969 it has been the world's most influential science centre. It had inspired small and large alike everywhere, from the excellent Science Centre at Richmond, Virginia, to the mammoth French Government project at La Villette. Unfortunately, not all of Oppenheimer's humanistic ideals have survived the journeys. His Exploratorium is essentially a place with participatory exhibits through which the public can explore science in an unconstrained and playful way. It was 'conceived to communicate a conviction that nature and people can be both understandable and full of newly discovered magic,' states the Statement of Broad Purposes.

It therefore provides experiential opportunities for learning that are difficult, if not impossible to achieve through school classrooms, books or television programmes. The Exploratorium is not a substitute for other vehicles for learning, but it provides a fascination with learning that cannot be found elsewhere and which facilitates traditional teaching at all levels.

There is no attempt to impose a structure on the learning experience. This may be a shortcoming in that it makes the Exploratorium less effective educationally than it might be otherwise.

Another novel 'experiential opportunity for learning' is provided in the Micrarium, a unique museum in the spa town of Buxton, England. It is the creation of its director, Stephen Carter, who from childhood has been absorbed by microscopy. The visitor views the world of the living small – for instance, endless activity in a wasps' nest, red blood cells in the gills of a live baby newt, small embryos developing inside their eggs – through an original system of projection microscopes made by the Carter family themselves (Tony Jones, 'Through a glass clearly,' *New Scientist*, 28 March 1985, pp. 38f).

Finally, I need to comment on the sad fact that there are almost no case studies of science exhibitions. This is a major defect. The arguments are that museum staff are bad at writing up their experiments, and are keen to bury their mistakes. Clearly, this is not good enough. Some museums see the value of case studies: for instance, the Laurence Hall of Science *Star Games* exhibition is written up. And the Natural History Museum's new exhibitions have been studied in fair detail since *Human Biology* opened in 1977. There is the advantage, also, that the Museum staff have some idea of what secondary schoolboys understand about evolution before they are taught the topic (Clarke, 1981; Griggs, 1983). Possibly of much greater importance is a long account of the Smithsonian Institute's exhibition on the dynamics of evolution (Wolf & Tymitz, 1981). This is a fine statement of what can go wrong in trying to make a popular presentation of science.

CHAPTER 3

SCIENCE FICTION: TRUTH AND REALITY

The use of fictional techniques to explore the dimensions of scientific belief-systems is fully within the great heritage of free and independent inquiry within the scientific field, as is the use of scientific methods to define and explore human problems in fiction. Problems of high complexity are a fact of human existence. To turn away from them in search of lost innocence is to beg the question of whether extension of human knowledge is limited by 'the cold equations,' or by randomly distributed probabilities, or by some universal order beyond our present level of consciousness. Any or all of these hypotheses may correspond to 'truth.' Therein lies reality.

Richard Weholt, 1984

I first met the leading figures in British Science Fiction (SF) when as science editor of *Illustrated*, a million-plus readers weekly — now, with *Picture Post*, sadly gone — I wrote what was probably the first picture story on the young British SF writers of the day. The photos were taken in Arthur C. Clarke's first-floor flat in north London. Like many of those there, I was a member of the British Interplanetary Society, which was concerned with preparing society for the not-so-distant future, not in a fantasy but in a seriously technical sense.

Of course we had read Jules Verne and H. G. Wells, and seen such classic films as *A Trip to the Moon* made in 1902 by Georges Meliès, based on a story by Verne. We were, whether technically qualified as Clarke was or not, beginning to shape our abilities and aspirations to the development of a new aesthetic form, primarily literary. SF as a genre is, basically, a product of this century. It began as a literary form, but moved very soon into other media, particularly film, so that we are probably more familiar with SF through its audiovisual form. During recent decades it has found expression in other creative media, such as the graphic arts.

SF is *fiction* based upon an approach expressing the physical laws of nature as defined by the hard sciences (*les sciences exactes et naturelles*). This is the basic constraint which determines how the writer handles his theme. It is the science laboratory and the scientific method (sic) in the service of an art form. This definition of SF is valid for all the creative media.

SF is often confused with Fantasy, an older literary form, whose basic constraint is the 'irrational' (myth and superstition) as distinct from the 'rational' in SF. There is a sub-form called Science Fantasy, which ignores the basic constraints of hard science, and which will be reinforced by the development of the new field of 'psychophysics'.

There is a clear British SF tradition, exemplified in the work of H. G. Wells, Olaf Stapledon, Aldous Huxley and Arthur C. Clarke. In the background, between the two world wars, there was the influence of such radicals as the biologist, J. B. S. Haldane, the geneticist, Lancelot Hogben, and the crystallographer, J. D. Bernal, who through their pioneer insistence on the social function of science helped to initiate a post-1945 generation into a wider understanding of the relationship of science to social behaviour and the need to harness science in the interests of survival and national development.

Their philosophical outlook — whether accepted, modified, or rejected — could not be ignored. It excited passionate debate in academic circles. In 1929, Bernal, age 28, had made clear his 'vision of the future' in his first book, *The World, The Flesh and the Devil*, sub-titled, 'An inquiry into the three enemies of the rational soul'. In this he drew a new world picture based on science to replace that of religion. It was essentially a demand for a planned society in which scientific method could be introduced into all areas of human activity. Given that, he believed there could be tremendous acceleration in man's understanding of himself and his universe, and release from war, famine and disease. The three enemies of the rational soul would retreat in defeat; these were the World (the forces of Nature), the Flesh (man's body and its physical limitations), and the Devil (man's passions, stupidity and ignorance).

I mention this book because although it was not a novel, it was a visionary approach based upon extrapolations from existing scientific knowledge. He made some remarkable forecasts. For instance, the conquest of space, reinforced by necessity. 'Ultimately it would seem impossible that it should not be solved. Our opponent is here simply the curvature of space-time — a mere matter of acquiring sufficient acceleration on our own part.' He proposed the use of the rocket, and envisaged a form of space sailing which used the repulsive effect of the sun's rays instead of wind. He described in some detail a three-dimensional, gravitationless way of living in space in a spherical shell, ten miles or so in diameter, made up in the main of the substance of one or more smaller asteroids, rings of Saturn, or other planetary detritus. In the 1970s, some astronomers conceived of a large spherical settlement, with 50,000 people, orbiting the Earth: they called it the Bernal Sphere.

Bernal was returning to the early days of modern science, when such speculation was common. Newton, for example, at the end of his *Opticks*, had many queries which were, in fact, speculations. Twentieth-century scientists would not normally engage in such behaviour. They were trained not to say anything for which they had no evidence, and certainly not to make their presentation personal. I suggest that SF writers became the twentieth century's speculators: and under the influence of John W. Campbell, Jr, the remarkable editor of the monthly, *Astounding SF* – today, Analog/Science Fiction/Science Fact, this marriage of hard science and fiction was made secure. The dominating romantic pap adventures of intergalactic heroes saving empires and beautiful women from BEMs (bug-eyed monsters) gave way to scientific and technical credibility. In its March 1944 issue, *Astounding* published a short story, 'Deadline', by Steve Cartmill. It described a successful attempt to destroy an atomic bomb which was to be used by an evil power. What was most interesting was its statement that U-235 had been separated from non-fissionable isotopes and was to be detonated in a working bomb, the details of which were given.

In the spring of that year, agents from the Manhattan Project's security division interviewed the author and John Campbell, suspecting a leak from the top-level secret project, although they knew that the SF bomb bore no resemblance to that being built and would not work. They wanted to know where the author and editor had got their idea from.

In the September 1984 issue of *Analog*, Albert I. Berger provides an answer to that question. Under the provisions of the Freedom of Information Act he was able after April 1983 to read the reports and memoranda prepared by the Manhattan Project Military Intelligence agents who had conducted the investigation. He saw, also, the comments of their superiors, and of one official of the Press section of the Office of Censorship.

Berger concludes that Campbell and Cartmill had created a problem by 'naming what was intended to be unnameable: the near-term practical possibility of an atomic bomb'. Of course, there had not been any leak. Campbell had shown how SF could put 'scattered bits of scientific knowledge together into a specific, concrete idea or device, and speculate on what that idea or device's impact might be on the world at large.' SF had become a problem-posing and problem-solving medium.

The rival monthly, *Galaxy*, edited by Horace L. Gold, insisted also on 'accurate science', which included soft science. For him, the universe consisted of probabilities: for Campbell, only strict causality based on natural laws was acceptable. Both magazines published some highly original fiction. The American influence on the British SF writers' world was great: especially as the 'strength' of SF – that is, the publications and finance – lay then, as it does still, in the USA. That is why so many British SF writers adopted a mid-Atlantic style.

SF films continue to be popular. They have made us aware that there are many strange creatures in our galaxy. Three of the big money-making films of the early 1980s are *Star Wars*, *Close Encounters of the Third Kind* and *ET*. As with the written word, there are hard and soft SF films: the former is about gadgets, and the latter about the human implications of meeting beings from 'somewhere out there'. Soft SF films are usually pessimistic, for instance, a race of lizard-like beings lands on Earth determined to take over the planet.

It is not my intention here to provide a potted history of SF literature or films. I am concerned with two questions: 'Does SF influence science?'; and 'What impact has SF on popular understanding of science?'

The answer to the first is, 'Yes, it does.' In his, *The Spaceflight Revolution*[1] William Sims Bainbridge suggests Wernher von Braun[2] talked Hitler into developing the V1s and V2s not because he saw them purely as military weapons, but because this was the only way in which he could ensure that his spaceflight researches would continue. Similarly, with the Russian physicist, Konstantin E. Tsiolkovsky, who was the first to suggest the possibility of building a space station. Sputnik 1, the first man-made satellite, was launched in October 1957 to celebrate the one-hundredth anniversary of his birth. He wrote, also, an SF novel, *Outside the Earth*.

I was in Washington on that historic October day at an evening reception being given by the Russians for scientists who had been attending an international conference. At one stage, there was a sudden hush and an American announced, 'I learn that our Russian colleagues have succeeded in placing an earth satellite in orbit.' There was a short silence, then much applause and too much vodka. It was clear the Americans were not happy about being beaten in the race, and they had to wait four months, until 31 January 1958, before von Braun's group was able to place the first American satellite in orbit.

There are many reputable scientists who are SF enthusiasts, and some who are spare-time SF writers. For instance, the American electrical engineer, who working at Bell Telephone Laboratories, helped to develop communication satellites. Like von Braun he was an SF reader, and like Tsiolkovsky wrote SF, but under an assumed name. There is the controversial English astronomer, Fred Hoyle, who continues to write SF openly despite the sneers of many of his contemporaries. And the younger American astronomer, Carl Sagan, another controversial figure, who is an SF reader.

Of course, there are full-time, professional SF authors trained as scientists. Arthur C. Clarke is a qualified engineer; the American Jerry Pournelle worked on the American and Apollo space programmes, and has degrees in engineering, psychology and political science. Many writers have no scientific qualifications, but they do consult widely when constructing their 'imaginary universes'; fine examples are Brian Aldiss with *Helliconia Spring* and Harry Harrison with *West of Eden*. John Brunner's massive *Stand on Zanzibar* is often cited as *the* SF novel on overpopulation, containing much vital technological/sociological detail. Within SF we find many different kinds of science, from the 'imaginary future technologies that are extrapolated (like Jules Verne's submarine) from what we know already' (Nicholls, P., 1982) to ' "imaginary" science, which tends to be much sillier than the first sort. . . Three of the commonest examples are time machines, hyperspace travelling, and the idea of alternate universes. Yet modern physics now gives some warrant for ideas as strange as these and even stranger. We can no longer afford to dismiss "imaginary" science quite as contemptuously or patronisingly. . . .'

And there is controversial science: those areas of speculation that are rejected by a majority of the scientific community, but pursued by an extremely well publicized minority; examples are flying saucers, telekinesis, ESP and vanished civilizations such as Atlantis.

What is the effect of the presentation of science in SF on popular understanding of science? Directly, none. It is not an educational medium, although the editorials and specialist science articles in the SF monthlies, such as *Analog*, are usually first class in their scientific accuracy. What SF probably does is to prepare the public mind for a future that may lie ahead. It may act as a kind of vaccine against future shock. It could be that the film *Star Wars*, made by the SF buff George Lucas, made it easier for President Reagan to put across his military defence proposals, which were immediately christened as the Star Wars strategy by the media. And it may well be that the agonizing about *in vitro* fertilization, testtube babies and cloning is more apparent in intellectual circles than with people generally because these have long been the clichés of SF. It is important to differentiate between 'pop' SF, significant in reader influence but for the most part non-rigorous scientifically, and 'inside track' material, found in such limited circulation magazines as *Science Fiction Review*.

There are a number of SF courses in American universities. They are designed usually as part of a programme to bring the sciences and the humanities together. Some fit deliberately into an English curriculum, and are concerned with giving SF respectability by treating it purely historically. Courses of this type are not normally concerned with aesthetic and ethical values. They are concerned with presenting SF as a literary genre, beginning with, say, such stories of early moon flight adventures as Lucian's *The True History*, Francis Godwin's *The Man in the Moone*, and Cyrano de Bergerac's *Voyage to the Moon*.

Other courses will insist that SF is not merely escape writing, but more often is social commentary. That is true today, when contemporary SF illuminates the relationship between science and human values. For example, John Brunner's *Shock-Wave Rider* presents a world in which the ability to fit into a new environment and detach oneself from the present/past without psychological disorientation is essential for survival. And Ursula LeGuin's *The Dispossessed* dissects the varied ways in which technological, political, psychological and ethical issues are interrelated.

SF has much to say in providing alternative answers. As the industrialized countries move into the 'quaternary stage' (primary was training and agriculture; secondary, manufacturing; and tertiary, the wholesale and retail trade) we have need of 'early warning' systems, or technology assessment institutions. That is a function the SF writers perform so brilliantly. They are the historians of the future.

CHAPTER 4

THE EDUCATIONAL SYSTEM NO-GO

> But in physics I soon learned to scent out the paths that led to the depths, and to disregard everything else, all the many things that clutter up the mind, and divert it from the essential. The hitch in this was, of course, the fact that one had to cram all this stuff into one's mind for the examination, whether one liked it or not.
>
> Einstein

Schools have largely failed to inspire in students an understanding of science and mathematics. This is true for each country with a broad educational system. After years of study, the majority of school leavers move into adult life illiterate in science and innumerate. They are equipped neither with the appropriate skills, nor with the knowledge to form objective judgments. They cannot relate the S&T that dominates them with their daily lives. They cannot cope with such key issues as the development of atomic energy, the use of non-renewable resources, chemical pollution, over-population, surplus food supplies, genetic engineering, and nuclear weapons. In practice, they cannot function independently as citizens to assess alternative solutions on an informed and rational basis before making their democratic choice. They are potential fodder for the demagogue as they move into acceptance of pseudo-science, of astrology and horoscopes, and mystic sects.

It is not my purpose here to tread again a well-worn path of why this is: of how it is that in the USA, for instance, in the last few years great alarm has been expressed at the failure of the schools to produce a scientific literate population given the millions of dollars that have been spent in attempts to do so since the alarms of 1957 (the year of Sputnik 1).

The major influences on students are: the home, their peers, the media and economic pressures. School comes low down on the list. Clearly, the presentation of science at school does not carry over into the home. Some years ago I looked intimately into this aspect. As a temporary science teacher in a working-class area in London, I began to understand that the language of the classroom remained remote from the language used in the home. The authoritative and inflexible manner in which I was required for exam purposes to present the basic laws of science failed to excite the students. Finally, I gave up the formal approach and sought to relate, for example, Boyle's Law and Torricell's vacuum to everyday devices with which they were familiar. That succeeded with some: most, especially the girls, shrugged their shoulders: why worry about how the electrons perform to carry the voice through the telephone wires when all we need to do is to use the phone, they argued.

Although it is clear that the language of science is not part of working-class homes, yet working-class students see science as a means of providing social mobility. Science is regarded as 'middle-class', and a route to a university. Students from middle-class families have more positive attitudes to science than those from other socio-economic groups (Gardner, 1975).

A particular way in which schools have failed is in regard to girls and science. In statistical terms, females do lag behind males in their understanding of, and attitude toward, science. This is well documented in the UK, the USA and Australia. There may be genetic differences responsible, but I would assign more importance to sociological pressures and expectations. In a study of a group of gifted Australian secondary school pupils (Leder, 1979), girls showed a greater fear of success in mathematics and the physical sciences than did boys. They were aware that it was social pressures, rather than innate limitations, that held them back in these fields. The stereotype of femininity, learned very early in life, is at variance with the characteristics demanded by a scientific training (Walden, 1972). Girls are expected to conform to Society's expectations more strictly. This contributes to boys developing a freer cognitive style and greater proficiency in abstract reasoning, hence to potentially greater success in science (Musgrove, 1967).

The strength of the social factors allows biology to be seen as appropriate for girls to study, almost certainly because of its link in the popular mind with nursing and the other 'caring' professions. Only rarely exceptionally do women come to the top in physics and chemistry. By contrast, in the USSR women are 75 per cent of all doctors, and 35 per cent of all engineers. This is strong evidence that the different performance between the sexes arises from social rather than from genetic factors. Only recently have schools come to be aware of how powerful are such social factors, and how relatively powerless they are in face of such pressures.

Are schools attempting the impossible in seeking to produce a scientifically literate and numerate population? In general, yes.

Should they then give up the task? Certainly not. We require our specialists, and science is an essential part of our general culture. Schools are no more successful with other subjects. French, for example, is taught widely in schools in England, yet the English are among the world's worst linguists. And in October 1984, the Minister of Education, Sir Keith Joseph, was launching a campaign to improve the teaching and use of the English grammar!

In the classroom, science is taught in a social vacuum. Too many students do not regard the scientists, whose 'laws' and methods they are told about in class and imitate in the laboratory, as human beings. Boyle is 'a law;' Roentgen is 'X-rays'. Mme Curie is more real as she has been dramatized in film and on TV. What is lacking in the teaching is the historical background.

Science entered the school curriculum relatively late. For example, Winchester College, one of the oldest of British schools, founded in 1394, did not teach mathematics until 1820, and then only as an 'extra'. The teacher was 'the Writing Master, a despised functionary whose main duty was to clean the slates and mend the pens' (Firth, 1949). Science teaching at Winchester did not appear until 1870, and became an integral part of the curriculum only in 1920. In British schools generally, science teaching began about 1840, but its development depended on social and political, much more than on educational, considerations. Following the Revised Code of 1862, teachers were paid by results. That virtually put an end to classroom science, except in the Public Schools (where it had hardly begun), as it was not one of the subjects in which pupils were tested by school inspectors when grants to school managers were being determined. Social attitudes insisted that the Classics were the best training for the elite mind, and that science could be left to those most backward in classical knowledge (Devonshire Reports, 1872).

By contrast, in 1917 a committee of the influential British Association for the Advancement of Science set up to make recommendations on the future of science teaching reported that science had an essential part to play in a broad, general education: it should occupy at least one-sixth of teaching time for boys, and one-seventh for girls. This change of emphasis arose from an awareness of the contribution science was making to the prosecution of the First World War.

In the 1960s, in a number of industrialized Western countries there was a substantial drop in the proportion of pupils opting to continue to study science subjects, particularly the physical sciences, at those times when the school system required of them to make choices between subjects. This was well documented in the 1968 Dainton Report, which suggested that a contributory cause in the 'swing from science' was that for many young people, science, engineering and technology (SET) seemed impersonal and dangerous, and indifferent to human and social concerns. Scientists were seen as responsible for many of the world's troubles (Ashton and Meredith, 1969; George, 1974). These social and humanitarian concerns had a greater influence on girls than on boys in leading them to give up the study of the 'hard sciences'.

To meet this widespread retreat from science, the first major wave of school science curriculum reform in the 1960s was content-based. There was little concern with social issues. A US study of new directions in secondary school science (Hurd, 1969) identified future trends in the curriculum. It found that of 24 shifts of emphasis advocated not one was about social issues. The Council of Europe (1972) listed the aims of the more than 200 school physics courses in use in the late 1960s in seventeen European countries. Only seven of those courses had any social connotation: four were concerned with the philosophy of the human implications of physics, one with the education of an elite in France, one with the education of future science specialists in Britain, and only one with 'the development of a sense of social responsibility'.

By the mid-1970s it was becoming clear that despite the massive expenditure on science curriculum development over the previous decade the student antipathy towards science, typical of the 1960s, was being replaced by widespread disinterest. In 1976, Lord Bullock, in his presidential address to the Association for Science Education, urged science teachers to try to raise the level of science literacy in the whole school population, particularly among non-science specialists, by presenting science as a humane study, deeply concerned with humankind and society. This approach was a key part of a programme advocated by the American Association for the Advancement of Science ('Science – a Process Approach') as a way of teaching science more effectively in the prevailing unfavour-able-to-science social climate. Since the mid-1970s the focus of much curriculum development has followed this path, as new curricula sought to relate the teaching of scientific ideas and techniques to social and political issues.

Today, there are a number of courses concerned specifically with social relevance. For example, the British Association for Science Education has a range of publications which provides materials for teachers working on a modular science course with less academically motivated pupils in the 14–16 age range. The former Schools Council developed an integrated science project which viewed science as a human activity, closely linked with values and social responsibility. Among its objectives were to develop skill in understanding the significance, and the limitations, of science in relation to technological, social and economic issues; and to foster an attitude of concern for the application of scientific knowledge within the community. Teachers were invited to provide opportunity for pupil self-expression and development in class discussion, and for the pupils to consider the effects of science-based decisions on themselves and their community. Amongst the practical problems in teaching this project approach was the need to overcome traditional attitudes among teachers of the humanities (Silver, 1974). There was developed also at first degree level SISCON (Science in a social context) (William, 1978).

Although there are many examples of the ways in which attempts have been made to ensure that the science curriculum is responsive to social demands, attitude formation favourable to science has not been successful. Why? The reasons are many and complex, but science teachers themselves in the main have been unable to cope with the special demands made on them. For instance, in 'discovery-learning' they have used predetermined syllabuses which have given little opportunity for pupils to experience the thrill of real discovery. Science as taught is still compartmentalized, increasingly so as the pupil gets older. 'Nature study' in the early primary gives way to 'science' in the upper primary/lower secondary, and this then splits into the three traditional great divides – physics, chemistry and biology – before the public exam stage at 16+. Recently, there have been attempts to teach 'combined science' or 'integrated science', but these have not been successful, partly for reason of status, and partly because of the traditional demands of the university science faculties (Haggis and Adey, 1979).

Dr J. R. Ravetz contrasts the preoccupation of such science with 'facts' rather than with 'processes', which is the way in which scientists work. It has been true always that subject matter is an easier basis on which to plan and to teach, but there must be a fitting place for method-of-inquiry (Glass, 1970). The key factor, however, is that the real motivation for studying science for most pupils (and their parents) is to pass examinations. And that is true, also, for most teachers, who, regrettably, have no experience of being working scientists. Thus, the approach is over-theoretical, and largely responsible for the low standard of school technology.

Yet, the major trend in recent years in school science curricula is an emphasis on the social significance of science teaching. This is concerned not only with a broad cultural perspective of science, but also with contemporary science-related issues. Some have argued that the development of socially responsible attitudes in future citizens is the most important aim of school science education.[1]

Of particular significance internationally has been the development of the Commonwealth Association of Science, Technology and Mathematics Educators (CASTME). This Commonwealth professional association was born in the Caribbean as CASME

(The T was not added until much later). In March 1973, the Commonwealth Foundation sponsored a conference in Jamaica, in collaboration with the Guinness Awards for Scientific Achievement, on the theme 'Social significance in science and mathematics teaching', at which there were about 14 Commonwealth countries. To quote from the Conference report: 'Science and its promotion of technological development bring sociological consequences which cannot be ignored. Technological advance may bring social benefits, but it can also bring social disadvantages.' It was to consider the relationship between science education and its social context that CASME was founded. Since then there has been rapid growth in the interaction between science and mathematics education as well as an increasing awareness of the cultural context in which such education takes place. No longer may the teaching of science and mathematics be seen as value-free, independent of the society which sustains it and which it is designed to serve.[2]

There is still much to do before science teaching becomes effective in producing a literate and numerate population. In Britain, a particular handicap is the cuts in educational provision which are damaging to the quality of teaching of science, with its dependence on practical activity. Effective science teaching requires adequate material provision in the form of laboratories, apparatus and materials. The financial cuts mean a retreat to disastrous rote learning. Barriers between the classroom and the outside world have still to be broken down. Science education must be motivated by, and to a great extent be conducted in, the local environment and community. Students must be given greater opportunity to pursue investigations in science at all levels in a manner less constricted by considerations of timetables and examinations. The trend towards assessment of student achievement in schools will be helpful.

The advent of microcomputers adds a dimension to school science learning. It is a technique of value in its own right, and it can enhance the integration of school science with the world of work. It relates strongly to the out-of-school interests of many students (Fletcher, 1983). Above all, teachers matter. All is dependent on their supply and quality.

CHAPTER 5

THE THIRD WORLD (TW): ILLITERACY AND COMMUNITY

... in a man's lifetime (in the Cochabamba Valley), he will buy one suit, one white shirt, perhaps a hat and one pair of rubber boots. The only things which have to be purchased in the market are a small radio-record player, the batteries to run it, plaster religious figures, a bicycle, a picture of a popular ex-President, and some cutlery.

Caska, 1978

The concept of the Third World (TW) is a nonsense, designed to foster the view that there is a coherent group of countries with status sufficient to be regarded as ranking with the first and second worlds. As I believe it is a nonsense, why then do I continue to use the term? Because I dislike the concept of 'developing countries'. This ensures a patronizing approach, which assumes that economic development along first-world lines is the major route to national wellbeing, and which carries with it the implicit assumption of social and cultural backwardness. However, I continue with the term TW because for me it has the merit of an imposed mathematical neutrality. However, I am not wed to it.

I recognize that TW 'is a lumber-room of a term ... what is there in common between Saudi Arabia and the people's Republic of Vietnam, between Israel and Yemen, between Cuba and Brazil?' (Debray, 1969). The World Bank ranks countries in order of level of development, based on such indicators as: GNP *per capita*, food production, annual production growth, adult literacy and life-expectancy rates. Thus, Kampuchea ranks lowest and the United Arab Emirates highest. This is a simplistic approach. Apart from the impossibility of obtaining valid figures for most countries, the unequal *per capita* distribution of wealth, with its profound social implications, is covered up. Certainly, deep inequalities divide people and individuals: in 1980, the average *per capita* gross product in the industrialized countries was $10,660; in the low-income TW it was $250.[1]

But that is another book. My concern here is to consider what are the special problems in presentation of science and technology in the TW, and by this I mean in the impoverished nations in which live the majority of the Earth's people. I include countries such as India, Brazil and Mexico, in which there is great wealth with an affluent bourgeoisie, but in which most citizens are illiterate, hungry and poor. There are approximately 4.5 thousand million inhabitants of our planet, and sometime within the next 50 years that number will double – given no nuclear war or other catastrophe. The approximate one-third who live in the industrialized countries have a reasonable standard of wellbeing, enjoying averagely good health, sufficient food, and the estimated maximum lifespan. Almost all of the other two-thirds live with the fear of constant hunger and debilitating disease, and with a lifespan far below that of the other third. Depending on the criteria used, the World Bank estimates there are some 800m living in a state of absolute poverty, and the International Labour Organization about 1,100m poor.

In this complex mix of social problems, I propose to concentrate on just two aspects: the basic needs of health and food.

The World Health Organization (WHO) make clear that the aim of health for all by the year 2000 rests in great part on overcoming poverty and malnutrition, which are exacerbated by overpopulation: which in itself is a cause of disease, poverty and malnutrition. There is increase of population because larger families provide food winners, and there is the near certainty that at least half of the children will die in infancy or early childhood. Annual deaths are estimated at some 15m. They die mainly from diarrhoea diseases caused by contaminated water supplies and faulty hygiene, and malnutrition. The other major health problem for all – adults and children – is from parasitic disease: 500m threatened by amoebiasis; 1.2 billion by malaria, with about 175m infected each year; 70m by trypanosomiasis in Africa and South America; and about 200m by schistosomiasis. These are morbidity, not mortality, statistics. Actual annual deaths are very much smaller. Thus, the problem is of living with chronic, debilitating and often incapacitating disease, a burden on an impoverished community. The numbers are large: some 15m lepers and some 20m blind from trachoma. The crippling loss of human energy and productivity is significant in economic terms.

Much can be done to counter these diseases. In the last century, drinking water contaminated by human faeces was the greatest single cause of disease in the Western world. Typhoid fever, cholera and dysentery were rampant until sanitary engineering with good plumbing took over. And it was only 50 years ago that the industrialized world had available new techniques for treating and preventing infection on a large scale, so that the killer diseases of tuberculosis and rheumatic fever, and crippling tertiary syphilis have been conquered. Preventive medicine is the key, yet in TW nations the emphasis is still on curative medicine. Observe the contrast: China practises preventive medicine, Brazil does not: China's *per capita* GNP is one-fifth that of Brazil, but the average expectation of life in China is several years more than in Brazil.

So far as parasitic diseases are concerned, there is great basic ignorance about the pharmacological and immunological aspects. Much research is needed, and now is the time to engage in this. Cellular biologists can cultivate malarial parasites, immunologists are active in work on surface antigens, and molecular biologists are about to clone the genes responsible for the surface agents with which malaria parasites protect themselves against the infected human host. A malaria vaccine is on the way (but nobody can say when); and if a vaccine can be produced against the trypanosome infection in humans and farm animals, agriculture would have available in Africa alone a vast fertile area at present almost uninhabitable. Resources made available for parasitology research should provide new frontiers of social and economic benefit. Agriculture can be transformed by genetic manipulation, which will improve the stress tolerance and disease resistance of crops and grasslands. This would make available some 40 per cent of the world's uncultivated but potentially productive land.

How is this transformation to be secured? It requires action on an international, coordinated scale in close collaboration and consultation with the countries affected. A conclusion to be drawn from this positive use of Science and Technology (S&T) is that population growth might be halted as parents develop confidence that their children will not suffer an early death. This would be linked with the upswing in basic health standards and in food. But two things are needed: one, communal stress on economic self-reliance (that is, not waiting for the government to do everything), and two, the development of confidence and endogenous expertise. Neither can be secured without the popular presentation of S& T, for self-reliance is meaningless without the understanding and motivated support of the people.

Most people in the TW live in the countryside. This means that their lifestyles are traditional, relatively unaffected by industrial civilization. To be self-reliant they must insist on asserting their cultural identity, to develop their cultural heritage. For instance, in countries such as India, where there are rich cultural traditions, rural people drink water out of a clay pot in every house; this practice must be reinforced so that it is not replaced by the imported metal container, which in time kills the local technology. The shaping of clay by a variety of techniques – unbaked or baked to become terracotta – has been, and is, a medium to serve art, religion and everyday needs. A pot, also, is made in an extremely scientific manner. It represents linear links with traditions going back some 4000 years.

I give the clay pot as one example of the approach needed in popular presentation. It presents S&T as part of general culture, for, as distinct from Western society, there is no dichotomy between endogenous technology and its application.

But the imitative pressures of urban society are oppressive. A warning is necessary. Imitation of first world institutions needs to be most carefully considered. TW countries have tended to extrapolate from happenings in industrialized countries, often their former colonial rulers, and mechanically to apply solutions designed for basically different – but seemingly similar – situations. Examples are the Academies of Science and the Associations for the Advancement of Science which have come into existence in a number of TW countries. As with the transfer of technology, the transfer of institutions can be most harmful. There is a view that the creation of such institutions is an indication of national development. It is usually a status symbol conferring a false dignity on office holders: possibly an indication of flattery to the country's leader or ruling party that all goes well. There may be, also, the influence of the concept of 'less developed' or 'underveloped' or 'developing': that is, a TW country may be stating, 'Look, see how we are advancing . . .' in using technologies which are culture-bound to another environment. It is true that there are no distinct techniques which without reservation can be applied to meet poverty and economic backwardness. That is why 'indiscriminate' borrowing may flourish.

Many TW countries and regions are creating associations of S&T writers. I would advise them to wait. Perhaps an entirely new form of association is needed, and will develop. What is required is not the presentation of Western-dominated, laboratory science, but the presentation of basic information in, say, coping with health problems and developing agriculture, always using local resources.

The written word is of little use. Most rural people are illiterate. 'In many communities of limited literacy, the habitual response is, "We can't speak, we have no education." There is a belief in educational inferiority and there seems to be almost a compulsion for self-denigration before the mystique of education' (Low, 1974). According to UNESCO, 41 in every 100 – and in some cases, 60 – are illiterate. The 123m children of primary school age have no opportunity to attend school, and of 814m adult illiterates, 60 per cent are women.

In these circumstances, a technique I am much in favour of is the medium of popular theatre, so called because it is aimed at the whole community, not just those who are educated. Its forms include drama, puppetry, song and dance. Performances deal with local problems and situations with which people can identify. They are given without admission charge in a public place to ensure maximum attendance, and in the local language. But it is more than entertainment. It aims at securing what good popular presentation of S&T always dreams of – helping people to get interested in problems of concern to them, and then wanting to do something about solving them.

Experience in the use of the popular theatre has shown that if there is to be action, then people need to talk about the problem in common, to agree communally on what to do, and if necessary to seek further information and skills: for example, on how to make a new type of latrine, or how to grow vegetables. Popular theatre fails if it is not combined with such extension work as discussions, demonstrations and practical activities.

A great advantage is that there are no special or difficult skills to be learned. Rural life in the TW is full of songs, dances and stories, and the popular theatre builds on the fact that people sing while they work, dance at parties, and act out stories around the fire.

... many people in these communities, because they are not influenced or inhibited by extensive schooling, have a great oral tradition of story telling, versifying, singing and so on. They can be witty and colourful and the language is sometimes richer than the homogenized textbook variety. Making use of the talent can change attitudes toward the educational mystique.

Low, 1974

The actors who, as in the Laedza Batanani project in Botswana, are linked with the University's adult education extension programme involving agricultural, health and community development staff, use the popular theatre to explore and expose problems they want people to talk about. They improvise their own songs, plays and dances. All they need is a little time for preparation and practice, but not major rehearsals and routine memorizing of lines. Spontaneity, more or less, is the keynote in providing a show, which can be organized speedily.[2]

Each year the popular theatre campaigns are usually one-week programmes in which the team of actors tours the major villages in a district, with one day in each village, providing a one-and-a-half hour show, including drama, puppetry, dancing, singing and drumbeat poetry. The aim is to stimulate community participation, discussion and action, so after each performance the actors and other extension workers involved divide the audience into groups and organize discussion of the problems.

A District Extension Team is responsible for planning and organizing the campaign. They work out the detailed timetable and select problems they consider to be appropriate to be included along with those suggested by the community. Choice is made usually of such small tasks as a clean-up campaign, which can be done easily, and the result of which is clearly visible. Those tasks are done with local rather than with government resources and action. Problems selected are those whose solution can be aided without difficulty by regular extension work.

In 1976, for example, the Laedza Batanani team chose poor diet as their topic. The solution needed an integrated approach involving improved nutrition, home gardens, and new cookery methods. Other issues presented included 'cattle theft, inflation and unemployment, the effect on community and family life of migrant labour and the drift to towns, the conflict between modern and traditional practices, school leavers, and family and health problems' (Kidd et al., 1978). The District Team is responsible, also, for working out the costs of the campaign. Normally, this is not expensive. Many costs, such as transport and salaries, are met by the university department, and equipment costs are minimal, usually for the stage backdrop (a piece of painted canvas) and a puppet stage.

TW popular theatre presents much of value to the industrialized world. Its value lies in its role of social transformation, 'expanding participation and self-confidence and providing a mirror for critical analysis, and a stimulus for discussion and action' (Kidd et al., 1978). In Latin America, it is used as a form in 'the struggle for the transformation of society' (TW popular theatre newsletter 1982); in Bangladesh as a tool working 'closely with popular organizations committed to the struggle against the social, economic, political and cultural oppression of the people'.[3]

I am not concerned here with the obvious fact that the popular theatre has the ability to transform its presentations into direct political statements. The Catholic Church in Latin America is well aware of the impossibility of avoiding such statements, despite the negative voice of Rome. I am convinced that appropriate technological solutions for TW countries are not available yet, and that such solutions will have to be derived by making use of and developing new criteria and approaches. Popular theatre is an important way of having people come to grips with reality, of ensuring self-reliance is not concerned with romanticizing a cultural heritage, but with developing and implementing realistic and enlightened policies.

What of the other main forms of mass communication – radio and the press?

Radio is of importance because for the majority of people the spoken word is still the most important form of communication. Most people live in rural areas and have been brought up in an oral tradition, learning about their village and the people who live there through songs and stories, many of which will have been passed down by word of mouth for generations.

Against this background, radio is the only form of mass communication which is likely to have any impact on these communities. Although television undoubtedly would have an appeal, the sheer expense of the equipment needed to make and receive programmes places TV way beyond the local means.

According to the latest International Broadcasting and Audience research figures published by the BBC, in most African countries it is impossible to measure how many TV sets there are because they are so scarce. Radio sets, however, are common. In Nigeria, for instance, there are 17 radio sets per 100 of the population, and in Equatorial Guinea 27 per 100 of the population. This is not impressive when compared with Great Britain, where almost everyone has a radio set, but they are impressive figures in countries where it is usual for a village to use one set communally. People gather to listen to the news bulletin or a particular programme.

In the Pacific and Caribbean regions, where village and tribal links are not so strong, many more people own radios personally: for example, in Fiji and Jamaica more people own radio sets than in Portugal (in Fiji 61, and in Jamaica 40, per 100 of the population).

The great advantages of radio over other forms of communication are flexibility and comparatively low cost. Transistor radios are within the means of most people who earn a wage. Also, it is quite cheap to make a programme. One person with a portable tape recorder can make perfectly adequate programmes.

In countries, where most people have little or no access to formal education, neither newspapers nor books will have an impact, and both these media are comparatively expensive to produce and to distribute in remote areas where transport and communication lines are unreliable. The radio is used instead.

The Gambia, in West Africa, uses radio extensively to inform and educate in a wide variety of subjects, including technology and science. The Gambia, as with many African countries, has many different local languages, and this is an added reason why printed matter is difficult and expensive to produce. This presents little problem for radio. The national radio station divides its output during the day into four main sections, and carries material in all the local languages, and in English.

Programme makers can be very mobile, collecting material from remote areas and involving local people in programming. This generates much interest in the output of a radio station. In the Gambia, for instance, two most successful programmes are one for farmers, which includes interviews with Gambian farmers, and a self-help development magazine, called *Tesito*, which also includes interviews with local people running such schemes.

This approach to programming is common in the tropics. Fiji Broadcasting divides the day up to allow programming in Hindi, English and Fijian. And it will send programme makers out into the communities with portable equipment.

This ability to control programming effectively is the main reason why radio is not going to be superseded by TV in the near future. Fiji, for instance, could afford a TV service, but has chosen not to because it would not be able to afford to produce all its own programmes. The government see that Fijian culture would suffer if air time was taken up by cheap imported American films and videos.

Radio is a powerful medium. Its impact can be considerable because it makes great demands on its audience. As the listener has to supply his own mental images, a good deal of imagination is required, and there is considerable audience involvement. Listeners to the BBC World Service often identify strongly with the announcers: a retired BBC announcer, Peter King, has newspaper clippings from African papers to prove that people in Ghana have named their children after him!

By and large, the press in the TW cannot handle a science story if it has not been interpreted by the overseas press or news agency. This is becoming less true, as the regular presentation of science is coming to be accepted as a normal function of the leading third world newspapers or popular magazines.

By contrast the Chinese Science Journalists' Association (CSJA), founded in 1981, has about 800 members, all editors and reporters in the mass media. Two-thirds have a scientific or engineering background, and the remainder have been trained in journalism. The majority are college and university graduates.

What is needed urgently is to train the press and the scientists to get to know each other. An interesting pioneer example was when about 60 journalists, publicity experts and agricultural scientists met early in 1985 at the International Institute of Tropical Agriculture (IITA), in a three-day workshop on the role of the mass media in food production in Africa. It was designed, as IITA director-general Ermond H. Hartmans put it, to 'create new pathways between the research station and the farm' by establishing a partnership between science and journalism.

There is much good research carried out in universities, and in publicly-funded research institutions. Most TW countries have research departments dealing with such basic topics as agriculture, fisheries, forestry, and health. And some individual countries host international research institutes, for crops such as cocoa, palm oil, rice or rubber. There are good stories in each of these. Journalists need to be made aware of them.

Special mention must be made of the rural newspaper, with a content and style linked with the specific information needs of the community, much as is the English local paper, and read only locally. In Africa, such papers are recent: they began in Liberia in 1963 and in Niger in 1964, in the form of mimeographed bulletins concerned with literacy. In 1972, following a survey of 11 African countries which indicated the practicability of a rural press where printing facilities and basic journalism skills were available, and where there was an ongoing literacy or rural development campaign, Mali launched a rural paper, *Kibaru*, under the direction of the National Agency of Information and of the daily paper, *l'Essor*. This followed certain guidelines suggested by experts, and the paper soon had a circulation of 10,000 copies.

A UNESCO survey published in 1981 states that at present not one African rural newspaper is economically autonomous, and their future expansion cannot be left to chance (UNESCO, 1981). However, throughout Africa there do exist training and research institutes for rural journalists, and the papers do offer a field for the expression of science in relation to local needs. This should be linked with a national scientific and technical information service.

CHAPTER 6

THE SCIENCE CRITIC: A CASE FOR NEW CREATION

... the critic steps back from the object of perception in order to 'get closer to it' (focus, clarity, intelligibility are factors of direct access, of nearness to relevant phenomena). He establishes and organises distance in order to penetrate. He widens or narrows the aperture of vision so as to obtain a lucid grasp. This motion – we step back to come nearer, we narrow our eyes to see more fully – entails judgement . . .

Steiner, 1984

Our advanced technological society is rapidly making objects of most of us and subtly programming us into conformity to the logic of its system. To the degree that this happens, we are also becoming submerged in a new 'culture of silence'.

Shaull, 1983

The old order is dying: a new order begins to emerge. But what will emerge will depend on our re-assessment of strategy and tactics in so far as our scientific efforts are concerned. In this rapidly changing period, we must seek to develop the science critic, for he is a being necessary to help us move forward to the new level of integration.

I first made published reference to the need for a science critic in 1974 (see 'Popularization of Science' in the British Association supplement to *Nature*, vol. 250, no. 5469, p. 752). June Goodfield in her *Reflections on Science and the Media* (1981, AAAS, Washington, DC), believes 'present circumstances require a new group of people, outside the scientific profession yet looking at it critically.' She quotes Dr Louis Thomas on the need for 'a highly professional, analytic, meditative scrutiny of how (scientific) work is done; why; whether good, bad, or trivial; and what is its inner meaning.' It is not the same job as the daily reporting of science. For

Goodfield it is ' "a critic" in the accepted, old-fashioned meaning of the term' (p. 95).

I do not think we can discuss this only with scientists. By and large they are idea-bound, seeking to further their own special needs, unable to see clearly the needs of their scientific neighbours. They cannot provide regular, informed comment across a broad area of linked interests. And perhaps it is not their business to see the whole picture.

This should be one of the tasks of the science critic – to see the whole picture. But first I attempt to define a science critic by what he might do.

Herbert Read, in his *Contemporary Art*, pointed out that there is no unity in the modern art movement anywhere in the world, but rather a diversity that reflects the fragmented nature of our society. He sees two predominant tendencies that may in the end prove complementary; these are the search for formal harmony, for an equilibrium that compensates for our spiritual chaos, and a desire to express the chaos itself in images of uncompromising reality. The situation is somewhat similar in the sciences.

The first quality required by the science critic is to have an overall view. And this is more likely today than it was a generation ago.

In the book I have quoted, Read also points out,

The whole history of art demonstrates . . . that the particular emotions and particular outlook of a period always seek to find their particular forms. Similarities exist betweeen the form of similar phases of history, but as history never repeats itself exactly, such forms of art remain similarities, and are never identical. Criticism is the science of their classification. It is part of the business of the art critic to reach for such similarities . . .

This classification of similarities, not only through past history but also in terms of living history is to be a function of the science critic. This is most difficult. But the historiography of science – following the basic work of men such as Sarton, Koyré, Butterfield and Needham – is being expressed in the new chairs and depart-ments of the history of science. Of course, 'living history' is really a contradiction in terms. As we are entangled in the present it is not possible to assume easily an assured objectivity.

The science critic must also see the future. But this is difficult when history lies before us and we must perforce see its unfold-ing in terms of an incomplete present.

Of course, the science critic will be concerned with the integrity of science. Statements in a survey made in 1964 by the Committee on Science in the Promotion of Human Welfare of the American Association for the Advancement of Science on the effects of the rapid expansion of science are still valid. One conclusion was that 'under pressure of insistent social demands there have been serious erosions in the integrity of science. This situation is dangerous both to science and to society.' The erosion of integrity of science was defined as compromising the ' . . . internal factors of science – the methods, procedures, and processes which scientists use to discover and to discuss the properties of the natural world which have given science its great success . . . .' The Committee did not use the term 'integrity' to refer to the behaviour of scientists, which is dependent on 'the system of discourse in which scientists must operate'. What the Committee was saying was that the integrity of science could not be ensured by the scientists alone:

> the continued strength of science cannot be taken for granted
> . . . Scientists are human beings, and science is a part of culture
> . . . Can the very success of science and its closer interaction
> with the rest of our culture lay it open to the influence of new
> and possible alien points of view which derive from other
> sectors of society: military, business or political?

Further, the Committee pointed out that

> in a number of instances individual scientists, independent
> scientific committees, and scientific advisory groups to the
> government have stated that a particular hazard is *negligible*,
> or *acceptable* or *unacceptable* without making it clear that
> the conclusion is not a scientific conclusion but a social
> judgment.

Their comment was, 'To arrogate to science that which belongs to the judgment of society or to the conscience of the individual inevitably weakens the integrity of science.'

I have dealt throughout this book with the need to replace the concept of popularization with that of the public understanding of science and the public appreciation of its impact. Here is another function for the science critic. He must seek to bring what we see in science around us into some relationship with the non-science things we see.

This is a subtle and difficult task. It is no mere mechanical adding together of A and B. What is required is rather something more in keeping with the reply by the German playwright Bertolt Brecht to the following question: 'Ought not the actor try to make the man he is representing understandable?' Brecht replied, 'Not so much the man as what takes place. What I mean is, if I choose to see Richard II, I don't want to feel myself to be Richard II, but to glimpse this phenomenon in all its strangeness and incomprehensibility.'

The science critic should help the non-scientist to penetrate more deeply so that the latter, too, might be able to enjoy the poetry of scientific experience. But to do this the science critic must have a warm sympathy for his fellow-beings. The popular presentation of science requires that it should be intelligible to people; to ensure this the presenter must take over the people's own forms of expression and enrich them. Science must not only be made popular, but it must be seen to be becoming popular in terms of public acceptance.

I summarize what the functions of a science critic might be. First, to see the whole picture; second, to see the future because of what he knows of the past; third, to classify the similarities in scientific experience; fourth, to uphold the integrity of science; fifth, to interpret science; sixth, to communicate science, so that people understand its poetry and cease to have fear of it.

Thus far I have attempted to make out a case for a new type of communicator to cope with a situation in which the politicians, and we who elect them, continue to think in idea-tight compartments, although the beliefs, values, and institutions which nourish us are subject to rapid change. The new type of communicator, a poet and a visionary, can help us to meet the flight from reason originating from the false belief – and one that the popular communicators have failed to counter – of the threat of a blind, self-motivating, all-possessing, juggernaut of science and technology.

Where is this generalist I call the science critic to come from? Clearly not the scientist who prepares an annual review of scientific progress, for he is too narrow.[1] Nor is he the information officer who has a clearly defined task and is mission-oriented. Nor is he the science writer, although he is more likely to emerge from this category of communicator than from the others, mainly because of his imposed breadth of interest.[2]

I believe we shall have to create our science critic in another way. A curriculum might include the following: a course in general science to get an understanding of the linking concepts now being formulated; a course in the history and philosophy of science and technology; a course in the meaning and significance of the arts; a course in the psychology of communication and the nature of creativity; a course in the techniques of communication.

The aim is to provide a solution to what the science fiction writer Arthur C. Clarke describes in his *Profiles of the Future* as 'the great problem of finding a single person who combines sound scientific knowledge – or at least the *feel* for science – with a really flexible imagination.' Or, as the French poet Guillaume Apollinaire put it:

> For the theatre should not be a copy of reality
> It is right that the dramatist should use
> All the mirages at his disposal . . .
> It is right that he should let crowds speak inanimate objects
> If he so pleases
> And that he should no longer reckon with time
> Or space
> His universe is the play
> Within which he is God the Creator
> Who disposes at will
> Of sounds gestures movements masses colours
> Not merely in order
> To photograph what is called a slice of life
> But to bring forth life itself in all its truth . . .

My kind of science critic will, I suspect, aim at this too: to be his own creator so that he can bring forth life itself in all its truth.

APPENDIX 1

PERFORMERS AND AWARDS

To say that a journalist's job is to record facts is like saying that an architect's job is to lay bricks. True, but missing the point. A journalist's real function – at any rate his required talent – is the creation of interest. A good journalist takes a dull, or a specialist situation, and makes the readers want to know more about it. By doing so he both sells newspapers and educates people. It is a noble, dignified and useful calling.

Nicholas Tomalin, *Sunday Times*, London, 17 February 1963

(i) NATIONAL AND INTERNATIONAL ASSOCIATIONS OF SCIENCE WRITERS

Science writers are a product of this century. Science writing as a fulltime occupation was born between the two world wars. It grew to adolescence immediately following the second war, nurtured by the growing importance of the application of science in that war: in, for example, the production of the successful death-dealing atom bomb, the spectacular success of radar, the significance of operational research, and the effectiveness of early rocket technology as shown in the V1 and V2 weapons.

Much of what the public in any country learn about science is provided by the science writer (and I use the term to cover the science communicator in the different media). He is recognized today as performing a key role in acting as 'an interpreter' of science.

Because of the way in which science writing was born, most science writers are not trained scientists, either in the natural or social sciences, but are primarily good reporters. This is becoming less true so far as science is concerned; the number of science graduates in journalism is increasing. Although most science writers cover a wide range of disciplines, a growing number are becoming specialists – experts, in fact – in fields such as the environment, nuclear energy, space and the oceans. As such they are concerned not only with making clear what the research project means in purely scientific terms (e.g. in explaining the structure of DNA), but also in assessing the social significance of the research.

Each major industrialized country has an organization, or organizations, of science journalists.

(a) The United States

The firstborn association was conceived in 1934 in a bar of a Philadelphia hotel during the annual meeting of the American Philosophical Society (Perlman, 1974). The founder-fathers were three journalists, pioneers in the communication of science. They were David Dietz (*Scripps-Howard Newspapers*), William L. Lawrence (*New York Times*), and Robert D. Potter (*New York Herald Tribune*). The National Association of Science Writers (NASW) became fullborn later that year in Washington during a meeting of the National Academy of Sciences. Almost all fulltime science writers in the country became members: there were 12 charter members.

In the *Scripps-Howard News* in 1964, David Dietz tells the story somewhat differently, but with the same cast.

> One pleasant afternoon in April of 1934, William L. Laurence and I were strolling down Chestnut Street in Philadelphia, returning to the Benjamin Franklin Hotel from a session of the American Philosophical Society. 'Bill,' I said, 'Roosevelt has organized everybody and his brother into some kind of an association under the N.R.A. Why don't we for the National Association of Science Editors?'
>
> Lawrence thought the idea was excellent, but since at that time he did not possess the title of science editor, suggested we call it the National Association of Science Writers. In the hotel lobby we bumped into Robert D. Potter who was then on the staff of Science Service. He was equally enthusiastic over the idea.
>
> The following week our little group of science writers was in Washington at the annual meeting of the National Academy of Sciences. Preliminary organization of the NASW took place at a dinner in the Cosmos Club on April 25. The first formal meeting was held in Cleveland on June 13. Our group was in Cleveland that week to report the annual meeting of the American Medical Association. The NASW had exactly 12 members.

Dietz, who died aged 87 in December 1983, was elected the first president, as the first American newspaperman to have the title of science editor. He has been one of four reporters who, in 1922 at the annual meeting of the American Association for the Advancement of Science for the first time covered a scientific meeting seriously on a national scale. Encouraged by B. W. Scripps, founder of the *Press*, he was, also, the first American newspaperman to visit Europe with an exclusive assignment to write about science by covering the 1923 meeting of the British Association for the Advancement of Science. He met such distinguished figures as Sir Oliver Lodge, Lord (then Sir Ernest) Rutherford and Sir Arthur Eddington.

Today, there are about 600 active members – writers, editors and producers employed by newspapers, magazines, radio and television stations. There are another 500 plus in business, academic and government public relations, making a total of about 1200. The NASW is an influential body, with a fine record of achievement in creating a corpus of science writers with their own ethos and ethic, providing a link between scientists and the public. Its aim as defined in its constitution is to 'foster the dissemination of accurate information regarding science through all media normally devoted to informing the public.'

The NASW performs a unique function in liaison activities with policy-making bodies, such as the AAAS, of which it is a member society with nominated representatives on relevant committees. One of the two annual meetings of the NASW is organized to coincide with the Winter meeting of the AAAS, and the other with the June meeting of the American Medical Association.

The NASW influences procedures for the release of news by scientific bodies and government agencies: for example, after it protested officially to the Food and Drug Administration about a rule limiting the reporting of critical information about new medicines the rule was modified. The NASW plays a part in 'educating' the science writer and bringing him into contact with the scientist through background briefings, usually a day of lectures and discussions on new areas of research. It sponsors luncheons and dinners at meetings of scientific societies.

Research into aspects of the dissemination of science news is supported by the NASW. In 1956, 1957 and 1959, the Rockefeller Foundation provided substantial grants for basic studies on the impact of science news on the public. The results appeared in *Science, the News and the Public* and *Satellites, Science and the Public*. Other studies have sought the opinions of editors on science news and its place in the media, and of scientists on the adequacy of the coverage.

Special aids for members are provided by the Vocational

Committee, a job placement service, and the Freelance Committee, which prepares a specialist guidebook and an index-by-speciality directory of freelance members. There is a quarterly *Newsletter*, a quarterly *Clipsheet* which brings together science stories that have appeared in the press, a Handbook of 'Science News Conventions,' first published in 1968, giving guidelines for press arrangements at scientific meetings, and an annual membership list.

In 1959, the NASW established an independent, non-profit foundation, the Council for the Advancement of Science Writing, with a grant of $60,000 from the Alfred P. Sloan Foundation. This emerged from the nationwide surveys already mentioned, which had revealed a deep interest by the public in science but a lack of specific knowledge. The aim of the CASW is to improve public understanding through more effective use of the mass media. Activities include on-the-job training programmes for general reporters on papers that do not have science writers to give them some experience and facility in covering science, and opportunities for science writers to engage in scientific research in laboratories during the summer vacations.

At the Massachusetts Institute of Technology (MIT) in the 1983/84 academic year, there were eight Vannevar Bush Fellows, attending the first of the science journalist programs, directed by Victor K. McElheny, a seasoned science reporter. The Fellowships were funded by the Alfred P. Sloan and Andrew M. Mellon Foundations, and each Fellow received a stipend of $16,000. Their age range was from the middle-20s to the late-30s. There were five women and three men, covering a wide range of media experience. Although the eight agreed in general that 'the fellowship had been a remarkable experience,' there were inevitably certain 'start up' difficulties: for instance, they felt that they were too isolated from the faculty researchers.

Richard Saltus, one of the fellows, felt that the time he spent in

a molecular biology lab was probably the most valuable single part of the program. Getting in was the most difficult part, partly because the research was politically sensitive (developing gene transfer techniques that have implications for gene therapy) and partly because many of the researchers simply didn't like the idea of a reporter around. I spent many weeks talking my way in. Richard Mulligan, the head of the lab, was elusive but eventually agreeable. His graduate students and post-docs were skittish, and it wasn't until I had talked to nearly all of the 14 researchers over a period of weeks that they voted to let me in (and only after I had agreed to ground rules that would have been unacceptable under ordinary circumstances).

Thus, he had to allow a review of anything he wrote, 'which made sense only because the chief value was to be the experience rather than the product' (Richard Saltus, 'First Bush Fellows Review Year's Activity', *NASW Newsletter*, vol. 32, no. 3, July 1984, p. 9).

The 1984/85 Fellows – four women and four men – studied energy, environmental hazards assessment, artificial intelligence, information theory, and the nature of scientific evidence. There are other programs at Stanford, Drexel, Duke, Johns Hopkins and California (at Santa Cruz) universities.

(b) Great Britain

The Association of British Science Writers (ABSW) is the second-born. Its birth arose from an initiative – one of many – by J. G. Crowther, who in the war years, about 1942, had set up a Society for Visiting Scientists (SVS), formally established in 1944, to help the numerous scientists from many countries who came to London from Nazi-dominated Europe. The SVS had provided a centre where journalists could meet scientists, and an environment to foster understanding relations between 'those who were best able to make their needs, and the needs of science, known and understood through the press and other media of communication to the public' (Crowther, 1970).

In 1947, Crowther called a meeting at the SVS to discuss the formation of a science writers' association. Those present, who agreed to form the ABSW, were Ritchie Calder (*News Chronicle*), J. G. Crowther (*Manchester Guardian*) who was elected chairman, William Dick (*Discovery*), A. W. Haslett (*The Times*), and Maurice Goldsmith (*Reynolds News*) who was elected secretary/treasurer. They understood the social role of science, and set about attempting to raise standards of science writing in the mass media. The aims included:

1. The establishment of science writing as a definite profession with appropriate status. 2. Co-operation to the advantage of members, for example in the passing-on of requests for articles and books which cannot be accepted because of other engagements, to those in a position to carry them out. 3. The improvement of the opportunities of members to receive science news from official institutions and laboratories, learned societies, industrial laboratories, etc. 4. The increase in space in the press for science. 5. The improvement of rates of remuneration for science writing. 6. The organization of meetings to exchange views on matters of common interest,

and attempt to form agreed views on the problems that affect science writers.

(Crowther, 1970)

We recognized that not any or all of these aims could be realized without great effort. Today, the five pioneers have become 264 (210 full members, 34 Associate, 15 Academic, and 5 Life), and there is no newspaper or magazine concerned with science and technology without a member of the ABSW. It is recognized as the voice of science writers in the country, and there are regular meetings with the leading scientific policy makers, such as the president and council of the Royal Society, and the chairman and members of the Research Councils.

Three of Britain's most prestigious science bodies – the Royal Society, founded in 1660, the Royal Institution, 1799, and the British Association for the Advancement of Science, 1831, set up in November 1985 a joint committee on the public understanding of science. Its purpose is to keep under review and promote activities aimed at increasing the levels of scientific and technological understanding of the public.

The first chairman is Sir George Porter, president of the Royal Society, recently director of the Royal Institution, and 1985/86 president of the BAAS. Members include representatives of the scientific community, the media, industry and commerce, engineering, education (formal and non-formal), public bodies, and publishers.

The new body arose from a proposal by an *ad hoc* Group of the RS which in 1985 reported on the public understanding of science. The Chairman was Sir (then Dr) Walter Bodwer.

In 1985, also, the Royal Society began in collaboration with the Association of British Science Writers a series of press briefings, which take place before every major discussion meeting at the RS. The subjects covered have been on the frontiers of new knowledge dealing with, for instance, oncogenes, free radicals, evolution of DNA sequences, and the electron microscopy of surfaces. At the briefings, the leading exponents in the field are providing many leads for features, topic ideas and names of contacts. They are building, also, a special relationship between the scientists and the journalists. By and large, the six aims proposed almost 40 years ago have been met. Unfortunately, what we lack still is an environment in which 'a scientific temper' can flourish.

(c) France

The Association des écrivains scientifiques de France (l'AESF) is the third born. It arose from an initiative by two senior UNESCO staff, then in the Department of Natural Sciences, concerned with the communication of science. They were François le Lionnais and Maurice Goldsmith. M. Louis de Broglie, secrétaire perpetuel of the Académie des Sciences, agreed to be the président d'honneur, and in June 1950 the AESF came into existence.

In his Inaugural Lecture in January 1951, M. de Broglie said it was essential to keep public opinion in touch with scientific developments and their repercussions on thoughts, life and the future. But that the problem was very delicate because there was

a conflict between the legitimate and irrepressible desire to make known the discoveries of science and the scientist's conscientious scruples lest he obscure the truth or make claims which are not absolutely borne out by his scientific research . . . . By bringing together scientists and qualified science writers, the Association is bound to enhance popular scientific writing, which through the competence of its authors will instruct without distortion and raise the intellectual level of the reader.

He urged that encouragement be given only to 'works of unquestionable merit, works that will draw people to admire science for the right reasons. If it succeeds in doing that, it will have rendered a great service both to science and to French thought' (Louis de Broglie, 'Scientists and the Problem of Popularizing Science', *New Perspectives in Physics*, Oliver & Boyd, Edinburgh & London, 1962, pp. 184f. trs. by A. J. Pomerans).

Under its first chairman, Le Lionnais, engineer, mathematician and distinguished Resistance fighter, the Association developed a distinctive style which has secured for it a special intimacy with the media, the scientific community, government and industry. In May 1985 there were 281 members (246 active and 35 honorary).

In 1958, the AESF began a series of discussions on the nature of 'la vulgarisation scientifique.' Let me digress. 'Science popularization' is the usual translation, but this is a crude, limiting expression. 'Vulgarisation' has as a root the Latin noun *vulgus*, which means not only the people, the great multitude, but comes also from the verb meaning to make common to all, to communicate, to make accessible to all. The English word *vulgar* has two distinct meanings: it may be 'the common language of a country,' or 'a person not reckoned as belonging to good society.'

Le Lionnais defined the term 'vulgarisation scientifique' as meaning all activity concerned with explaining and spreading scientific and technical knowledge, but with two conditions: one, that these are undertaken outside of any official educational programme, and, two, the explanations are not designed for specialists. For him, a *vulgarisateur des sciences* is a term of social merit. In fact, when the AESF was set up, we wanted to include the word *vulgarisateurs* in the official name. But there was opposition to this from such eminents as Academician Jules Romains. The debate continues.

> It is clear that the word *vulgariser* sounds bad, not because of its Latin origin, but through its evident relationship with 'vulgarity.' Another word must be found: nobler, more capable of arousing enthusiasm, to express the indispensable contact between the scientist and his fellow citizens engaged in such very different activities,

commented physicist, Louis Leprince-Ringuet, member of the Académie Française and of the Académie des Sciences, in 1983 (Leprince-Ringuet, 1983).

For almost 30 years, the AESF has organized a biennial series of meetings on aspects of *la vulgarisation scientifique* (v.s.). These are probably the best discussions of their kind anywhere. They bring together the eminents and the practitioners from all fields, and probe all aspects of this basic social complex. Thus, at the first in February 1958, the main questions posed were: 'What do we understand by the term v.s?' 'Is v.s. work of national interest?' In May 1960, the main theme was, 'Can we *vulgarise* all of science?' and the questions were: 'What can we *vulgarise* in science?' 'Who is *vulgarisation* aimed at?' 'What difficulties does v.s. encounter, and how can we overcome them?' 'Do we have the right to *vulgarise* all of science?' In 1962, the discussion was on, 'Who should *vulgarise* science?'; in 1964, the theme was, 'Sensationalism in v.s.'; in 1966, 'Education and v.s.'; and in 1968, 'The language of v.s.'

Other AESF activities include a number of regular publications, and the organization of a Service of Consultations on Science by Telephone (SCST), which is provided without charge for science writers. A list is available of scientists and technologists in a variety of disciplines who have agreed to answer questions from *bona fide* journalists. The AESF awards, also, the Jean Rostand Prize – annual until 1985, when due to rising costs it became biennial – for the best book on popular science written by an author under forty years of age.

(d) Other national associations

There are now national bodies in almost all industrialized countries, and in a very few third world countries. There is a strong European regional grouping.
The European list is as follows:

Austria
Klub für Bildungs und Wissenschaftjournalisten (founded 1971) c/o ibf, Reichsratsstrasse 17, 1010 Vienna.

Belgium
Groupement des journalistes professionels scientifiques (1967) c/o 31 Avenue Bel Air, 1181, Brussels.

Denmark
Danske Videnskabsjournalisters Klub (1976) Henrik Hertz Vej 13, 2920 Charlottenlund.

Federal Republic of Germany
Technisch Literarische Gesellschaft e.V. Teli (1929) Journalisten-Vereinigung für Technisch-Wissenschaftliche Publizistik c/o Klaus Goschmann, AUMA, Lindenstrasse 8, 5000 Cologne 1.

France
Association des écrivains scientifiques de France (1950) c/o M. Charles Penel, 129 rue de l'Abbé-Groult, 75015 Paris.
Association des Journalistes scientifiques de la presse d'information (1971), 29 rue de Louvre, 75002 Paris.

Ireland
Irish Association of Science & Technology Journalists (1976) 303 Martello Estate, Portmarnock, Co. Dublin.

Italy
Unione Giornalisti Scientifici Italiani (1966)
c/o Federazione Nazionale Stampa Italiana, Associazone Lombarda, Viale Montesanto 7, 20124 Milan.

Netherlands
Nederlandse Verenigung van Wetenschapsjournalisten (1985) c/o Huub EGGEN, Hardeenkosmos, PO Box 108, 1270 AC Huizen.

Norway
Norsk Forening for Forskningsjournalister (1983) c/o Per Torbo, R Kul, NRK, 0340 Oslo 3.

Philippines
Science and Technology Journalists of the Philippines Inc. (STJPI)
President – Leo Deocadiz, *Business Day*
807 E. de los Santos Avenue, Quezon City, Metro Manila.
Formed in 1983, STJPI has a popularizing programme, including
workshops and the weekly Thursdays at 10.00 a.m. Isaahan sa
N.P.C. (Tea at the National Press Club) for informal discussions
with leading scientists.

Spain
Asociación Española de Periodisma Científico
c/o Centro Iberamericano de Cooperación, Ciudad Universitaria,
Madrid 3. President – D. Manuel Calvo Hernando.

Sweden
Svenska Forenigen for Vetenskapsjournalistik (1977) P.O. Box
14131, S–104 05 Stockholm.

Switzerland
Schweizer Klub de Wissenschaftsjournalisten (1974) Club Suisse
des Journalistes scientifiques c/o Tages-Anzeiger, Werdstrasse
21, CH-8021 Zürich.

United Kingdom
Association of British Science Writers (1947). The Executive
Secretary of ABSW is Ursula Laver, Information Officer of the
BAAS.

The Biochemical Society, through its executive secretary,
provides the services of the society, including a well equipped
meetings room.

The administrative address of the ABSW is c/o New Scientist,
1–19 Commonwealth House, New Oxford Street, London W1A
1NG.

The *European Union of Scientific Journalists' Associations
(EUSJA)* was founded on 8 March 1971, as a union of existing
national associations. The permanent secretariat is provided by the
European Community (c/o Dr Ernest Bock, EEC-DG XII, rue de
la Loi 200, B-1049 Brussels, Belgium). English is the official
language. An occasional Newsletter is produced and distributed to
European science and technology writers.

The *Israel Association of Science & Medical Journalists* is an
Associate Member of the EUSJA, and its representative is Kapai
Pines, c/o MADA (Hebrew Science Journal), P. O. Box 81, Jeru-
salem, 91007.

The *International Science Writers' Association (ISWA)* is based
on individual not corporate membership. It is an organization

formed in 1967 'in response to the increasingly international scope of science popularization and technical communications'. The aim is to provide 'an ever wider circle of contacts'. It has had a difficult existence since the idea was expressed in 1965 at a meeting in London, at which a committee was formed including Dennis Flanagan (*Scientific American*), John Maddox (*Nature*), W. A. Osman (then ABSW chairman), and G. Rattray Taylor (freelance). The first plenary meeting was held in Montreal, funded locally by the *Montreal Star*, in May 1967, coincident with the world's fair, Expo 67. The first full meeting of the Council was held in the home of Fred Poland. The current president is James Cornell, of the Harvard-Smithsonian Center for Astrophysics, Cambridge, Mass. 02138. The secretary-treasurer is Howard Lewis, 7310 Broxbourn Court, Bethesdall, Maryland 20917, who is, also, editor of the informative occasional *Newsletter*.

ISWA, which now has 111 members in 21 countries, is much concerned with stimulating international symposia. Cornell represented the NASW at the 1977 Congress of Ibero-American Science Journalism held in Madrid, under the sponsorship of the Spanish Institute of Hispanic Culture. He comments,

> Unlike similar meetings in the United States, where the main concerns often seem to be job opportunities and the location of the open bar, gatherings of Hispanic journalists are highly politicized, concerned as much with international rivalries, the impact of multi-national corporations, and the role of the media in shaping national destinies as with the problems of science writing. This, in part, stems from the nature of Latin American journalism.
>
> In those countries with controlled presses, the journalists tend to echo government policies and goals. By contrast, in other countries where urbanization, industrialization and technological development are all proceeding at a breakneck pace, the science journalist has a position akin to the American space writers of the 1960s, recognized as powerful social spokesmen who are promoting national aspirations and articulating national self-perception.
>
> Cornell, 1977

There had been a previous international meeting of science writers from the Hispanic world in Caracas in 1975, at which Cornell proposed the creation of Inter-American Science Writers' Association to link the journalists of North and South America. The proposal was not adopted formally, 'probably because many nations still regard it as somewhat unseemly to join the United

States in any formal organization,' wrote Cornell. On the last day of the Caracas meeting, the Venezuelan delegation proposed the formation of a World Union of Science Journalists, but that has not moved forward at all.

The Ibero-American Congress in Madrid in 1977 was organized by the *Associación Ibero-americana da periodístas especializados y técnicos (AIPET)*. This body came into formal existence on 21 October 1980. The president is M. Jorge Cometta Manzoni, and the headquarters address in Santiago des Estero 286, 6 piso-oficina 1, 1075 Buenos Aires, Argentina.

There is a *Union des journalistes africains*, founded in March 1977, whose secretary-general is M. Cheick Mouctari Diarra, c/o Ministère de l'Information, Boîte postale 141, Bamako, Mali.

The *Chinese Association for Science Writing (CASW)*, founded in August 1979, has a membership of 7,000, of whom 3,400 are scientists, scientific workers, teachers, doctors, and cultural workers; 600 are engineers and technicians in industry and transport; 1,400 are in journalism and the press; 100 are government officials and officers in the armed forces. Most of them write popular science books and articles in their spare time. Only a few are fulltime science writers.

Xie Chu, editor of *Aerospace Knowledge Magazine*, Beijing, as a member of the Standing Council of the CASW, wrote that

> jointly with several other national organizations it initiated in 1981 popular science awards on a grand scale across the nation. My article on Chinese aviation history, The First Aircraft Designer and Aviator in China, won the national award for excellence in popular science writing.
>
> Recently, publication of popular science books has been booming. In 1978, 1175 were published, and in 1982 over 2000. At present, there are 120 Chinese popular science magazines on the mainland, and 54 popular science newspapers, the total circulation exceeding 17 million copies. In addition to these, a number of national and local newspapers, broadcasting and TV stations, have included special columns, pages and programs about science. . . .
> *Aerospace Knowledge*, a monthly magazine, issues 200,000 copies every month at home and abroad.

(ii) AWARDS – INTERNATIONAL AND NATIONAL

I like these award projects: they stimulate, they excite, they encourage, and they provide a demonstrable prize – usually financial. I regret I have never gained one, but I live in hope.

I have initiated a few, such as the Guinness Awards for Scientific Achievement in the Service of the Community, which ran for 10 years, rewarding achievers in the anglophone and francophone third world countries; the annual CASTME Awards for Science Teachers; and the recently begun Commonwealth Professional Associations' Awards for Innovation (in the natural and social sciences).

The awards for science communicators I detail below are known to me. There are others. I would appreciate further information.

International

The *Kalinga Prize* for the popularization of science is awarded annually by UNESCO. It was established directly as a result of my initiative in 1951. It was at a lunch in Paris with Mr B. Patnaik, of the state of Orissa, India, founder and president of the Kalinga Foundation Trust, that I suggested he could help in the communication of science at a popular level by providing funds for an annual award. He agreed, and provided a grant of one thousand pounds sterling to be given each year, with the condition that the award winner is required to travel in India.

My aim was to encourage the grassroots writers, to give them a necessary lift so that they would regard their difficult task, especially in the third world, as worthwhile. Unfortunately, at UNESCO my head of department thought otherwise. He argued that it was necessary to establish the award by giving it the imprimatur of a distinguished person for the first few years. In the list of laureates there are very few fulltime professional science communicators. I am sad: in this special context elitism is snobbism and is harmful. In any event, I have good reason to state that UNESCO would like to hand over the Kalinga Prize to another body: they find it too expensive and too awkward to administer. There is a great sigh of thanks-giving when each year, finally, a name or names, emerge to be rewarded.

Tyler Prize for Environmental Achievement. Established in 1973, it is the largest achievement award presented by an American institution and it is the largest ecology-energy prize in the world. Prizes to 1984 have totalled over $1m, ranging from $150,000 to $200,000 annually.

Citizens of all nations are invited to nominate individuals or institutions of any nation who have benefited humanity in the fields of ecology or energy. Persons eligible to make nominations include, but are not limited to: scientists in fields of biology, geology, chemistry; and environmental, petroleum and chemical engineering; National academies of science, engineering or their equivalent, and their members; Research institutions and their members. Nominees can be associated with any field of science. Nominated institutions can be universities, foundations, corporations or other types of organizations.

(Details from Dr Jerome B. Walker, Executive Director, The Tyler Prize, University of Southern California, Los Angeles, California 90089–4019, USA.)

McLuhan Teleglobe Canada Award, established in 1983, to honour the work of the communications philosopher, Marshall McLuhan. It is a biennial award funded by Teleglobe Canada, and administered by the Canadian Commission for UNESCO. It is open for nominations by national commissions or recognized organizations representing the member states of UNESCO. Nationals of any country, either an individual or a group, are eligible for an award of C$50,000 and a commemorative medal, for 'any work or action that will have contributed in an exceptional manner to furthering a better understanding of the influence exerted by communication media and technology on society in general and in particular on its cultural, artistic and scientific activities.'

National

Belgium
Glaxo Prize for Medical Journalists. Biennial award of BFr 200,000. Sponsored by the Association of Pharmaceutical Industry.

Canada
The Canadian Science Writers' Association (CSWA) has an annual awards ceremony, with prizes given by sponsors such as Upjohn, the Federal Department of Fisheries and Oceans, Ontario Hydro, Ortho Pharmaceutical, and Control Data Canada.

There is a rather complicated awards process as they cover 14 categories. Thus:

Magazines – Science & Technology, Science & Health, Science & Natural Resources.

Newspapers – with three categories as for Magazines.

Magazine and Newspaper writers may compete for one award for Science & Society.

Radio – News (items of 10 minutes or less), Features (10–30 minutes), Documentary (30 minutes or over).

Television – as for Radio.

Junior Award – is given for Magazine or Newspaper writers with less than two years' experience.

France

Prix Jean Rostand founded by the Association des écrivains scientifiques de France. Annual prize of Ffr 5000 for the best popular science book.

Italy

Glaxo Prize of L8m is awarded occasionally by the Glaxo Group of Verona under the auspices of the Unione Giornalisti Italiani Scientifici.
Unione Petrolifera Italiana prize for writing on scientific and technological problems in energy.

Sweden

Nils Gustav Rosen Prize for Science Journalism, equivalent to about US $6000, given annually by the Council for the Planning and Coordination of Research.

USA

AAAS-Westinghouse Science Journalism Awards presents annually five $1000 prizes – for writing in dailies with circulation over 100,000; under 100,000; general-circulation magazines; radio; and television broadcasting. (Details from Carol Rogers, American Association for the Advancement of Science, 1515 Massachusetts Avenue, N.W., Washington, D.C. 20005.)
Science-in-Society Journalism Awards made annually by the National Association of Science Writers for outstanding writing about science and its impact on the quality of life. The NASW gives these awards *without subsidy from any professional or commercial interest* to encourage the kind of critical, probing article that would not receive an award from an interest group. There are separate prizes of $1000, engraved medallions, and Certificates of Recognition for writing in each of three categories: Newspapers, Magazines, Television and Radio. Publishers and broadcasters of awarded material also receive Certificates of Recognition.
Nate Haseltine Fellowships in Science Writing are awarded by the Council for the Advancement of Science Writing in memory of a distinguished medical writer. They are funded by a grant from the

American Medical Association. In 1984, there were six Fellowships awarded, totalling $9000.

CASW Internships: two available annually for general-assignment reporters, journalism school graduates and others interested in becoming science or medical writers. A grant in 1984/85 from the Henry J. Kaiser Family Foundation to the CASW provided each intern with $325 per week for 36 weeks. Interns spent 18 weeks at the University of California – San Francisco, and then 18 weeks at University of California – Berkeley, working for the university news bureaux.. (Details from CASW, 618 N. Elmwood, Oak Park, Ill, 60302.)

Science Writing Awards in Physics and Astronomy given by the American Institute of Physics annually: prizes of $1500 each to a professional science writer and a scientist.

Grady-Stack Award for Interpreting Chemistry for the Public was established in 1955 by the American Chemical Society 'to recognize, encourage, and stimulate reporting directly to the public, which materially increases the public's knowledge and understanding of chemistry, chemical engineering and related fields.' Annual award consists of $3000, a gold medal, and a bronze replica of the medal.

Patrick M. Mcgrady, Sr. Memorial Scholarship Awards in Science Writing of the American Tentative Society (ATS): Awarded annually for an essay about a scientist embodying the ATS founding principle that the essence of science and commonsense is to regard present knowledge as subject to growth, addition and revision, therefore, tentative. The top award is $2500. ATS president is Alton Blakeslee, former Associated Press science editor.

Mass Media Science and Engineering Fellows: Each year the American Association for the Advancement of Science (AAAS) invites outstanding students in the natural and social sciences and engineering, preferably at graduate level, to apply for the Mass Media Science and Engineering Fellow Program. Fellows will work as reporters, researchers, production assistants and script consultants for ten weeks during the summer at newspapers, news magazines, radio and TV stations. They will have an opportunity to take part in news-making, to increase their understanding of editorial decision-making, and to develop skills in conveying to the public a better understanding and appreciation of science and technology.

The AAAS pay a weekly stipend of $300 and travel expenses. Minorities and handicapped people are especially encouraged to reply.

AAAS, 1333 Eighth Street, NW, Washington, DC 20005.

Great Britain
Glaxo Science Writers Fellowships: Glaxo Holdings in conjunction with the ABSW provide four annual Fellowships of £1,250 each for (1) the best article or series in a national or regional newspaper on a science subject; (2) in a trade or technical journal; (3) the best broadcast script and/or radio or TV programme; (4) the best entry in any medium on 'Improving Human Health in the 1980s'. (Details from: ABSW c/o Group PR Department, Glaxo Holdings, 6–12 Clarges Street, London W1Y 8DH, England.)

United Nations Educational Scientific and Cultural Organization. Revised regulations for the awarding of the Kalinga Prize

1 The Prize
The Kalinga Prize for the popularization of science was established by the United Nations Educational, Scientific and Cultural Organization (UNESCO) in 1951. It is an annual international award of one thousand pounds sterling, based on a grant to UNESCO from Mr B. Patnaik, of the state of Orissa, India, founder and President of the Kalinga Foundation Trust.

2 Candidates
The winner of the Prize must have had a distinguished career as a writer, editor, lecturer, radio/television programme director or film producer which has enabled him to help to interpret science, research and technology to the public. He is expected to have a knowledge of the role of science, technology and general research in the improvement of public welfare, the enrichment of the cultural heritage of nations and the solution of the problems of humanity. He should also be acquainted with the scientific activities of the United Nations, UNESCO and the other Specialized Agencies. He should preferably be proficient in English.

3 Gift facilities
Under the terms of the gift, the Kalinga Prize will enable the recipient to travel to India where he will be the guest of Mr B. Patnaik and of the Kalinga Foundation Trust. He will be provided appropriate facilities to familiarize himself with Indian life and culture, Indian research and educational institutions, and the development of India's industry and economy. He will also be invited to visit Indian universities and attend meetings of Indian scientific societies, particularly those of the Indian Science Congress Association.

While in India, the recipient will be asked to deliver lectures in English and take part in meetings, with a view to giving an interpretation to India of recent progress in science and technology or the social, cultural and educational consequences of modern science. Upon his return to his country, he is expected similarly to make India and its scientific achievements known by means of articles, books, lectures, radio/television programmes or films.

4 The Jury
The Kalinga Prize-winner is named by the Director of UNESCO on the recommendation of a jury of four members designated by him. Three members of the Jury from different countries of the world shall be designated on the basis of equitable geographical distribution and the fourth on the recommendation of the Kalinga Foundation Trust.

5 Presentation and selection
Every year the Director-General of UNESCO shall invite the National Commissions of Member States to nominate one candidate each, on the recommendation of the national association for the advancement of science or other science associations, or national associations of science writers or scientific journalists. Nominations or applications from individuals will not be accepted. Nominations shall be sent to the Director-General of UNESCO by 31 January of each year. Files shall include, in five copies each complete biodata on the candidate and a list of his published works and shall also be accompanied by five copies of his principal publications.

UNESCO
Laureates of the Kalinga Prize for the Popularization of Science

1952	Louis de Broglie	France
1953	Julian Huxley	United Kingdom
1954	Waldemar Kaempffert	United States of America
1955	Augusto Pi Suner	Venezuela
1956	George Gamow	United States of America
1957	Bertrand Russell	United Kingdom
1958	Karl von Frisch	Federal Republic of Germany
1959	Jean Rostand	France
1960	Ritchie Calder	United Kingdom
1961	Arthur C. Clarke	United Kingdom
1962	Gerard Piel	United States of America
1963	Jagjit Singh	India
1964	Warren Weaver	United States of America

1965	Eugene Rabinowitch		United States of America
1966	Paul Couderc		France
1967	Fred Hoyle		United Kingdom
1968	Gavin de Beer		United Kingdom
1969	Konrad Lorenz		Austria
1970	Margaret Mead		United States of America
1971	Pierre Auger		France
	Philip H. Abelson		
1972	Nigel Calder	ex aequo	United States of America
			United Kingdom
1973	Nil		
1974	José Reis	ex aequo	Brazil
	Luis Estrada		Mexico
1975	Nil		
1976	George Porter	ex aequo	United Kingdom
	Alexander I. Oparin		USSR
1977	Fernand Seguin		Canada
1978	Haimar von Ditfurth		Federal Republic of Germany
1979	Sergei Kapitza		USSR
1980	Aristides Bastidas		Venezuela
1981	David F. Attenborough	ex aequo	United Kingdom
	Dennis Flanagan		United States of America
1982	Oswaldo Frota-Pessoa		Brazil
1983	Abdullah Al-Muti Sharafuddin		Bangladesh
1984	Yves Coppens	ex aequo	France
	Igor Petryanov		USSR
1985	Sir Peter Medawar		United Kingdom

BRITISH NEWSPAPERS (DAILY/WEEKLY) AND MAGAZINES (WEEKLY) WITH SCIENCE WRITERS ON STAFF (SELECTED FROM MEMBERSHIP OF THE ABSW AS AT OCTOBER 1984)

Name	Function	Newspaper
Ronald Bedford	Science Editor	Daily Mirror
Adrian Berry	Science Correspondent	Daily Telegraph
Alan Burns	Medical Editor	Sunday Mirror
Alan Cane	Technology Editor	Financial Times
John Delin	Science Correspondent	Sunday Telegraph
Clare Dover	Medical Correspondent	Daily Express
David Fishlock	Science Editor	Financial Times
Oliver Gillie	Medical Correspondent	Sunday Times
Peter Large	Technology Correspondent	The Guardian
Robin McKie	Science Correspondent	The Observer
Peter Marsh	Technical Writer	Financial Times
Anthony Osman	Science Editor	Sunday Times Magazine
Matt Ridley	Science Correspondent	The Economist
Bryan Silcock	Science Correspondent	Sunday Times
Arthur Smith	Science Correspondent	Daily Mirror
Anthony Tucker	Science Correspondent	The Guardian
Jon Turney	Science Correspondent	Times Higher Education Supplement
Elaine Williams	Technology Writer	Financial Times
Michael Cross	Technology News Editor	New Scientist

Kos Herman	Science Policy Editor	*New Scientist*
Michael Kenward	Editor	*New Scientist*
John Maddox	Editor	*Nature*
Peter Newmark	Deputy Editor	*Nature*

NEWSPAPERS WITH WEEKLY SCIENCE SECTIONS

Newspaper	Section Title	Day of week	Date started	% own news	% other sources*	Circula-tion**
The Boston Globe	Sci-Tech	Monday	4/83	98	2	514,817
Chicago Tribune	Tomorrow	Sunday	4/10/83	33	67	1,116,403
The Cincinnati Post	Newscience	Friday	9/23/83	50	50	135,585
Columbus Citizen-Journal	Medic One	Monday	8/1/83	75	25	121,676
The Columbus Dispatch	Discovery	Sunday	9/4/83	50	50	352,842
The Commercial Appeal Memphis, TN	Future Currents	Sunday	11/28/82	33	67	280,880
The Dallas Morning News	Discoveries	Monday	8/1/83	70	30	328,332
Detroit Free Press	Science/New Tech	Tuesday	5/4/82	50	50	635,114
The Detroit News	Science/Computers	Thursday	5/83	95	5	650,683
The Evening Gazette Worcester, MA	Today/Science-Health	Thursday	9/83	60	40	85,273
The Globe and Mail Toronto, ONT	Science/Medicine	Monday	6/84	75	25	310,689
The Journal-news Rockland County, NY	Discovery	Thursday	10/82	25–30	70–75	43,772
The Miami Herald	Science/Medicine	Wednesday	1981	40	60	424,939
The New York Times	Science Times	Tuesday	11/78	99	1	910,538
Newsday Long Island, NY	Discovery	Tuesday	10/2/84			525,216
The Oregonian Portland, OR	Science	Thursday	11/3/83	Over 50	Under 50	289,600
The Plain Dealer, Cleveland, OH	Science-Health	Tuesday	9/83	60	40	493,329
San Jose Mercury News	Science & Medicine	Tuesday	11/16/82	33	67	290,109
The Washington Post	Health	Wednesday	1/9/85			718,842

*Figures are estimates supplied by editors
**Figures are daily averages from 1984 *Editor & Publisher Yearbook* Sections appearing on Sunday list Sunday circulation

EXAMPLES OF PRESENTATION

And the end men looked for cometh not,
And a path is there where no man thought,
So hath it fallen here.

Euripides, *The Bacchae*

In this twentieth century began the great 'leap into the cosmos' (Weiskopf, 1968). Relativity and quantum mechanics modified the classical theories to ensure the great revolution which has given us a place in our solar system, and new insights into our ignorance of the workings of Nature.

In the latter part of the century there has come the equally important revolution in biology based upon the unravelling of the genetic code. This is expressed in genetic engineering and biotechnology. And since the second war, 1939–45, science itself has come to be recognized clearly as having basic social impact.

These developments are expressed in the examples of popular presentation I have selected. I am obviously limited by space: that is why such important communicators in their day as James Jeans and Arthur Eddington do not appear. But those I have selected unveil and set before our eyes with simplicity and harmony the natural phenomena which shape us, and which in their effects on social and individual behaviour and our institutions, in their turn have affected science.

I continue in this chapter my 'autobiographical' approach by presenting some samples selected from my own writings, or from joint works, the product of blended intimacy. My reason for so doing does not arise, I am persuaded, from a false immodesty, or from an unwanted embarrassment at the possibility of commenting on some displeasing writing by a contemporary, or from singling out another for high praise. It arises from my desire to present a continuity and development in my writing which have led from straight reporting of science to seeking to express the social impact of science and technology.

My first book, *The Scientist and You* (1959) was edited by me. I fashioned the contents and organized the authors, all of whom were leaders in their fields. The distinguished scientist and advisor to governments, Sir Henry Tizard (1885–1959), wrote in a Foreword:

FOREWORD
Sir Henry Tizard G.C.B., A.F.C., F.R.S.

Sir Henry Tizard has been concerned for most of his life with higher education in science and technology, and with the application of science to national affairs, both civil and military. He has been Rector of the Imperial College of Science and Technology, London, Chairman of the Aeronautical Research Committee, Chairman of the Advisory Council on Scientific Policy, President of the British Association for the Advancement of Science, and a Vice-President and Foreign Secretary of the Royal Society.

This book is an experiment. Its purpose is to give boys and girls a wider vision of the influence of science on society than they are likely to get in the ordinary course of their school work. It consists of a series of essays by well-known men on some of the victories of applied science – victories which have had a far more beneficial effect on humanity than any victory won in war, but which, because they have been fought over a long period, and are still incomplete, are seldom recognized and honoured. The essays represent a selection only of such victories. Many more might be written. For instance the essay on the *Conquest of Space* deals with only space beyond the earth, not with the conquest of space on the earth. The book might have been entitled *Some Decisive Battles of the World*, except that this would have deceived the ordinary reader into thinking he would learn about a different kind of battle. *The Scientist and You* will not deceive anyone, and may induce some adults, who have not had a scientific education, to profit by what has been written primarily for a younger generation.

The book concludes with a section on the Careers which are open to young scientists today. Their number is legion. I open my copy of *The Times* today and find nearly a full page devoted to advertisements for chemists and physicists in a variety of industries; for professors and lecturers at universities at home and abroad; for pharmacologists, entomologists, medical officers, development and design engineers; instrument, electrical, electronic and civil

engineers; and so on. And that is not all, by any means. The Civil
Service, the greatest administrative service of any country, is crying
out for recruits with a scientific education; the defence of the
country depends vitally on the quality of the scientists and engineers devoted to it; and the schools are terribly short of science
teachers.

Indeed there is no profession that would not gain by having
scientifically educated men within its ranks. As the Duke of Edinburgh remarked at Uppingham School, the time is approaching
when no man can be called properly educated who has not
combined a real study of science with an education in the arts. But
no man can study deeply all the sciences and all the arts. It is
enough to be able to appreciate all the many triumphs of human
endeavour, and to decry none.

The demand for scientists was almost negligible in England at
the beginning of the century. I have consulted a copy of *The Times*
issued on my fifteenth birthday. There were no advertisements for
scientists. I have read another copy, which appeared when I took
my degree in chemistry at Oxford University. The only advertise-
ment for a scientist was for a Director of Research at the Cancer
Hospital, Fulham. There were indeed some industries even then
which had started research laboratories. One of these was the
famous firm of Brunner, Mond & Co., which kept a sharp look-
out for promising university graduates. Its research laboratory was
very small; the number of university graduates employed there
never exceeded six up to the outbreak of war in 1914. That does
not mean that only six had been recruited, because many men
passed through the laboratories to other positions in the company,
and subsequently became leaders of the chemical industry. Imperial
Chemical Industries now employ nearly 2,000 graduates on
research and development.

The schools of science at universities were then few in number
and ill-endowed. When Sir Norman Lockyer was President of the
British Association in 1903 his remarks on the neglect of science
roused a number of representatives of universities and university
colleges, and many other educational authorities, to put their views
to the Prime Minister and Chancellor of the Exchequer, and to ask
for an additional sum of £200,000 a year from the Exchequer. The
Prime Minister was then Mr. A. J. Balfour, who was a highly
educated man, judged by the Duke of Edinburgh's standard. He
had many close friends in universities and in the scientific world,
and was himself President of the British Association in 1904. So
he received the deputation with the greatest sympathy. He
acknowledged that England was far behind Germany in the application of science in industry. But he said that there was little point
in substantially increasing the number of students of science so

long as industry preferred to employ practical men who had been brought up to business in the old way at an early age. Can one say that he was wrong? The Chancellor of the Exchequer (Mr. Austen Chamberlain) added that it would be a misfortune if it were thought that it was the duty of the State to take upon itself the whole or main cost of the higher education of the country! In any case, he said, the country was not so well off that the Treasury could afford to be generous without having to place fresh burdens on the taxpayer. Income tax was then 11d. in the pound. Today the State is providing 34 million pounds for the general support of universities, of which it is estimated that some 20 millions goes to schools of science and technology; about 25 millions on technical colleges; and at least a further 10 millions for scientific research in universities – and the taxpayers appear to carry these additional burdens quite cheerfully.

The great change that has occurred within 50 years in public opinion and in the thoughts and actions of men in high positions, has been mainly accelerated by two devastating wars, which have made the powers of science obvious to all. Many people have said that war stops science, that it 'grinds up the seed-corn of scientific progress', and that the only good effect of it is to promote the application of science. How they can continue to say this after the experience of the last war passes my imagination. One cannot promote the application of science without advancing science itself; the one reacts on the other. It is true that in war much of the ground covered by science remains fallow for the time, but it then becomes more fertile to new ideas. The progress of science since the war has been remarkable; still more striking has been the rate of application of science. I fancy that no one would be more astonished at the sight of large nuclear power-stations than the great Lord Rutherford, from whose scientific work they have all evolved.

It may rightly be argued that this progress has been bought at too great a cost of destruction and human misery, and that the world would have been happier with a slower rate of advance in peace. But it is useless to let our minds dwell on the mistakes of the past; we must take things as they are and look upon their bright side. And the bright side is that we can now see our way clearly to abolish poverty and disease all over the world. It is only a matter of time: and the time it takes will depend on the number of scientists ready to take their part in the advancement of knowledge and its application to human affairs. I am inclined to think that we are now, or soon will be, on the steepest part of the curve of what Macaulay called 'the long progress from poverty and barbarism to the highest degree of opulence and civilization'; and that the slope of the curve will be determined in future as much by the wisdom of men in guiding social and economic affairs as by science itself. But whether I am right or wrong, we may agree that we live in most exhilarating times, and that old men whose life is nearly done have some cause to envy young scientists who are on the threshold of their active careers.

It is well to realize how short a time relative to human history has elapsed since the Scientific Revolution, which is still far from its zenith. There is much evidence now that men capable of making and using tools, of drawing deductions from observations and passing them on by example or word of mouth, existed on the earth half a million years ago. They provided themselves with food and clothing by collecting, hunting and fishing. Slowly, throughout the ages, they developed their primitive handicrafts. Slowly the priests and witch-doctors amongst them acquired and jealously guarded knowledge. And then, about 10,000 years ago, agriculture was invented. It became possible to grow food and to keep domestic animals, and thus to enable a man and his wife to provide in average years more than enough food and clothing to satisfy their family. A small proportion of the population was gradually set free from the never-ending task of finding food and could concentrate on improving the useful and ornamental arts. This was the first great social revolution in the history of the world; it took some 5000 years to spread over a considerable portion of the earth's surface, and culminated in the invention of writing, about 5000–6000 years ago. We conveniently take this date as the start of civilization. There is clear evidence that by then well-organized villages in Mesopotamia held more than 10,000 inhabitants, including skilled artisans, priests, administrators and clerks. The way became clear for the steady development of useful and ornamental arts. Many things remain from this period to astonish and delight us with their beauty. Long before the Christian era, say the editors of the *History of Technology*, 'human dexterity had already so marvellously developed that it has never since been excelled'.

Wars and plagues caused the centre of civilization to shift from place to place, moving gradually westward so far as we are concerned, until it came to Greece and then to Rome. In Athens civilization reached intellectual, artistic and literary heights which were not surpassed for 2000 years. The Greeks were conquered in the end by the more practical Romans, who, after many struggles, held sway over the western world for centuries. The Romans were great soldiers, organizers, lawyers and engineers. The distinguished Professor R. J. Forbes of Amsterdam has written that 'their practical knowledge of machinery was not surpassed until the eighteenth century'. But they were not great innovators; they adopted and improved the techniques of the East and spread them over Western Europe by conquest. Most people must be familiar with the remains in England of Roman cities, villas, roads, baths and walls; every year something is discovered to add to our knowledge of the past; but far fewer people realize the lasting influence of the Romans on our habits, laws, methods of administration, and even, up to the twentieth century, on the art of medicine.

When the Romans at last quitted Britain and Western Europe after 400 years of rule, they left a void behind. For the next 1000 years the history of the British people is really a history of invasion, and of wars and quarrels between men of power and influence. Society remained practically static. A Roman revisiting the country after 1000 years of sleep would have been astonished at the primitive state of the roads, at the absence of sanitation in the cities and of baths in the better homes, and at the paucity of fine buildings. He would have commented on the slow advance of technology; the materials available to the artisan were the same as they had been in his time, and the workmanship no better. Science, or natural philosophy, had been at a stand-still since the time of the Greeks. The common people lived, laboured, fought, suffered and died as their fathers had done and were sullenly resigned to remain in that state of life unto which it had pleased God to call them.

All this was completely changed by the Scientific Revolution, which is at last becoming to be recognized as the greatest, the most far-reaching and the most beneficent event in the history of civilization. There are few people now living on the earth who have not felt its impact. There is practically no department of human thought and action on which it has not impinged. It has swept away age-long dogmas and prejudices from the mind of man. It has profoundly influenced philosophy and religion. It is the ultimate cause of all the great increase in human welfare and health in our time. Its influence on the conduct of human affairs becomes greater every year. All the social changes in the world that are taking place today, and that are the preoccupation of statesmen, are due directly or indirectly to the advance of science. That some of them involve unprecedented dangers to mankind no one will deny; but I for one believe that they can and will be avoided. And all this has been due in the first place to a few men, who, tired of perpetual wars, political strife, and the wranglings of theologians, set themselves to acquire a knowledge of nature by patient experiment and logical processes of thought.

We date the Scientific Revolution in this country from the foundation of the Royal Society of London, 'for Improving Natural Knowledge' nearly 300 years ago. But the scientific movement was stirring all over Western Europe at about the same period. Italy must be given the honour of being the birthplace of the Revolution; England was the home of its adolescence. Galileo certainly laid the foundations on which Newton erected a majestic structure, and probably also helped to inspire Harvey's researches on the circulation of the blood. But all revolutions smoulder for long in the minds of men before they burst into flame. The early traces of the scientific revolution can be detected as far back as the thirteenth century, when the invention of paper (certainly) and gunpowder (less certainly) was brought to Europe from China, and when Roger Bacon was preaching to the unconverted the value of experimental inquiry. It smouldered away quietly in the fourteenth century, and more briskly in the fifteenth, fanned as it was by the needs of the great sea explorers, and by the tales and specimens that they brought back to their countries. At the end of the fifteenth century that great genius Leonardo da Vinci was engaged on an astonishing series of investigations which had a local influence only, because he published nothing, and his voluminous and nearly illegible notes were not made available to the world at large for over 250 years. By the end of the sixteenth century many entirely new ideas were put forward and discussed. One has only to read the works of Francis Bacon to realize this. His prophecies of things to come are of extraordinary interest, but he must have derived them from the conversation of others. John Evelyn recorded in one of his letters that Francis Bacon '(although never to be mentioned without honour and admiration) was used to tell all that came to hand without much examination'. There can be no doubt, however, of his influence on the scientific movement in the seventeenth century.

For the last 300 years agriculture, industry and medicine have advanced with science, at first side by side, until at last in the twentieth century they have been closely knit together. The progress has not been regular of course. The great triumphs of medicine, for example, have been mainly achieved in the last half century. Their effects can be illustrated very simply by the statement that when I was born the expectation of life of a baby boy was 42 years. It is now 65 years, and for some unaccountable reason a newly born baby girl will live on the average some five years longer! The tide of science itself seemed to pause towards the end of the nineteenth century. I remember Sir Joseph Thomson saying that when he was young many physicists thought that the main principles of physics had been firmly established, and that all that remained was to determine more accurately a few physical constants. That was just before the discovery of the electron and of X-rays and radioactivity, which ushered in what has been called the Golden Age of Physics. There is no such pessimism in the scientific world today. Indeed the more knowledge accumulates, the more there seems to discover and to understand. The Frontiers of Science seem to us to be endless indeed.

The young scientist of today need have no fear that no one will need him when he finishes his formal education. His problem will rather be to choose out of countless opportunities that one which will give the greatest scope for his energy, ability and tastes. I cannot promise him a life of opulence, nor one of great influence on national affairs – though times are changing in that respect too. I cannot promise him complete freedom from the shocks that flesh is heir to: but I can confidently predict that his life will never cease to be interesting, and that when it draws to a close he will be able to comfort himself with the thought that he has added something to the stock of knowledge and experience on which the future of the human race depends.

I followed in 1961 with *The Young Scientist's Companion*, a book of information and advice, in which I described one of the most exciting moments of my life – when I met the Periodic Table of the Elements.

THE STORY OF THE PERIODIC TABLE
Maurice Goldsmith

One of the most exciting moments of my life was when, for the first time, I met the Periodic Table of the Elements. As I began to understand it, the separateness of Physics, Chemistry and Biology – the major disciplines of science with which I had gained some superficial familiarity at school – stood revealed as formal hangovers from a Victorian age, convenient labels retained for examination purposes. When I saw Science as one, it was a moment of revelation.

The Periodic Table is a shining example of how scientists have seized truth by anticipation. I read it as a guide to what we are, what we came from, what we may become, what our links are with the inanimate, and with the stuff of the Universe in other worlds in the vastness of space. These are vast claims but not exaggerated. Let me begin to substantiate them.

An element is a material that cannot be broken down into a simpler one by chemical means. The origins of the word are not clear, but T. H. Savory points out that it is a convenient translation of the Greek, *ta stoikeia*, meaning 'the simplest constituent part'. It is in this metaphorical sense that Mark Antony uses the word when he says of Brutus:

His life was gentle and the elements
So mix'd in him that Nature might stand up
And say to all the world, 'This was a man!'

Empedocles (c. 490–430 B.C.), the ancient Greek philosopher, spoke of the elements Fire, Air, Earth and Water as material substances in their own right. Aristotle accepted this, but the terms came to be used to mean the quality that a body had: for example, hot substances contained much of the element Fire. Another Greek Anaxagoras (c. 500–428 B.C.), used the term to denote 'simple' substances: for example, sand and salt were 'simple' substances because they could be separated out from a mixture.

But it was Robert Boyle (1627–91) who first gave the word 'element' its modern meaning when, in *The Skeptical Chymist*, he wrote: 'No body is a true principle of element . . . which is not perfectly homogenous but is further resolvable into any number of distinct substances how small soever.' He believed the elements were ultimately made up of particles of various sizes and kinds, into which, however, they could not be resolved in any known way. In fact, the techniques available in his day were unable to produce elements as defined by him. Chemistry was struggling to emerge from alchemy, in which astrology provided the theoretical basis. Thus, the metals were linked with the planets.

A common rhyme was:

> Like the planets up in heaven,
> Metals also number seven;
> Copper, iron, silver, gold,
> Tin and lead, to smelt and mould
> Cosmos gave us; listen further:
> Fiery sulphur was their father,
> And their mother – mercury;
> That, my son, is known to me.

There was one metal for each heavenly body. Gold, the perfect metal, was obviously linked with the Sun, and its alchemical symbol was the circle, the mark of mathematical perfection (see Fig. 1). 'Gold is not strictly speaking a metal: gold is light,' is the comment legend reports the Arab philosopher Averroes (1126–1198) to have made as he buried a beam of sunshine under a pillar in the great mosque of Cordova. Silver was linked with the Moon, quicksilver with Mercury, copper with Venus, iron with Mars, tin with Jupiter, and lead with Saturn. Even today when we use the word 'base' in reference to certain metals, we use it in the mediaeval sense of 'sinful', that is, those metals which can never be pure and virtuous like gold.

The word 'chemistry' is in fact derived from this ancient link with metals. It is taken from the Arabic *Al-chemy*. The prefix is the definite article, which came to be omitted. But the Arabs took the word from the Greeks who spoke of the Art of Khem, meaning the skill in working with metals of the people who lived in the Black Land (or Khem). This was Egypt, where after the autumn floods of the Nile a fine black mud was deposited in the soil. The

Egyptians were metal workers, whose great skill impressed their fourth century Greek conquerors.

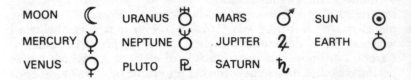

It was Lavoisier who began to make clear that an element is a substance that cannot be split up by any means known at that time into something simpler. But of course, as techniques improved the number of accepted 'elements' began to be reduced: for example, caustic potash was regarded as an element until Davy in 1807 decomposed it by electrolysis.

The real story that led up to the Periodic Table and laid the foundations of modern chemistry was the investigation into the nature of combustion. We are not concerned here with giving you a potted history of chemistry. There are some good books which do this admirably. Let us take a more detailed look now at the Periodic Table.

The Periodic Table is the inspired work of the great Russian chemist, Dmitri Mendeleyev (1833–1907), 'the Copernicus of the atomic system', a happy phrase used by J. D. Bernal. By the second decade of the nineteenth century some 50 elements were already known. Scientists began to look around for some relationships which could allow of classification. The obvious road of choice was to use atomic weights. (We define 'atomic weight' later in this chapter.) In 1829, the German chemist, Johann Wolfgang Döbereiner (1780–1849), put forward the idea of triads. He found by inspection that of three elements with similar properties listed one above the other, the atomic weight of the middle element was halfway between the other two, and its chemical properties were also a mean of the elements immediately above and below. For example: *Ca*lcium, atomic weight 40.1, *Str*ontium 87.6, and *Ba*rium 137.4

$$(\frac{40.1+137.4}{2} = 88.7);$$

*Li*thium 6.9, Sodium (*Na*) 23.0, Potassium (*K*) 39.1 $(\frac{6.9+39.1}{2} = 23).$

In 1862, the French chemist, A. E. B. de Chancourtois (1819–1886), introduced the idea of plotting the elements along a spiral according to their atomic weights. This revealed that elements with similar properties were grouped together, and these corresponding pairs differed by 16 (atomic weight of Oxygen) in their atomic weights.

About 100 years ago, Professor John Newlands (1838–98), reported to the Chemical Society in London, that when he listed the elements, starting with *Li*thium, in order of their atomic weights, the first seven elements – *Li*thium, *Be*ryllium, *B*oron, *C*arbon, *N*itrogen, *O*xygen, *F*luorine – stood at the head of columns in which every eighth element fell, and these elements in each column all had similar properties. Like a conductor of music, because his arrangement was similar to the octave of the musical scale, with its eight-note interval, he called it 'the law of octaves'.

The great advance was made almost simultaneously – as so often happens in scientific discovery – by the German, Julius Lothar Meyor (1830–95), and by Mendeleyev. On 4 June 1889, the Russian scientist delivered the Faraday Lecture in London at the invitation of the Chemical Society.

Mendeleyev also pointed out 'how far the periodic law contributes to enlarge our range of vision. Before the promulgation of this law the chemical elements were mere fragmentary, incidental facts in Nature; there was no special reason to expect the discovery of new elements, and the new ones which were discovered from time to time appeared to be possessed of quite novel properties. The law of periodicity first enabled us to perceive undiscovered elements at a distance which formerly was inaccessible to chemical vision; and long ere they were discovered new elements appeared before our eyes possessed of a number of well-defined properties.'

Here Mendeleyev gave three examples of the way in which he had been able to use the Periodic Table to *predict* the existence of three new elements. This he was able to do because he was bold enough to leave 'boxes' for elements that might one day be discovered, and he forecast what they would look like, what their weight would be, and their chemical behaviour. He was fortunate in that in his own lifetime his predictions were found to be true. He himself said: 'When, in 1871, I wrote a paper on the application of the periodic law to the determination of the properties of the as yet undiscovered elements, I did not think that I should live to see the verification of this consequence of the law, but such was to be the case.' He expected the three elements to have properties similar to those of *B*oron, *Al*uminium, and *Si*licon, so he named them in advance eka-*B*oron, eka-*Al*uminium, and eka-*Si*licon. (*Eka* is Sanskrit for one: thus, eka-*B*oron is *B*oron plus one.)

In 1875, the Académie des Sciences in Paris heard that the young scientist, Lecoq de Boisbaudran (1838–1912), had discovered a new element in a mineral zinc sulphide, which he had named *gallium*: in honour of France (Gallia). When Mendeleyev heard about this in St. Petersburg, he wrote to Paris: 'Gallium is the eka-Aluminium I predicted. Its atomic weight is about 68, its specific gravity about 5.9. Investigate and see if this is not correct.' And it was so: successful first prediction. In 1879, the Swede, Lars F. Nilson (1840–99), discovered a new element in a rare mineral, found only in the Scandinavian peninsula. He called it *Scandium*. It turned out to be eka-Boron: successful second prediction. And when, in 1886, a German chemist, Clemens Alexander Winkler (1838–1904), discovered another new element in a silver ore, he named it *Germanium*. It turned out to be eka-Silicon, predicted by Mendeleyev 15 years before.

The genius of Mendeleyev was clear. He had shown that the chemical properties of the elements were a periodic function of their atomic weights. Now a powerful tool was available to the scientist: it revealed the relationships between the elements, and indicated where new elements remained to be found. But Mendeleyev had not said the final word. There were inconsistencies in his Table: for example, certain elements were not in the order of their atomic weights. Perhaps there was some other fact on which to base the organization of the elements?

In 1913, a brilliant young physicist, H. G. J. Moseley (1887–1915) – one of the remarkable men who worked with Lord Rutherford – made the next step forward. He measured the wavelength of X-rays from elements of successively increasing atomic weight. He found that as the atomic number (that is, the position of the element in the Periodic Table) increased, the wavelength increased also. This did not happen in every case when he used atomic weights instead of atomic numbers.

Moseley wrote. 'We have here a proof that there is in the atom a fundamental quality, which increases by regular steps as we pass from one element to the next. This quantity can only be the charge on the central positive nucleus of the existence of which we already have definite proof.' The inconsistencies left by Mendeleyev were now ironed out. Now it was seen that the chemical properties of the elements are a function of their atomic numbers.

Before we can go any further, we must have some definitions. An element is made up of atoms of one kind only. For example a piece of Copper (*Cu*) is made up of atoms which are all identical in chemical terms; the element Carbon is made up of atoms with a distinct structure which we call that of the atom of Carbon. When these atoms of the *same* or any kind combine they form *molecules*. For example, two Hydrogen atoms combine to form a Hydrogen molecule. And most atoms can also combine with atoms of *different* kinds to form *compounds* of molecules. For example, an atom of Carbon will combine with two atoms of Oxygen to form carbon dioxide (CO_2). Or, two atoms of Hydrogen will combine with one atom of Oxygen to form water (H_2O). Notice that we have had to detail specific numbers of atoms; also, that in combining the elements themselves may change, and the properties of the products may be amazingly different from those of the elements. For example, Hydrogen, a highly flammable gas, combines with another gas, Oxygen, to form the liquid water. Or, the highly poisonous metal Sodium (Na) combines with the equally poisonous gas Chlorine to form essential-to-life salt (NaCl).

To understand the elements we must know about the atoms that make them up. The Periodic Table of the Elements is also the Table indicating the way in which the different atoms combine to form molecules.

There are very many good books and films on the atom, so we shall not go into great detail here. An atom may be considered as made up of a central nucleus, and revolving around it are electrons. (It all seems rather like our solar system, with the nucleus representing the sun and the electrons the plancts.) The nucleus is made up of protons and neutrons, each of which has unit mass on the atomic scale. The proton has a positive electric charge, the neutron has no electric charge, and the electron has a negative charge. The number of protons in the atomic nucleus is equal to the number of electrons revolving around that nucleus. This means that the positive electric charge is balanced by the negative. The atom is therefore electrically neutral. The atomic number states the number of protons (therefore, electrons) in an atom of that element. For example, the atomic number of Carbon is 6: this means that there are six positively charged protons in the nucleus, and six negatively charged electrons whirling around. It is the electron charge which Moseley was able to measure.

The atomic weight is the relative *weight* of an atom of the element. At one stage this was compared with Hydrogen as unity, then with Oxygen as 16. Thus, as the atom of Helium has a weight of 4, the Oxygen atom is four times as heavy. The atomic *number* and the atomic *weight* describe the make-up of the nucleus of the element. (There is a modification which we explain below when we deal with isotopes.) For example, Oxygen has an atomic number of 8, therefore it has 8 protons: its atomic weight is 16, therefore it has 8 neutrons. (Remember the nucleus is made up of protons and neutrons.) Why do not the electrons matters in discussing mass? Because the electron has insignificant weight. The *matter* of the atom, that is, its *mass*, is contained in the nucleus, which is extremely small. To take an example, the size of a Carbon atom is about one-one hundred millionth of an inch. A pencil dot on a piece of paper contains more carbon atoms than there are people on earth. The diameter of an atom is about ten thousand times its nucleus, yet the latter has no less than 99.9 per cent of the mass of the atom. The atoms which make up matter are mostly empty space.

In the *Hydrogen* atom the nucleus consists only of a single proton, that is, a single charge of positive electricity of unit mass. It has no neutron, but one electron in orbit. The nuclei of heavier elements may be considered as built up by adding protons and neutrons. But the nucleus is only part of the atom: electrons are needed to counter-balance the positive electric charge of the proton. For example, the *Helium* atom has two electrons balancing two protons. But we cannot add electrons simply at will. The electrons travel around the nucleus in orbits or shells, and these hold only a fixed number of electrons. In the shell nearest the nucleus there can be only two electrons; in the next shell, eight. The general formula for the number of electrons in each shell is $2n^2$, where n increases by whole numbers from one shell to the next. For example, in the innermost, known as the K shell, $n = 1$ and the number of electrons in the shell is $2 \times 1^2 = 2$. In the next, known as the L shell, $n = 2$ and the number of electrons is $2 \times 2^2 = 8$. In the next, known as the M shell, it is $2 \times 3^2 = 18$; and in the next one, the N shell, $2 \times 4^2 = 32$. Beyond this, we are not clear at the moment: the formula does not seem to apply. We can now look at the Periodic Table in another way; that is, in terms of the way in which the electrons are arranged in orbits. It is this which determines the chemical properties of an element.

Chemical combination depends on the number and arrangement of the electrons in the outer shells, and has nothing to do with the nucleus nor the electrons in the inner shells. For example: Sodium (*Na*) has an atomic number of 11, that is, there are 11 protons and to counter-balance them 11 electrons. There are two electrons in the first shell, and eight in the second. That means there is one electron free in the third shell. In the case of the two elements *Hydrogen* and *Helium*, for example, there is only a single occupied electron shell. In the gases *Helium*, *Neon*, *Argon*, *Krypton*, *Xenon*, and *Radon*, all the outer shells are not filled, though they are stable shells of eight electrons. The inner forces are so well balanced that no compounds can be formed with other elements.

We pointed out above that it is not strictly true to say that an element is a substance all of whose atoms are identical. The reason is because some elements have *Isotopes* (from the Greek, *iso*, occupying the same or equal, *topes*, place), and this statement is only true of pure isotopes. You cannot distinguish the isotopes of an element by inspecting the Periodic Table. The isotopes have the same number of protons and electrons, but the number of neutrons in the nucleus differs. This means that the atomic weight of the isotopes is different: for example *Uranium-238* has 92 protons and 146 neutrons. But the isotope *Uranium-235* – used as a fissile material in nuclear energy – has the same number of protons, and only 143 neutrons.

The Chlorine atom has 17 electrons – two in the inner shell, eight in the second shell, and seven in the outer shell. That is, there is one electron short of the stable number of eight in the outer shell. The Sodium (Na) atom has a free electron and a 'match' is easily arranged: the result, a compound forming salt. In the process, a single electron from the outermost shell has gone from the Sodium (Na) atom. The atom now has more protons than electrons: that is, it is positively charged electrically. As the Chlorine atom now has an additional electron it is negatively charged. Atoms whose outer electrons have become detached are said to be *ionized*. The word is taken from a Greek word meaning 'to go', because these charged atoms can be made to move under the influence of an electrical potential difference. In solution or a melt it is their movement which constitutes the electric current. Electricity is a flow of electrons.

There is another way in which atoms can combine to form molecules: the electrons are not transferred, but atoms join together and share electrons. For example, the Oxygen atom has eight electrons – two in the inner shell and six in the outer, leaving places for two electrons. A Hydrogen atom has one electron. To form water, two Hydrogen atoms (more precisely, their electrons) and an atom of Oxygen (more precisely, its outer six electrons) join together and share their electrons. This is known as electron pair bonds, or co-valences.

You will have noticed that when mentioning the names of the elements in this chapter, we have italicized the letters that make up the symbol (for example, *O*xygen), or written this in parenthesis after the element when the letters are not contained in the name (for example, Lead (*Pb*)). This gives us a clue as to the way in which the symbol is derived.

It was the Swedish chemist, Baron Jöns Jakov Berzelius (1779–1848), who, in 1814, made the proposal that the initial letter of the Latin name of the element be used as the chemical symbol: for example, *K*alium (K for potassium). When the same letter begins the name of different elements, the second distinctive letter of the name is added: for example, *Au*rum (Au for gold) to avoid confusion with Ag for silver. But as Latin ceased to be the language of scholars and scientists began to write in their own native tongues (roughly during the seventeenth century) non-Latin names for the elements appeared: the same system was used. These symbols were clear and each one stood for a definite atomic weight.

There are nine elements whose symbols are based on old Latin names: they are sodium (*Na*trium), potassium (*Ka*lium), iron (*Fe*rrum), copper (*Cu*prum), silver (*Ar*gentum), tin (*Stan*num), antimony (*Stibium*), gold (*Aurum*), and lead (*Plumbum*). The symbol for tungsten is W (from Wolfram), and for mercury it is Hg (from Hydrargyrum).

What is the origin of the names of the elements? This is a question you should attempt to answer yourself. It will plunge you into a fascinating aspect of history and lead you to understand how inter-related are our social ideas with our scientific. We shall take a few elements and have a closer – but brief – look at them in this way.

First, Carbon (from the Latin, *carbo*, coal). This element has been used by man through history as charcoal, graphite, or diamond. He first made its acquaintance when he began to use fire. The word graphite is derived from the Greek *grapho*, I write, because it leaves a greyish black mark on a rough surface. For centuries graphite was confused with molybdenite because both are like lead (Greek *molybdos* means, resembling lead). Pure Carbon forms the diamond, which is the hardest substance known to man. The word diamond comes from the Greek *adamos*, meaning invincible.

Second, Mercury (*hg*). The fact that this metal is found naturally in a liquid state gives it its name. The Latin word *hydrargyrum* comes from the Greek words *hyder argyros*, meaning liquid silver. The first written record we have of it is by Aristotle. It was used extensively by the alchemists.

Third, Oxygen. For centuries it was recognized that there was some active part in the air. The Chinese called it yin; the Italian artist, Leonardo da Vinci (1452–1519) stated that the air is not completely consumed in respiration or combustion and said there were two gases in the air. George Ernst Stahl (1660–1734), Professor of Chemistry at Halle University, in 1697 stated that bodies that burned possessed a property of combustibility, that he called *phlogiston* (from the Greek, *phlox* or flame): and bodies that did not burn did not contain this property, therefore they were de-phlogisticated.

In 1731, Stephen Hales (1677–1761), the Vicar of Teddington, heated saltpetre (potassium nitrate), and made Oxygen, but did not realise that this gas he had prepared was also found in the air. The original discovery is attributed to the English clergyman Joseph Priestley (1733–1804). In 1775, he informed the Royal Society of how he filled a test tube with mercuric oxide and inverted it over a pan of mercury. He heated the red powder with the sun's rays by focusing them on the test tube with a burning glass 12 inches in diameter. A gas was given off, and a heavy silver-white metallic liquid was left. A burning candle placed in the gas shone more brilliantly: a mouse placed in a jar of the gas rushed about excitedly. But the Swedish chemist, Karl Wilhelm Scheele

(1742–86), had also prepared oxygen about the same time. His account was not published until 1777, so Priestley obtained first credit for the discovery. Scheele called his gas *empyreal* or *fire air*; Priestley called it *dephlogisticated air*, because it did not contain the property of combustibility but absorbed it from around. He found that in burning and in breathing it was the dephlogisticated air that was used up. Both Scheele and Priestley understood that air was a mixture of two gases, one of which was 'combustible' and the other 'useless'. But they did not understand the real nature of combustion. The limiting factor for them was the phlogiston theory of combustion. This is a most interesting example of the way in which the accepted view seems to put the mind in blinkers. But, the phlogiston theory was an important advance on man's previous level of understanding in its day; and neither Scheele nor Priestley, brilliant investigators though they were, could step out of their day, although their researches led to the abandonment of the phlogiston theory and laid the foundation of modern chemistry.

It was left to the brilliant Frenchman, Antoine Laurent Lavoisier (1743–94), to effect the great revolution that was to lead chemistry away finally from alchemy and make it a qualitative science. When he heard of Priestley's experiments, he realized that the theory of phlogiston did not correspond to the experimental facts. The new gas alone was responsible for combustion. He called it *oxygen* or *acid-producer* (from the Greek, *oxus*, sharp or acid; *gennao*, I produce), a word he coined himself because he thought it to be an essential component of acids. What a great change now occurred in chemistry, as revolutionary as the industrial and political changes sweeping Europe. The elements had to be looked at anew. For example; tin could no longer be regarded as made up of two elements, phlogiston and tin rust. Water turned out to be not an element, but a compound.

Let us now leap 120 years and come to our fourth element, *Helium*. It takes its name from the Greek *helios*, the sun. The story of the discovery of this inert gas is as full of excitement as a good detective novel. In 1868, there was a total eclipse of the sun visible in India, and the French astronomer, Pierre Jules César Janssen (1824–1907), was able there to make the first-ever spectroscopic examination of the light from the sun's chromosphere. In the solar spectrum Janssen discovered a bright yellow line, which was at first thought to be the normal sodium line. But he soon showed that this was something new.

The story is now taken up by two British scientists, Edward Frankland (1825–99) and Sir Norman Lockyer (1836–1920). Investigating the red gaseous layer around the sun, they discovered a black line in the yellow portion of the spectrum. It belonged to no known element on earth. Was it a new element? On the sun?

Entirely unknown on earth? The scientists had to answer yes to these questions. They named it *He*lium. It was 27 years before this element was found on earth, and then only as a result of experimental work which had nothing to do with the search for it. But that often happens in science. The poser was where to fit *He*lium into the Periodic Table. It had no natural place where it could go.

About 20 years later, the English physicist, Lord Rayleigh (1842–1919), was doing a series of experiments at the Cavendish Laboratory at Cambridge, to determine the density of certain gases. The fun began when he weighed Nitrogen. He obtained it from the air, which is made up of four parts Nitrogen to one part Oxygen. He also obtained it from ammonia. He found that a litre of Nitrogen from the air weighed 1.2572 grams at standard temperature and pressure: but obtained from ammonia it weighed 1.2506 grams. This was something for the conscientious scientist to worry about. Was it an error? Rayleigh checked. Again a difference – this time of 1/1000th of a gram. How could Nitrogen have two weights? Early in 1894, Rayleigh read a paper at the Royal Society in London on his work with nitrogen. To him came Sir William Ramsay (1852–1916), a great chemist, with the suggestion that there must be some impurity in the Nitrogen from the air, possibly an unknown gas. He asked for permission to work on the problem.

Then from the past – a century back – came a clue that was to help solve the mystery. Rayleigh was advised to have a look at some work on the composition of the air done by the eccentric Henry Cavendish (1731–1810) and reported by him in 1785. As Rayleigh read, these words of Cavendish leaped up at him: 'From this experiment, I have come to the conclusion that the phlogisticated air (that is, nitrogen) is not homogenous. Part of it acts in a different way from the larger, principal part. It is probable that the phlogisticated air is not one uniform substance but is a mixture of two different substances.'

Ramsay, too, had been working in his own laboratory. On 13 August 1894 he and Rayleigh announced that they had discovered a new element – one of the component parts of the air around us. Like *He*lium, it did not fit into any place in the Periodic Table. It was a gas, without colour, odour, or taste. It kept to itself chemically, not mixing in any way. He named it *Ar*gon (from the Greek word *Lazy*).

Ramsay now began to check a report that when a rare mineral *cleveite* was heated in sulphuric acid, it gave off a gas, which was believed to be Nitrogen. He was asked whether this might not also contain Argon. His work led him to conclude that besides Argon there must be another new element, which he called *Kr*ypton (from the Greek *kryptos*, hidden). He sent a sample of his new gas to

Sir William Crookes (1832–1919), an expert in the use of the spectroscope. Within 24 hours, Crookes sent Ramsay a telegram. It read: 'Krypton is helium come and see for yourself.' In this remarkable way – from the sun to the earth via the discovery of another new element, Argon – was the gas *Helium* finally tracked down.

You will recollect that Ramsay suspected that a *new* column of elements might be found to be available for the Periodic Table. He now set out to find these other elements – and did so. He found the others also in the air: *Neon* (from the Greek *neos*, new), and *Xenon* (Greek *xenos*, stranger).

Thus, here we were in the year 1898, with Mendeleyev's Table confirmed in its key importance. Everything around us – and living creatures themselves – was made up of about 80 elements, each element itself made up of identical, indivisible atoms: a clear, sound, ordered world. Then the storm broke: doubt and uncertainty came in, and the end of the story is not yet clear. The atom turned out to be divisible!

Now we are ready to sum up a modern Periodic Table of the Elements, which is based on atomic numbers and atomic structure. You will find one on the endpapers [sic]. What does it tell us? It tells the *weight* and *structure* of each element and enables us to see which elements are similar to each other. You see it is made up of a large number of boxes. These are organized in a rather special way. There are, by the way, various forms of the Periodic Table available, and they differ somewhat in the way in which the data is presented. But the basic pattern is, of course, the same. In each group, all the elements resemble each other. They are arranged in seven horizontal rows and in 18 vertical columns. The rows are called *periods*, and the columns, *groups*. You will see that the first period has only two elements – Hydrogen and *Helium*; the second and third periods, eight elements in each. (In these two periods are most of the elements you will meet in school chemistry.)

In the fourth, fifth, and sixth periods, 18 elements in each: but in the sixth period, 14 (the Lanthanide series) inserted between box 56 and box 72, making a total of 32 in the period. The seventh period contains three boxes with elements in them, and 13 elements (the Actinide series) are added to box 89 – all those of number 93 and over are man-made elements not otherwise occurring in nature. In all, there are 102 known elements today.

At the top of each box there is a figure, e.g. $\boxed{\begin{smallmatrix} I \\ H \end{smallmatrix}}$. You can see

that these figures run in ascending order along each period. These figures are the *atomic numbers* of the elements. Each box is occupied by only one chemical element. (But see the explanation above on isotopes.) We can therefore define an element as a substance with only one atomic number. Look at each box again and you will see that there is a second set of figures which appears below the symbol for the element, e.g.

This number is the *atomic weight* of the element. You will see that the elements are arranged in order of atomic number, and these *more or less* (but take a look at Argon and Potassium (*K*), and Cobalt and Nickel) are in order of atomic weight, starting with Hydrogen, the lightest element, and ending with the heaviest.

Now let us look at the columns or Groups. You will see that they are numbered with Roman numerals running from I to VIII, and that those from I to VII have either the letter A or B attached to them. There is also a group which may be headed O, or Inert Gases.

Groups IA, IIA, and IIIB are side by side on the left: here are to be found the light metals (and hydrogen, which is a very interesting element to discuss in relation to metals with which it has more similarities than are apparent at first). The non-metals are to be found on the right-hand side: they run roughly from Carbon diagonally through Phosphorus, Sulphur, Selenium, Iodine, to Radon. Groups IA, IIA, IIIA contain the strongly-active metals; Groups VB to VIIB the strongly-active non-metals.

All the metals in IA are found near to each other in nature, except for Fr, which is man-made. We call these the *Alkali Metals*. In IIA are the *Alkaline-Earth Metals*, which also resemble each other and form one family. In IIIB there are: Scandium, Yttrium, a box with 15 *Lanthanides* (or Rare-Earth elements), and 13 *Actinides* (or Transactinium Earths). Group VIII occupies the middle of the Table, with columns 8, 9 and 10. They are all metals, and the neighbours of each metal resemble each other: for example, Iron (*Fe*), Cobalt and Nickel, which are always found together in nature; *Ru*thenium, *Rh*odium, and Palla*d*ium (the light platinum metals) and Osmium, *I*ridium, and Pla*t*inum (the heavy platinum metals) also resemble each other. These nine metals, grouped in three, may be referred to as *The Triads*.

Now come the heavy metals in columns 11, 12, 13 and 14. They include the well-known 'friends', which we have worked for centuries: Copper (*Cu*), Zinc, Tin (*Sn*), and Lead (*Pb*). In column 15, we begin to part with the metals. In the top box there is the gas Nitrogen; in the boxes below, Phosphorus and Arsenic, a semi-metal Antimony (*Sb*), and a metal Bismuth.

Now we find the non-metals, so-called because they do not

possess the properties of the metals. In column 16, there are Oxygen, Sulphur, Selenium, Tellurium, and Polonium. In column 17, there are volatile substances: the gases, Hydrogen, Fluorine, Chlorine; the liquid, Bromine; the volatile solid, Iodine; and Astatine, probably a solid, a synthetic product of uranium fission, which does not now exist naturally. The elements in this column (except Hydrogen) are known collectively as the Halogens (from the Greek hals, salt, gennao, I beget) because they form salts with alkalis.

Finally in column 18, there are the Rare or Noble Gases. They remain separate, aloof. They do not combine with anything, but are to be found in the free state in nature. In the top box there is Helium, then below are Neon, Argon, Krypton, Xenon, and Radon.

The arrangement of the Table also corresponds to the electron arrangement. All the elements in Group IA have one electron in the outermost shell; in Group IIA, two electrons; in IIIA, three, and so on for the A sub-groups. (This does not apply to the B sub-groups.) Group O (the inert gases) have eight electrons filling their outermost shells.

The atomic structure is given by the small figures on the right; this assumes that an atom has shells, each of which can hold a specific number of electrons only, but these shells need not always be filled. The atoms of the elements in Period 1 have a first shell which can hold only two electrons. Helium is an inert gas because the shell is filled, whereas Hydrogen has only one electron. In Period 2, the first shell has two electrons, and the second eight. Neon is inert because both shells are filled.

Table 14 gives the shells for each Period with the maximum number of electrons required to fill each shell.

All the elements in Group I have one electron in their outer shells. This makes clear that they have a valence, or combining number, of 1. This corresponds to the number of electrons they lose when they form compounds. Elements in Group II have two electrons in their outer shells; in Group III, three electrons, and so on up to seven in Group VII. Thus, the position of an element in the Table is determined by the arrangement of its electrons, and this will also give information about the element's chemical properties.

A good example of how the Table can be used as a research tool was given earlier in this chapter when we described Mendeleyev's prediction of three new elements. In recent years new products have been made possible through similar prediction.

Thus we see all the elements arranged in related groups: within each group the characteristics of the elements change gradually and in order as their atomic weights increase. From seeming chaos, an order was found – and Mendeleyev expressed this in the Periodic Table.

In the following year, in *The Young Physicist's Companion*, I emphasized 'the fantastic nature of reality'.

THE FANTASTIC NATURE OF REALITY
Maurice Goldsmith

Science forces men to create new ideas, new theories. Their aim is to break down the wall of contradictions which frequently blocks the way of scientific progress. All the essential ideas in science were born in a dramatic conflict between reality and our attempts at understanding.

Albert Einstein & Leopold Infield:
The Evolution of Physics

There are many books which tell the story of Physics, the science concerned with the study of matter in its simplest forms or organization. My aim is to concentrate on 20th century physics, the part of the science which is only slowly being written into the standard textbook. I regard this book, therefore, as a necessary *complement* to the academic school book. Because it highlights certain developments it presents a perspective which can be obtained normally by teacher and student only by reading the current scientific literature. Although the book is intended for the young person who is proposing to become a physicist, I believe it is also meaningful for the science teacher and might be read with pleasure by the interested non-specialist, whether in the sixth form or in the adult world.

Why should I be writing this book? I am not a qualified physicist. I am academically a sociologist. But when I graduated in the 30s, with a good honours degree, I – with many others – found it difficult to obtain employment. I began to ask why, and this led to my interest in science which has dominated my life ever since. I have taken courses in physics, spent hours with the great physicists, visited laboratories and research institutions throughout most of the world, and taught physics as part of general science in adult education and in secondary schools in England. It is this *rapport* with physics that I seek to communicate in this book. I would like the reader to share with me a unique experience, one that is as exciting and pervasive as first love. An experience that is therefore enriching.

Let the reader, for instance, take a brief look with me at the fantastic and try to get it into realistic perspective. Take an apparently simple thing like a pin. Look at its head. It weighs about eight milligrams. Its volume is one cubic millimetre. Yet within this tiny head there are one hundred million million million atoms of iron. Within each one of these atoms lie vast worlds of mass and energy. Each atom of iron is made up of a central nucleus around which spin 26 electrons in a variety of orbits. But that is not all. If we could strip off the electrons, so that our pinhead was made up of nothing but iron nuclei, it would weigh more than 10 Queen Mary ocean liners put together.

I'll go still further into this newfound realm of reality. Near Geneva, a proton-synchroton, a giant atom-smashing machine, is producing energies of 25,000 million electron volts. It cost several million dollars to build, and is helping us penetrate into the mysteries of matter that lie behind the nucleus of the atom. It is revealing particles with a lifetime of one thousand millionth of a second. Behind our real world – our pinhead – lie such staggering notions as these, which today we accept as manifestations of natural laws.

Or go from the infinitely small to the gigantic. At the National Radio Astronomy Observatory at Green Bank, West Virginia, there is a radio telescope, some 85 feet in diameter, that recently was beamed alternately on Tau Ceti, a star in the constellation of Cetus 11.8 light years distant from us (about 70 million million miles away), and on the star Epsilon Eridani, in the constellation of Eridanus, 10.9 light years away. The instrument was an essential part of Project Ozma, named after the wizard of Oz. The project was planned to help us find out whether living forms, if they exist or those star systems, might not be sending messages to us.

It's all a far cry from yesterday's orderly text books. Today the fantastic and the real are merging. A new world of 'fantastic realism' is emerging. Arthur Clarke, a leading science fiction writer, expressed this very thing – unconsciously – when we were seated side by side a few years ago at an historic meeting in London of the Royal Society. The subject was Space Travel. At one moment when an astronomer was speaking on how to use television relays in space to obtain closeups of the surface of the sun, Mr. Clarke made a gesture of impatience. He turned to me and said: 'I must leave.' 'Why?' I whispered. 'Because,' he replied, 'this is too unreal for me. The only reality left is in science fiction.'

What was really bothering him was that the frontier between our visible universe and our invisible universe has become so much easier to cross. The alchemist of old tried to transmute base elements into gold and in so doing hoped to transmute himself as a human being into something more divine. He did not succeed. Today we are really at a period when we can actually begin to do so. As the scientist learns to manipulate matter and energy, he secures a privileged position in his understanding of the universe. He is beginning to have access to realities of space and time that are normally hidden from us. Men like Newton and Einstein have driven their minds into regions untouched by normal human experience. So did other men of genius who were not scientists. They, too, have been crossing the frontiers into the invisible universe.

What does all this add up to? To the fact that nuclear physics today is opening up new doors into the infinity of the unknown. The human sciences have not yet grasped this. Let me draw an analogy. Light is both a form of energy and a wavelength. On the one side, radiowaves and infra-red, in the middle, the narrow band of visible light; and then, ultra-violet, X-rays, and gamma rays. Might there not exist a similar spread of human consciousness? We live with, say, the narrow band of present consciousness. Below this, there is infra, or subconsciousness. Why should there not also exist an infinite unknown of the ultra-consciousness? There is a vast new world here. In the same way as the forces which bind the atom's nucleus differ from the known forces of gravity and electro-magnetism, so may the properties of the mind in the ultraconscious differ vastly from those of our present experience.

To meet the needs of our new horizons, our thinking must embrace every aspect of reality. And that must include the so-called fantastic, which we must now study as a manifestation of natural laws.

I first met Nobel laureate, Louis de Broglie, shy genius, when he agreed to become involved with the Association des écrivains scientifiques de France (French Association of Science Writers) as honorary president in 1949. And I was concerned with his receiving the initial Kalinga Prize for Science Writing in 1952. He was a sensitive writer, concerned with those who had a certain level of education and broad literacy, as shown in this introduction to his book dealing with quantum theory (*The Revolution in Physics, a non-mathematical survey of quanta*, London 1954: Routledge & Kegan Paul, translated from the French, *La Physique Nouvelle et les Quanta* by Ralph W. Niemeyer.)

THE IMPORTANCE OF QUANTA (1936)
Louis de Broglie

1 Why it is necessary to know about quanta

Among those who will glance at the cover of this little book, many no doubt will be frightened away by the sight of that mysterious world: quanta. The general public does have some vague ideas — yes, often very vague — about the theory of relativity concerning which there has been a great deal of talk in the last several years;

but the said general public has, I believe, few ideas – even vague ones – about the quantum theory. It must be said, this is excusable, for quanta are a very mysterious thing. In my case, I was about twenty when I began to work with them and I have now been pondering over them a quarter of a century; very well, I must humbly confess that if in the course of these meditations I have come to understand some of their aspects a little better, I do not yet know exactly what is hidden behind the mask which covers their true face. Nevertheless, it seems to me one thing can be asserted: despite the importance and the extent of the progress accomplished by physics in the last centuries, as long as the physicists were unaware of the existence of quanta, they were unable to comprehend anything of the profound nature of physical phenomena for, without quanta, there would be neither light nor matter and, if one may paraphrase the Gospels, it can be said that without them 'was not anything made that was made.'

We can then imagine the essential shift in direction which the course of the development of our human science underwent the day when quanta, surreptitiously, were introduced into it. On that day, the vast and grandiose edifice of classical physics found itself shaken to its very foundations, without anyone very clearly realizing it at first. In the history of the intellectual world there have been few upheavals comparable to this.

Only today are we beginning to be able to measure the extent of the revolution which has come to pass. Faithful to the Cartesian ideal, classical physics showed us the universe as being analogous to an immense mechanism which was capable of being described with complete precision by the localization of its parts in space and by their change in the course of time, a mechanism whose evolution could in principle be forecast with rigorous exactness when one possessed a certain amount of data on its initial state. But such a conception rested on several implicit hypotheses which were admitted almost without our being aware of them. One of these hypotheses was that the framework of space and time in which we seek almost instinctively to localize all of our sensations is a perfectly rigid and fixed framework where each physical event can, in principle, be rigorously localized independently of all the dynamic processes which are going on around it. Thereupon, all the evolutions of the physical world are necessarily represented by modifications of local states of space in the course of time, and that is why in classical science dynamic quantities such as energy and momentum appeared as derived quantities constructed with the aid of the concept of velocity, kinematics thus serving as the basis for dynamics. From the point of view of quantum physics, it is entirely otherwise. The existence of the quantum of action, to which we must so often return in the course of this work, does imply a kind of incompatibility between the point of view of a localization in space and time and the point of view of a dynamic evolution; each of these points of view is capable of being used in a description of the actual world, but it is not possible to adopt both simultaneously in all their rigor. Exact localization in time and space is a sort of static idealization which excludes all evolution and all dynamism; the idea of a state of motion taken in all its purity is on the contrary a dynamic idealization contradictory in principle with the concepts of position and instant. A description of the physical world in the quantum theory can only be made by using to a greater or lesser degree one or the other of these two contradictory images: there results then a sort of compromise and the famous Uncertainty Relations of Heisenberg tell us in what measure this compromise is possible. Among other consequences, it follows from the new ideas that kinematics is no longer a science having a physical meaning. In classical mechanics it was permissible to study displacements in space by themselves and to define thereby velocities and accelerations without bothering about the manner in which these displacements are materially realized: from this abstract study of motion, one then progressed to dynamics by introducing several new physical principles. In quantum mechanics a similar division of the subject is no longer admissible in principle since the spatio-temporal localization which is the basis of kinematics is acceptable only to a degree which depends on the dynamic conditions of motion. We shall see later why it is nevertheless perfectly legitimate to make use of kinematics when studying large scale phenomena; but for phenomena on the atomic scale where quanta play a predominant role, it can be said that kinematics, defined as the study of motion made independently of all dynamic considerations, completely loses its signficance.

Another implicit hypothesis underlying classical physics is the possibility of making negligible by appropriate precautions the perturbations which are exerted on the course of natural phenomena by the scientist who, in order to study them with precision, observes them and measures them. In other words, it is assumed that in well-conducted experiments the perturbation in question can be made as small as one wishes. This hypothesis is always sensibly realized for large-scale phenomena, but it ceases to be so in the atomic world. Indeed, as the fine and profound analyses of Heisenberg and Bohr have shown, it follows from the existence of the quantum of action that all attempts to measure a characteristic quantity of a given system has the effect of changing in an unknown fashion the other quantities associated with this system. More precisely, any measurement of a quantity which permits the exact localization of a system in space and time has the effect of changing in an unknown fashion a quantity which is conjugate with the first and which serves to specify the dynamic state of the system. In particular, it is impossible to measure at the same time and with precision any two conjugate quantities. We can then understand in what sense it can be said that the existence of the quantum of action makes a spatio-temporal localization of the parts of a system incompatible with attributing a well-defined dynamic state to that system, since, in order to localize the parts of a system, it is necessary to know exactly a series of quantities, the knowledge of which excludes that of the conjugate quantities relative to the dynamic state, and conversely. The existence of quanta imposes a lower limit of a very definite kind on the perturbations which a physicist can exert on the systems he is studying. Thus, one of the hypotheses which served implicitly as a basis of classical physics is contradicted and the consequences of this fact are considerable.

From the above it follows that we are never able to know with precision more than half the quantities the knowledge of which would be necessary for an exact description of a system in accordance with classical ideas. The value of a quantity characterizing a system is indeed that much more uncertain when that of its conjugate quantity is more exactly known. From this springs an important difference between the old and the new physics in so far as the determinism of natural phenomena is concerned. In the old physics, the simultaneous knowledge of the quantities fixing the position of the parts of a system and of the conjugate dynamic quantities permitted, at least in principle, a rigorous calculation of the state of the system at a later instant. Knowing with precision the values x_0, y_0 . . . of the quantities characterizing a system at an instant t_0, we could predict without ambiguity what values x, y . . . would be found for these quantities if they were determined at a later instant t. This resulted from the form of the basic equations of the physical and mechanical theories and from the mathematical properties of these equations. This possibility of a rigorous forecasting of future phenomena starting from present phenomena, a possibility implying that the future is in some manner contained in the present and adds nothing to it, constituted what has been called the determinism of natural phenomena. But this possibility of a rigorous forecasting requires the exact knowledge at the same instant of time of the variables of spatial localization and of the conjugate dynamic variables; now, it is precisely this knowledge which quantum physics considers impossible. Hence there has resulted a considerable change in the manner in which physicists (or at least a great many of them) today conceive the power of forecasting in physical theories and of the concatenation of natural phenomena. Having determined the values of the quantities characterizing a system at an instant t_0 with the uncertainties with which they are necessarily affected in quantum theory, the physicist can not predict exactly what the value of these quantities will be at a later instant; he can only state what the probability is that a determination of these quantities at a latest instant t will furnish certain values. The bond between successive results of measurements, which convey for the physicist the quantitative aspect of phenomena, is no longer a causal bond consistent with the classical deterministic scheme, but rather a bond of probability which alone is compatible with the uncertainties that, as we have explained above, derive from the very existence of the quantum of action. And here is an essential modification of our conception of physical laws, a modification of which all the philosophical consequences are still far, we believe, from being fully realized.

Two ideas of considerable general application have come out of the recent evolution of theoretical physics: that of complementarity in the sense of Bohr and that of the limitation of concepts. Bohr was the first to observe that in the new quantum physics, in the form which the development of wave mechanics has imparted to it, the ideas of corpuscles and waves, of spatio-temporal localization and well-defined dynamic states are 'complementary'; by that he means that a complete description of observable phenomena requires that these concepts be employed in turn, but that these concepts nevertheless are in a sense irreconcilable, the images that they furnish never being simultaneously applicable *in toto* to a description of reality. For example, a great number of observed facts in atomic physics can be treated simply by invoking only the idea of corpuscles so that the use of this idea can be considered indispensable to the physicist; likewise, the idea of waves is equally indispensable for the description of a great number of phenomena. If one of these two ideas were rigorously applied to reality, it would completely exclude the other. But it is found that both of them in fact are useful in a certain degree for the description of phenomena and that, despite their contradictory character, they should be used alternately depending on the situation. It is the same for the ideas of a spatio-temporal localization and of a well-defined dynamic state: they too are 'complementary' as are the ideas of corpuscles and waves to which they are moreover closely allied as we shall later see. It can be asked how it is these two contradictory images never happen to collide head-on. We have already indicated the reason for this: the two complementary images can not collide head-on because it is impossible to determine simultaneously all the details which would permit making these two images entirely precise and this impossibility, which is expressed in analytical language by the Uncertainty Relations of Heisenberg, rests squarely on the existence of the quantum of action. In this way the important role played by the discovery of quanta in the evolution of contemporary theoretical physics appears in all its clarity.

To complementarity in the sense of Bohr, the limitation of concepts is closely allied. Simple images, such as that of a corpuscle, of a wave, of a point in space, of a well-defined state of motion, are, in short, abstractions, idealizations. In a great number of cases, these idealizations are found to be approximately realized in nature, but they nevertheless have a limited application; the validity of each of these idealizations is limited by the validity of the 'complementary' idealization. Thus corpuscles can be said to exist since a great number of phenomena can be interpreted by invoking their existence. Nevertheless, in other phenomena the corpuscular aspect is more or less hidden and it is an undulatory aspect that is revealed. These more or less schematic idealizations which our mind constructs are capable of representing certain aspects of things, but they have their limits and can not incorporate into their rigid forms all the richness of reality.

We do not wish to continue too long this first survey of the new viewpoints, which has let us get a glimpse of the development of quantum physics. We shall have the occasion in the course of this volume to take up again all these questions one by one, completing them and investigating them thoroughly. What we have said will suffice to show the reader how deep the interest of the quantum theory is: it has not only stimulated atomic physics, which is the most really alive and zealous branch of the physical sciences, but it has also incontestably broadened our horizons and introduced a number of new ways of thinking, deep traces of which will without doubt remain in the future expansion of human thought. For this reason, quantum physics should interest not only specialists: it merits the attention of all cultivated men.

2 Classical mechanics and physics are approximations

We now should like to examine quickly what value the whole of classical mechanics and physics retains in the eyes of a quantum physicist. Of course, they retain practically all their value in the domain of facts for which they were created and in which they have been verified. The discovery of quanta does not prevent the laws of the fall of heavy bodies or those of geometrical optics from remaining valid. Each time a law has been verified in an incontestable manner to a certain degree of approximation (all verification carries with it a certain degree of approximation), we have a definitely acquired result which no later speculation is able to undo. If it were not thus, no science would be possible. But it can very well be that, in the light of new experimental facts or of new theoretical conceptions, we are led to consider previously verified laws as being only approximate, that is to assume that, if the precision of the verification were indefinitely increased, the laws would not be more exactly verified. This has happened many times in the course of the history of science. Thus the laws of geometrical optics – for example, the rectilinear propagation of light – although having been verified with precision and at first regarded as rigorously true, were seen to be only approximations that day when the phenomena of diffraction and the wave character of light were discovered. It is just by this process of successive approximations that science is capable of progressing without contradicting itself. The structures that it has solidly built are not overthrown by subsequent progress, but rather incorporated into a broader structure.

It is in this way that classical mechanics and physics can be regarded as entering into the framework of quantum physics. Classical mechanics and physics were built up in order to give an account of phenomena which operate on an ordinary, human scale and they are also valid for larger scales, the astronomical ones. But, if one descends to the atomic scale, the existence of quanta is going to limit their validity. Why should this be so? Because the value of the quantum of action, measured by Planck's famous constant, is extraordinarily small in relation to our usual units, i.e., in relation to the quantities that are found on our scale. The perturbations introduced by the existence of quanta, in particular the uncertainties of Heisenberg, are a great deal too small in the usual conditions on our scale to be perceptible; they are indeed a great deal smaller than the inevitable experimental errors which affect the verification of the classical laws.

In the light of the quantum theories, classical mechanics and physics seem then not to be rigorously exact in principle, but their inexactness is completely masked in usual cases by experimental errors in such a way that they constitute excellent approximations for phenomena on our scale. Thus we meet again the customary process of scientific progress: well-established principles, well-verified laws are conserved, but they can be considered valid only as approximations for certain categories of facts.

Perhaps, in the presence of this validity of classical mechanics or physics for facts on our ordinary scale where quanta do not intervene, one might be tempted to say: 'In short, quanta do not have all the importance you attribute to them since, in all the immense domain where classical mechanics and physics are valid, a domain which covers in particular that of practical applications, quanta can be completely ignored.' Such a way of looking at this matter does not seem to us justified. First of all, in a domain so alive, so important, so full of future possibilities, as that of atomic and nuclear physics, quanta do play an essential role and it is totally impossible to interpret the phenomena without appeal to them. Then, in macroscopic physics, quanta, although hidden by virtue of their smallness through an unavoidable lack of precision in measurement are however there, and their existence entails in principle all the consequences we have enumerated: if in practice these consequences have no appreciable influence, that does not detract in any way from their general, philosophical application. The knowledge and study of the quantum of action is one of the essential bases of natural philosophy today.

Like many others, I was fascinated by the Curie family: Marie and Pierre, Nobel prizewinners, daughter Irène and husband Frédéric (the Joliot-Curies), also Nobel laureates, and each couple working as partners in highly original fields. In *Frédéric Joliot-Curie: a biography* (1976), I wrote:

CREATION – AND WARNING
by Maurice Goldsmith

The experiments undertaken by the Joliot-Curies led them to a series of pronouncements which soon convinced physicists that a team of distinction was at work in Paris. Despite this, they had still to achieve one great discovery, which everyone would declare without doubt to be theirs uniquely. The great discovery, which detached itself like a fruit perfectly ripened on the finely shaped tree of delicate, sensitive experiment they cultivated, was that of artificial radioactivity. It was to lead them to world fame: and Frédéric, in particular, into an enlarged universe of social striving in which the simple, direct happiness of scientific discovery was never again to be his.

At this period of the early 30s, the only subatomic particles known were electrons, protons and neutrons. There was also the photon, a 'particle' without mass associated with electromagnetic radiation. But to provide a satisfactory theory of the nucleus theoretical physicists were drawing attention to the need for further types of subatomic particles. One lead that was followed was in the study of a mysterious radiation which originated in space. Following the pioneer work of the Austrian physicist, V. F. Hess, who before the first world war had declared, as a result of his remarkable experiments with balloons in flight, that the unknown radiation originated in outer space, it was believed by the mid-20s that this radiation caused ionization in air. In 1925, the American, Robert Millikan, suggested the name 'cosmic rays' for this radiation. They were considered to be a form of electro-magnetic radiation, although it was possible that they might be particles of a special kind. They were more penetrating than either X-rays or gamma-rays. Another American, A. H. Compton, showed, at the beginning of the 30s, that they were charged particles because they were deflected by the earth's magnetic lines of force, and did not fall equally on all parts of the earth's surface.

From April 25 to May 8, 1932, the Joliot-Curies were at the scientific station, 3,500 metres high on the Jungfraujoch in Switzerland, doing some experiments on the effect of cosmic rays on the atomic nucleus. They had negative results, and came to the conclusion that neutrons were not a major constituent in cosmic radiation. Fred wrote to his mother, 'We are working hard here, and when the weather is good we ski on the glaciers. The countryside is imposing and impressive. The Institute hangs from the side of the mountain, and opposite we see the glaciers with their wide crevasses and the Jungfrau. It's a dream to work so comfortably in this fine Institute. However, we are sorry not to be with you, and to be deprived of Hélène and Pierre. We eat in the big hotel of the Jungfrau station and we are treated like royalty. The food is excellent, despite the height. We are the only workers in the Institute.'

In that same year, the American, Carl Anderson, after studying thousands of tracks obtained in a Wilson chamber divided into two by lead, found a remarkable particle to have been ejected: it had a curvature which indicated it had a mass equal to that of the electron, but it curved in an opposite direction. This was brilliantly confirmed shortly after by Patrick Blackett and G. P. Occhialini in England. It was the positive electron, or positron as Anderson named it, predicted in 1930 by the British theoretician, P. A. M. Dirac. He foresaw that the electron ought to be able to exist in either of two energy states, in one of which it was an electron with negative charge, and in the other with positive charge. (We recognize the positron now as an anti-particle, and might call it the anti-electron.)

Once again, the Joliot-Curies might have discovered it first. During their researches, they had noticed that among the electron tracks in their Wilson chamber some were curved in an opposite direction to that of other electrons coming from the same source. As the later work of Anderson and Blackett made clear, these were positrons. The Joliot-Curies were able to produce the first photograph showing the appearance of a pair, a negative electron and a positron, due to the action of a photon. This they called 'external conversion', in contrast to 'internal conversion' in which instead of emitting a photon an excited nucleus emitted an electron/positron pair. This work was a stepping stone to their great discovery.

Experiments by Joliot showed that the absorption of positive electrons in lead or aluminium gave birth to photons of energy around 500 KeV, and he was able to confirm part of Dirac's prediction. Joliot determined the energy of the photons emitted by measuring their absorption in lead of thickness sufficient to obtain fine precision, and to show in addition to the photons of 500 KeV those of 1,000 KeV.

From time to time, Irène had to leave the laboratory and rest, not at home but in the country or mountains: the sickness from which she was to die eventually was always with her. The letters she exchanged with Joliot contain many references to their research hopes and fears. On July 22, 1932, Fred wrote that he had only just been able to get tickets on the night train to Arcouest as it was almost fully booked. He adds, 'I have wrapped the source of Po in a thin sheet of silver and placed them in a glass tube. But we have trouble with the electrometer because the thread keeps on sticking. That's due to the blasted vaseline that you've used. It really is a most impressive gaffe, my dear. Savel has the same inconvenience with his apparatus.' A few days later, he repeated a question that their daughter, Hélène, then five, had asked about her new brother, Pierre. 'Tell me, papa, how much have you paid for the little brother?' 'I replied to her that it was Irène who had made him. Hélène was not astonished, but she asked if she, too, when she was taller, could also make one . . .' He went on, 'I am at the laboratory every day where I am clearing up many things. I have taught Lecoin how to use the directional counter, and I've been able to regulate the Hoffmann apparatus of Savel and ours which were making errors. We are doing measurements of Po.' Joliot then went on to describe the various domestic chores he had done.

The experiments the Joliot-Curies were engaged in early in 1933 showed that when fluorine, aluminium and sodium were bombarded by alpha-rays they emitted neutrons, and both positive and negative electrons.* This occurred also with borium and beryllium. All these elements when irradiated by alpha-rays emitted neutrons and, except for beryllium, also protons. Their statistics on the number of negatons (negative electrons) and of positrons (positive electrons) emitted by each of these elements following alpha-bombardment, and the measurement of the kinetic energies

---

* At one time they suggested the possibility of the capture by a proton in the nucleus of one of the electrons in the atom, converting the proton into a neutron. But they did not develop the idea at the time. It was not correct in their case, but it is of interest that the process, which is known as orbital electron capture, was discovered a few years later in other atoms by L. Alvarez at Berkeley, and is a fairly important process in nuclear physics.

of these electrons, revealed a great difference between beryllium and the other elements. Joliot had designed and constructed a Wilson chamber particularly well adapted for this field of research. He and his wife interpreted the emission of positrons by beryllium as due to an internal conversion of gamma-rays from the beryllium nucleus, the gamma-ray energy being converted into an electron/positron pair. On the other hand, the emission of positrons by boron, fluorine, sodium and aluminium was associated in some way with the neutron emission. They conclude that there were for these elements two types of nuclear reaction: one accompanied by the emission of a proton, and the other by a neutron and a positive electron. In both cases, the final product was the same stable element. One immediate consequence was a revision of the mass of the neutron. Chadwick had deduced that its mass was 1.006, thus less than that of the proton. The Joliot-Curies were able to make the first exact measurements of this mass, and they showed that it was greater than that of the proton.

At the Seventh Solvay Conference, held in Brussels from October 22–29, 1933, all the key figures in atomic physics were present – including, for the first time, the young American, Ernest Lawrence, inventor of the cyclotron. Langevin was chairman. On October 25, Fred wrote to his mother that they had been in the city for two days, but were working so hard they had not even found time to make social visits. The Joliot-Curies presented a lengthy paper on their recent work, which led to considerable discussion. Lise Meitner, the brilliant Austrian and the first woman to become a professor in the University of Berlin, increased their anxiety. She bombarded them with questions. She told them very plainly, 'That's not so'. She herself had carried out the same experiment and had never found anything but a proton to be emitted. If there was a positive electron, where did it come from? Although the young French couple were sure of the accuracy of their work, they were very upset by the doubts thrown on it. In fact, the majority of physicists there did not believe that their experiments had been accurate. After the meeting they felt rather depressed, although Niels Bohr took them aside and told them he thought their results were very important. But it was the encouragement of Wolfgang Pauli which stimulated and moved them most. They were not aware that Lawrence, too, was depressed at the highly critical way in which his report on deuterium experiments, which led him to conclude that the neutron and proton did not differ significantly in mass, was received. The impact of the Solvay meeting on the Joliots was also to affect Lawrence.

On their return to Paris, they decided that to justify their hypothesis it was necessary to show that the emission of neutrons and positrons always occurred simultaneously, whatever the energy of the bombarding alpha-rays. They began to irradiate aluminium with alpha rays and gradually decreased the energy. They noted that neutrons were no longer emitted. But positrons continued to appear for a certain time *even after the alpha rays had been removed.* They had demonstrated that their hypothesis was false. But their great discovery had been brought a step nearer.

'We were surprised to observe that when we progressively reduce the energy of the alpha rays, the emission of neutrons ceases altogether when a minimum velocity is reached while that of positive electrons continues and decreases only over a period of time, like the radiation of electrons from a naturally radioactive element. We repeated the experiment in a simplified form. We bombarded aluminium with alpha rays with their maximum velocity. Then, after a certain period of irradiation, we removed the source of alpha rays. We now observed that the sheet of aluminium continued to emit positive electrons over a period of several minutes. Everything became clear!' But not clear enough. They still had to be certain that the way in which the Geiger counter was behaving was due to the phenomenon they now understood, and not to a fault in the device.

And Joliot was right in wanting to have it checked. All the apparatus used was very primitive. It was home-made. Today's routine everywhere-in-use Geiger-Müller counter was then a novelty, and could not be bought commercially, so that if a laboratory wanted one, it had to make its own. Joliot's counter had a very small functioning zone of only a few volts. The amplifier which transformed the electric impulse, due to the passage of the electrons in the counter, into a signal which could activate an enumerator, included for the lower frequencies an old radio set made many years before by Joliot. Counter and amplifier worked with accumulators and batteries, and the enumerator could only accept impulses of up to 150 per minute. It was six in the evening, and he and Irène had an urgent invitation to dinner which they could not put off. He asked Wolfgang Gentner, a young German working with him, and an expert in the use of the counter, to check the movement. Gentner said, 'He explained everything very clearly to me, and left me to it. I did so, and left a note on Joliot's desk that I had found the Geiger counter in accurate working order.'

When he read this in the morning, Joliot was sure that he and his wife had discovered a new phenomenon – artificial radioactivity.* This has been stated in the plural, as Joliot would have

---

* That is the classic name used. But Joliot did not like it. He thought it badly chosen. The new radioactivity to him was just as spontaneous and natural as that of the normal radioactive elements. He would have liked the adjective 'artificial' applied to the radiosotopes produced, and not to their radioactivity.

insisted. But the original final observation was made by him alone working with his Wilson chamber. He understood the importance of the observation he had just made, and the urgent need as speedily as possible to derive all the consequences he foresaw. He went to fetch Irène from a nearby laboratory, to involve her in the experiments they would have to do to provide the physical and chemical proofs of the discovery.

As physicist he could be satisfied. But what about satisfying the chemist? That specialist would want physical proof of the existence of the new radioactive element; he would want to see it, isolated in a test tube. The Joliot-Curies set to work. Speedily, they bombarded aluminium and obtained radio-phosphorus. Its half-life is 3 minutes 15 seconds. Within three minutes, Joliot was able to identify it chemically: it was a record no chemist had ever dreamed could be set up. He felt 'a child's joy. I began to run and jump around in that vast basement which was empty at the time. I thought of the consequences which might follow from the discovery.'

That afternoon on Friday January 15, 1934, Pierre Biquard received an excited phone call. He ran from his nearby laboratory in the Rue Vauquelin to Joliot's sub-basement lab in the Rue Pierre Curie in a few minutes. He recalls that 'the apparatus he wanted to show me consisted of equipment scattered over several tables. Its newness and apparent disorder revealed to even the least observant observer that here was an experiment set up in haste, to reproduce as a demonstration a discovery made several hours before with Irène . . . Joliot summarized for me the experiments previously carried out in this field, taking great care – as he always did – to specify the part played by each one of them. Then he came to his own experiment. "I irradiate this target with alpha rays from my source: you can hear the Geiger counter crackling. I remove the source: the crackling ought to stop, but in fact it continues . . . ." At that moment the laboratory door opened, behind the experimenter . . . and Marie Curie and Paul Langevin came in.'

Joliot repeated the demonstration with the same precision and simplicity. All this took hardly more than half an hour. Biquard remembers that few words were exchanged, apart from several brief questions and the precise replies they received. 'I was not the only person to have retained an unforgettable memory of it. Joliot often later reminded me of that moment and spoke of his emotion, his pride and his joy at having been able to offer before these two scientists, to whom he was bound by so many ties, a fresh example of the vital character and ever-widening horizons of science.' Thanks to the excellent quality of the remarkable husband and wife team, Jean Perrin was able to present their paper to the Académie des Sciences on the following Monday. It was followed a fortnight later with another key paper.

Let me repeat what Frédéric and Irène Joliot-Curie had achieved. Physicists were agreed that when a particle – alpha-ray, proton, or neutron – entered a nucleus, the nucleus disintegrated to expel a particle different from the original one. It was supposed that as a result a stable nucleus would remain, and this does happen although on occasion it will be a rare isotope, such as oxygen-17 or carbon-13. However, the nuclei of fluorine, of sodium, and of aluminium, when acted upon by alpha-rays, may emit protons, in which case a stable nucleus results;* and may emit also neutrons, representing an instability for which an explanation was difficult.†

They bombarded a thin sheet of aluminium foil with alpha-rays from a polonium source. Using a Wilson expansion chamber, they observed the emission of protons from the transmutation of aluminium, and also many electrons. When the polonium source was covered with a thin layer of beryllium, both positive and negative electrons were emitted. They attributed these to internal conversion of the gamma-radiation produced in beryllium by the alpha-particle bombardment.

But their explanation could not fit the case where positive electrons were emitted by aluminium. They put forward an interpretation which met this; they suggested that when an alpha-particle was captured by a $^{27}_{13}$ Al nucleus, there could be emission either of a proton, or of a neutron and a positive electron. The resulting nucleus was in either case the stable $^{30}_{14}$ Si.

Whilst the Joliot-Curies were doing experiments to determine the minimum energy of alpha-rays which would produce positrons from aluminium, they noticed the emission of those particles was not instantaneous; the phenomenon occurred only after several minutes of irradiation, and continued for a short time after the irradiation had ceased.

Thus, when aluminium foil was irradiated for a few minutes, after it was withdrawn from the source it had an activity which decreased by half in 3 minutes 15 seconds. The radiation consisted of positive electrons. The decay of the activity was exponential: for irradiated boron and magnesium it was 14 minutes and 25 minutes respectively.

The conclusion was that they were dealing with new radioactive elements, and with a new type of radioactivity in which positive electrons were emitted. They found that the distribution of the electrons was in a continuous spectrum, and this confirmed them

---

* For example, $^{12}_{13}$Al + $^{4}_{2}$He = $^{30}_{14}$Si + $^{1}_{1}$H. The silicon is stable: the position is clear.

† Thus, $^{27}_{13}$Al + $^{4}_{2}$He = $^{30}_{15}$P + $^{1}_{0}$n. What was the unknown phosphorus of mass 30?

in their idea that the radioactive bodies produced were analogous to the natural radioactive elements.

It had never been possible to verify by chemical means the reality of the artificial transmutations, as the number of stable atoms produced in this way was too small to be revealed by any method of analysis. But radioactive elements could be identified by their radiation, despite the small number of atoms. The Joliot-Curies provided the first chemical proof for the transmutations caused. The method used was to dissolve a thin sheet of irradiated aluminium in hydrochloric acid. The hydrogen liberated was collected in a thin-walled tube which was found to carry the activity with it. As the decay of the activity was very speedy, the work had to be done in three minutes. This, and other chemical experiments, provided proof that the element formed by transmutation was different from the original element, and that the alpha-particle was captured in the nucleus.

The Joliot-Curies guessed that at Berkeley, 'chez Lawrence, around the cyclotron for a year already everything had been radioactive; but that had escaped them'. And when on January 15, 1934, their account of the discovery appeared in C.R., it ended with these words: 'To summarize, for the first time it has been possible to make certain atomic nuclei radioactive using an external source. This radioactivity can persist for a measurable time in the absence of the source which excites it. Long-lived radioactivity, analogous to that which we have observed, no doubt can be created by several nuclear reactions. For example, the nucleus $^{13}_{7}N$ which is radioactive on our hypothesis, could be obtained by bombarding carbon with deuterons, following the emission of a neutron.'

Lawrence read this on the afternoon of February 20. He thought it might be a message to him from Joliot. In his laboratory at Berkeley, the Geiger counter and the cyclotron worked on the same switch so that Lawrence had never been able to see whether his counters continued to count after the cyclotron was turned off. The reason for this was that by turning them both on and off simultaneously the experiments could be speeded up. Now, Lawrence ordered the wiring to be changed, so that they operated independently and had a carbon target put into the beam. 'Click . . . click . . . click went the Geiger counter,' said Livingston, a colleague of Lawrence, 'it was a sound that none who was there would ever forget.' Thus, the first nitrogen-13 ever known to exist in the world announced its presence. The staff had often bombarded carbon, but always before had switched off the Geiger counter with the beam. Now they stared at each other with an identical expression on their faces. 'We looked pretty silly,' said Thornton, another member of the team. 'We could have made the discovery any time.' Livingston, the most highly educated, interpreted most tersely, 'We felt like kicking each other's butts!'

The Joliot-Curies were now world famous. Their discovery led to the development of radio-chemistry with artificial substances. Marie and Pierre Curie had made experiments in radioactive chemistry with small quantities of naturally occurring radioactive substances. Now, all kinds of radioactive materials could be made, and applied widely in biology, medicine, industry and throughout the sciences and engineering.

Of the many letters of congratulations they received, none moved them as much as that from Rutherford. He wrote on January 29, expressing his delight at seeing the account of their experiments in producing a radioactive body by exposure to alpha-rays. 'I congratulate you both on a fine piece of work which I am sure will ultimately prove of importance. I am personally very much interested in your results as I have long thought that some such an effect should be observed under the right conditions. In the past I have tried a number of experiments using a sensitive electroscope to detect such effects but without any success. We also tried the effect of protons last year on the heavy elements but with negative results. With best wishes to you both for the further success of your investigations.' As PS. he added, 'We shall try and see whether similar effects appear in proton or diplon* bombardment.'

At about this time, there took place a most moving incident. Frédéric and Irène Joliot-Curie presented to Marie Curie the first chemically isolated artificially radioactive element. Frédéric recalled, 'Marie Curie had been witness to our research work, and I shall never forget the intense expression of joy which seized her when Irène and I showed the first artificial radio-element in a small test-tube. I can still see her taking between her fingers, burnt and scarred by radium, the small tube containing the feebly active material. To check what we had told her, she placed it near to a Geiger counter to hear the many clicks given off by the rate meter. This was without doubt the last great moment of satisfaction in her life. She died a few months later (July 4) of leukaemia.'

On November 14, 1935, Joliot received a telegram from Stockholm:

DD STOCKHOLM 201746 142339 NORTHERN CTF
J'AI L'HONNEUR DE VOUS INFORMER QUE L'ACADEMIE DES
SCIENCES DE SUEDE A DECERNE LE PRIX NOBEL DE CHIMIE DE
1935 A VOUS ET MADAME JOLIOT-CURIE LETTRE SUIT=

SIGNE PLEIJEL SECRETAIRE PERPETUEL

* The name he proposed for the deuteron.

'I have the honour to inform you that the Swedish Academy of Sciences has awarded the Nobel Prize for chemistry for 1935 to you and Madame Joliot-Curie letter follows.'

I was sad to hear of the death of John Davy in 1984 at the early age of 57. For many years he was the Science Correspondent of *The Observer*, a leading Sunday newspaper. I had helped to start him on his science writing career after his father, Charles Davy, *Observer* assistant editor, had sought my advice. The following article published in *The Observer* on 12 February 1961 shows the naturalness of his writing:

THE ORIGIN OF THE UNIVERSE
John Davy

Just as it was natural to think of the earth as the centre of the solar system, so for a long time there was a tendency to think of the sun as the centre of the starry system. But during this century the sun has been steadily demoted. It is now seen as a rather modest star situated towards the edge of a disc-shaped galaxy or array of stars. This galaxy is itself one member of a 'local' cluster of galaxies – and thousands of such clusters are scattered through the universe.

At the same time, our scale of cosmic distance has stretched vertiginously. By the turn of the century, the nearest star was known to be four 'light years' away, or 2,352,000 million miles. By 1930 it was agreed that the diameter of our galaxy is about 100,000 light years, while a neighbouring galaxy in Andromeda is nearly one million light years away. By 1936 the American, Edwin Hubble, was speaking of galaxies 500 million light years away, but in 1952 Walter Baade reinterpreted the observations and blithely multiplied the size of Hubble's universe by two. Subsequent corrections have again tripled this, and distances of several thousand million light years are now almost commonplace.

But the most startling discovery has been that all these distant galaxies appear to be hurtling away from us – and the farther away they are, the faster they hurtle. This is the starting point for all modern cosmologies.

The observation was based on a phenomenon which may be experienced on a railway platform. The whistle of an approaching express train sounds higher than that of a receding one – the pitch drops suddenly as the train roars through the station. This is because the sound waves are crowded together, so to speak, as the train approaches, and stretched out as it recedes.

The same effect applies to light waves. An approaching galaxy, blowing its luminous train whistle, should shift its light to a higher 'pitch' – it looks slightly more blue in colour. A receding galaxy looks redder. The effect is very slight, but it can be detected by measuring a shift in certain dark lines which appear when the light from the galaxy is analysed with a spectroscope.

By 1930 it had become clear that most of the galaxies showed a 'red shift', and were thus receding. But even more startling, the farther the galaxy, the greater the shift. Last summer, an object was photographed whose red shift is so great that it must be receding at nearly half the speed of light, and is about 4,500 million light years away.

If this relation between speed and distance holds still farther out, there could be galaxies receding from us so fast that their light never reaches us, and they can never be observed. Thus the speed of light now defines the ultimate frontier of the observable universe. It is within this frontier that the cosmologist must go to work.

The discovery of the receding galaxies led straight to the 'big bang' theory of the universe. If the galaxies are getting farther apart, they must originally have been closer together. By measuring the speed of expansion and reversing it, one can calculate a time in the past when all the galaxies were packed close together.

At this time, it was supposed, the universe may have consisted of a 'primeval atom' of incredibly concentrated energy. This atom then blew apart, and the universe has been expanding like a bomb ever since. In the future, it may simply disperse and run down like a clock – or it might start to contract again into another primeval atom.

Some scientists found this picture profoundly unsatisfactory. It raises unanswerable questions about the origin of the universe, the nature of the primeval atom, or the cause of the big bang. Why not simply abolish these questions?

The 'steady state' theory, first discussed half as a joke by Bondi and Gold, asks us to imagine a universe which has always existed and always will, and does not evolve in any way. It would look broadly the same anywhere in space and anywhere in time. This implies that the average number of galaxies in any part of space must stay the same. But astronomy shows that the galaxies are all rushing away from each other. Therefore, said the steady state theory, we must postulate a continuous creation of matter to fill up the gaps.

This, it was suggested, simply appears continuously in empty space in the form of hydrogen It condenses to form new galaxies, which develop just fast enough to balance the dispersion of existing ones, thus maintaining the average galaxy population for every part of the universe at the same time. Fred Hoyle showed that this daring idea, apparently contradicting classical laws of physics about the conservation of matter and energy, could be firmly based on a development of the relativity theory.

These ideas have gained momentum. Their great attraction was that they could be tested by experiment. If the universe is evolving, the galaxies would have been closer together in the past than they are to-day; but if the universe is in a steady state, it would look the same however far back in time we went.

Now astronomy is in the curious situation that it can undertake a kind of time travel and test these alternatives directly. The reason is that the stars and galaxies are so far away that light takes an enormous time to reach us. Light reaching us now left the more distant galaxies thousands of millions of years ago. We are thus not seeing them *as they are now*, but *as they were*. What is more, the stars that are farthest off in space are farthest off in time. Probing out into space with telescopes is like digging down a geological stratum of fossilised light. And a study of the most distant galaxies can indicate what the universe was like long ago.

If the big bang theory is right, the most distant galaxies we see should be closer together – since we are seeing them as they were nearer to the time of the bang. But if the 'steady state' theory is right, the distant galaxies should be distributed in the same way as the near ones. This was the most direct and obvious test suggested by the theory.

But a snag cropped up. It turned out that optical telescopes could not hope to see quite far enough out into space (or, if you prefer, back into time). The 200-inch telescope at Mount Palomar, California, the largest in the world, can just begin to probe the regions where there should be a detectable difference between an evolving and a 'steady state' universe. But the results are too marginal to be conclusive.

At this point, radio astronomy came into its own. Surveys of the sky with radio telescopes revealed a considerable number of 'radio stars'. These were point-like sources of radio waves in the sky, only a few of which seemed associated with objects visible through telescopes. But in 1952, Walter Baade turned the 200-inch telescope on the position of an intense radio star in the constellation of Cygnus which had been accurately pin-pointed by F. G. Smith at Cambridge. He succeeded in photographing a most peculiar object; it appeared to be a pair of galaxies in collision about 500 million light years away. Another theory holds that it is one galaxy splitting in two, but the important thing is that it emits extremely powerful radio waves. Optically it is almost invisible, but to the radio telescope it is as 'bright' as the sun.

It was immediately realised that if such a distant object could generate such powerful radio waves, similar objects at far greater distances would still be detectable by radio telescopes. In other words, the radio astronomers could probe much further into space and time, and settle the cosmological controversy.

This, in effect, is what the Cambridge team under Professor Martin Ryle are now confident that they have done. Since 1958, the team have been using a new and extremely powerful radio telescope. With this, they have made the most detailed survey and analysis of parts of the radio sky yet undertaken. On it, they have based the elaborate studies presented in the Royal Astronomical Society last week. The work is intricate, and the argument involved – but the gist of it is this:

If, like the radio telescope, we could perceive the radio waves from the sky, we would see many bright radio stars on a dimly luminous background of general radiation. We would ask, straight away: what are the radio stars, how far away are they, and what causes the background?

The first move was to direct optical telescopes at radio stars, and see if they are associated with any visible objects. Most of them, it turns out, are not. A few radio stars have been identified with visible objects inside our galaxy; a few more represent radio emission from nearby galaxies; and a rather larger number have been identified with extremely faint objects, some of which are probably colliding (or splitting) galaxies like the Cygnus source. These are a million times more powerful radio emitters than our own or nearby galaxies.

But there remain a very large number of radio stars for which no corresponding visible objects can be found. They could be nearby objects, which for some reason give out radio waves but little or no light. Or they could be very distant objects like that Cygnus source – in which case they might be used to settle the cosmological question. How are the two possibilities to be distinguished?

Suppose first, the Cambridge team said, that most of the radio stars are not far away, but are associated with our own galaxy. Suppose, too, that they are all powerful enough to be detected by the radio telescope. In this case, since we are situated towards the edge of our galaxy, we would expect to see more radio stars towards the galactic centre than towards the edge.

Our Cambridge study, therefore, was to analyse the distribution of the radio stars in the sky – and it was shown conclusively that they are distributed evenly.

Does this prove that the radio stars are beyond the galaxy? No – there is another possibility: they might be very weak radio emitters, so that the radio telescope detects only the nearest ones, *well inside* the galaxy. These would then represent a small sample of the total radio star population of the galaxy. The rest of this population, too faint to be distinguished individually, would contribute to the general luminous background.

If we assume the detectable radio stars are all very weak, they must all be very near, and hence packed rather densely round us. But the more distant, undetectable ones must be equally closely packed – so the galaxy as a whole must be very densely populated with these weak, radio emitting objects. This large population would contribute to the luminous radio background – and would make it very bright. In fact, measurements of the background show that it is nowhere near bright enough, and it turns out that the only way of reconciling the observed background with the observed number of radio stars is to suppose that most of these stars are *outside* the galaxy.

This argument was then extended beyond our galaxy, and it was proved that most radio stars are not only outside the galaxy, but are very far away. The majority of them, in fact, must be rare, very powerful sources, comparable to the Cygnus source, but much more distant. The sources now being detected at Cambridge, in fact, are providing a picture of the universe as it was some 8,000 million years ago – and the sources themselves are receding at something approaching nine-tenths of the speed of light.

The final step at Cambridge was to work out the cosmological implications of this. On the big bang theory, the radio star population should get denser as you go back in time – there should be more of the faintest, most distant sources. On the 'steady state' theory, the population density should stay the same. The Cambridge observations have shown now unmistakably that the density of weaker sources is at least three times, and probably ten times, higher than the 'steady state' theory predicts. Therefore, the universe must be evolving.

From an article in the *Observer*, 12 February 1961

I have written elsewhere in these pages on J. B. S. Haldane. His friend, John Maynard Smith, in an introduction to a collection of J. B. S. essays – *On Being the Right Size and other Essays* (Oxford University Press 1985) – edited by him, writes, 'The impact of those essays was such that, fifteen years later, when I decided to leave engineering and train as a biologist, I entered University College, London, where Haldane was at that time a professor, and became his student and later his colleague.'

STATISTICS
J. B. S. Haldane

Statistics is the branch of science, or perhaps of applied mathematics, dealing with large numbers of individuals. Originally the word meant the compiling of numbers about the state, for example censuses of population, production, and so on. Today it is applied to many other fields.

For example biologists use statistical methods. They shoot birds and count the numbers of various kinds of insects which they have eaten. They compare the average yields of different kinds of wheat, and the effect on them of rainfall, fertilizers, time of sowing, and other influences.

Modern physics is largely based on statistical mechanics. A liquid consists of millions of molecules in rapid motion. Any one of them near the surface may pick up extra speed as the result of collisions with its neighbours and fly away as vapour. One can predict very accurately what fraction of a large number will fly off in the next minute.

Astronomers also use statistics, because by now so many stars have been photographed that it is a hopeless task to catalogue them all. But we can count their frequency in different directions, compare the average speeds at which different kinds are moving, and so on.

There are two essentially different kinds of statistics, namely those based on a complete investigation of a group, and those based on a sample. Government statistics are generally supposed to be of the first kind. For example the census (which by the way ought to be taken in 1941, but will not be) is supposed to cover the whole British population, and probably does not miss one in ten thousand.

But statistics of this kind are often inaccurate. In some countries more births of girls than of boys have been registered because boys were reported as girls to avoid military service. British figures of unearned income are too low because, for example, a capitalist can invest in a new insurance company, drawing no dividend for some years, but be fairly sure that the shares will increase substantially in value. If he later sells out, he can pocket the profits without paying tax.

One might think, however, that it was always better to investigate a whole population than a sample. This is not necessarily so. Some official statistics are very dubious, notably the causes of death which are registered. A doctor sees that a man or woman is ill, does his best to diagnose the disease, and if they die, fills in a form accordingly.

Even the best doctors, with X-rays, bacteriological laboratories, and other such facilities, make some mistakes. The average doctor, with no such help, makes many more. This is probably specially so with young babies and old people.

Old people commonly have weak hearts, and whatever is the ultimate cause of their death, their hearts will often fail sufficiently to justify a doctor in ascribing their death to heart failure. Others die of pneumonia though the organ which first failed is not the lung.

If I were dictator (which heaven forbid) one of the reasons for my unpopularity, and I hope, for my violent removal, would be that I should insist on post-mortem examinations being made of everyone who died during a period of, say, a month. If I did, they would very likely be faked, as many things are under dictatorships, but we should perhaps gain more knowledge of the true causes of death.

For many purposes we can only study samples. For example a few towns have been studied intensively to find out what people actually eat, and from them rough calculations have been made to determine what fraction of the whole people are undernourished.

In the same way numbers of people have been carefully weighed, measured, and tested for colour-blindness, membership of different blood groups, and so on. If the set tested has really been chosen at random, we can calculate with great accuracy to the numbers in the whole population.

For example if out of 10,000 men tested, 250 were colourblind, then the odds are several hundred to one that in a population of a million the number of colour-blind men will be between 20,000 and 30,000, provided the first sample was taken at random.

But it may not have been a random sample. A large number were tested at an exhibition, but perhaps men who suspected they were colour-blind took the test more frequently than normal men. Or perhaps normal men took it, and colour-blind men did not, for fear of looking silly.

Such a test is only valid when it is done on a group selected in some other way, for example of all the children in a group of schools. And of course no amount of care in selecting your sample at random will get over other kinds of bias.

The Ministry of Information collects data on public opinion. Various people are asked their opinion on political questions, and are told that their names will be kept secret, and only total numbers given. Now I personally believe that this promise is kept. But it does not follow that the people who are asked the questions believe it too.

If they don't, some of them will give the answers which they think will please the Government. And so if more people say that they approve of official policies in January than in December, this may mean that the public is getting fonder of Churchill or more frightened of official spying. And there is no way of finding out which explanation is true.

Inquiries of this kind can hardly claim to be scientific. Observations of actual behaviour can. A comparison of the number of people who go to un-reinforced brick surface shelters as compared with basements or trenches would tell a very clear story of the small confidence which they inspire. Lenin was talking scientifically when he spoke about people voting with their feet.

In agricultural experiments great care is taken to avoid bias. Thus if you are testing three kinds of wheat in a field, it is useless to divide the field into three strips of equal area. One may be drier than others, or have better soil. The field must be divided into about thirty plots of equal era, ten sown with each kind of wheat.

It will probably be found that owing to differences in the soil, one or two plots of the best wheat will yield less than some of those of a poorer variety. There are, however, mathematical tests which allow one to say whether the observed higher yield of one kind of wheat over the others means anything, or is due to chance.

Unfortunately tests of this sort are rather rarely applied in official statistics. That is one reason why people say that you can prove anything by statistics. Actually you can prove a good deal. But it needs a lot of training to avoid pitfalls. In a later article I shall write about some of the methods which scientific statisticians use.

More about statistics

Scientists aim, so far as possible, at clear-cut experiments. A plant is grown in a solution of various minerals, including iron salts, and is healthy. The iron salts are left out; the leaves of a similar plant are yellow, and it dies. A sufficiently strong electric current always acts on a compass in its neighbourhood, a man always loses consciousness if he breathes nitrogen, and so on. But very often this cannot be done. Sometimes experiment is impossible. You can't do experimental astronomy. And an experimental study of human heredity is impossible in practice. Sometimes the experiment is not clear-cut. For only some, but not all, of the variation in the result is due to the experiment, and the remainder cannot be eliminated.

Here the statistician comes in. He may be asked two questions. The first is whether the experiment has had any effect at all. Suppose five rats fed on one diet weigh 6.5, 6.8, 6.9, 7.2, and 7.5 ounces and five similar ones fed on another diet weigh 7.1, 7.4, 7.7, 7.9, and 8.3 ounces (of course in scientific work grams would be used for weighing). Can we be sure that the diet has had an effect on the weight?

No, we can never be absolutely sure. But we can be quite sure enough for practical purposes. The statistician does not ask the question in this form. He asks what is the chance that two groups as different as these should have been picked out of the same population by mere luck. The statistical method for answering this question was devised by the late Mr. W. S. Gossett, who was employed by Messrs. Guinness, the brewers. The firm did not permit its employees to publish work under their own names, so he signed his papers 'Student', and many statisticians only learned his name when he died. In the particular case of the rats, the odds are about 40 to 1 against the difference between the two groups being due to chance. A good biologist would be fairly sure that the difference was due to the food, but he would repeat the experiment once or twice before publishing his result.

A second question asked of statisticians is 'How much of the variation in the quantity is determined by variation in another?' In the above case only some of the variation was due to the diet, because there was still a good deal of variation among rats on the same diet.

Human height is to some extent hereditary, though it also depends on diet, and probably on climate and other factors at present unknown. Suppose we measure a thousand boys of the same age, and their parents, we ask what is the difference between the average heights of a group of boys whose fathers are 5 feet 6 inches high and of a group whose fathers are 5 feet 10 inches high. The answer is 'about one inch'. An increase of an inch in the father's height means an average increase of about a quarter of an inch in the sons. The mother's influence is about the same. The degree of resemblance is measured by a number called the coefficient of correlation which, for parents and children, is about a half.

It is one thing to state a statistical result, and quite another to interpret it. The first statisticians who measured resemblances between parents and children put them down wholly to heredity. This is correct in some cases, particularly when the people concerned are taken from one class of a particular nation. For example there was probably little malnutrition among students of Oxford University in 1900.

But it is certainly untrue in other cases. The Japanese mostly have slanting eyes when compared with Europeans. They are also shorter. Both are often thought to be racial characters determined by heredity. If a Japanese couple emigrate, their children still have slanting eyes, so the eye shape may be regarded as a racial character fixed by heredity. But their average stature may be greatly increased, as Suski found after measuring some hundreds. The average height of Japanese boys of 15½ born in Japan is 5 feet 2.8 inches, of those born in America 5 feet 4.4 inches. The weight is increased from 99 pounds to 114 pounds. What is even more remarkable, the American-born Japanese at this age are slightly taller and heavier than American boys of European stock. This may be due to the fact that the whites lived in many different states, the Japanese mostly in California, where conditions seem to favour rapid growth. The few adult American-born Japanese measured averaged three inches taller than their parents.

Results like this make it rather hard to believe most of the stories of racial superiority and inferiority which are spread by conquerors, whether British or German. If a small fraction of the sum spent on the Chinese war had been devoted to feeding the people, the Japanese could probably be converted from a short to a tall race. Let us hope that a Japanese People's Government may soon try this interesting experiment.

This example shows the value and the limitations of statistics. It is a common gibe that 'you can prove anything by statistics'. All that you can prove is some fact such as that the children of tall parents are generally taller than the average. To find out why this is so, you must experiment yourself, or take advantage of a natural or social experiment such as the transplanting of Japanese to California.

But if you can't prove much by statistics alone you can prove nothing at all in some fields without them. Among a thousand badly undernourished children a few will be taller and even stronger than the average. If they are picked out they can be used for a 'sunshine story' about the adequacy of our food. The sunshine stories which we read in the press about shelters come into this category. We all know of cases where a house was destroyed and a shelter untouched. We also know of cases where the contrary happened. Only a statistical investigation could tell us which kinds of shelter are of any value. Such an investigation would have to be carefully done. If we merely compared the fraction of people sleeping in shelters who had been killed to the fraction of people sleeping in bed who had been killed, we should find that bedrooms were much safer than shelters. This would be unfair to the shelters, because few or no people sleep in shelters in country districts, while many do so in heavily bombed areas. For a fair comparison it would be necessary to compare bedrooms and shelters within a number of small districts, say within each of a hundred wards in different boroughs. I do not know what the result would be, though I think it would show that some kinds of shelter are worse than useless, whilst others give partial protection. Very inadequate surveys seem to show that the Anderson shelter cuts down the chance of death to about a quarter, while brick surface shelters without reinforcement are useless, as are many basements. If the Minister of Home Security has made an inquiry of this kind he should publish its results, if only to let people know which kind of shelter to choose. If he has not yet[1] made it, he should be replaced by someone who will apply scientific method to this vital problem.

These examples prove, I think, that statistical methods are indispensable. But they also show that they easily lead to false conclusions, and that everyone in the intellectual side of the Labour Movement should know something about the use and abuse of statistics.

An extract from the work of Jim Crowther is essential, for without him science writing would not have taken off in the form that it did in Britain. His books on British scientists provided a new historical, Marxist approach which shed new insights and reinforced the social responsibility movement so actively pursued between this century's two major wars.

BIOCHEMICAL PHILOSOPHER
J. G. Crowther

When Hopkins was young, before the end of the nineteenth century, chemistry in England was not in a flourishing condition. Organic chemistry was influenced by the needs of the not-very-large synthetic dyestuffs industry, and physical chemistry was meeting the demand for the more efficient management of chemical manufacturing processes. Under these circumstances, both organic and physical chemistry were concerned almost entirely with inanimate processes. It was not surprising, then, that British organic and physical chemists had little feeling and insight for the application of their sciences to the elucidation of the chemistry of living processes. Medicine provided a medium for the investigation of living processes, but it subordinated chemistry to its own limited field and ends. General biology subordinated chemistry to itself in an even cruder way. It propounded a hypothetical chemistry of its own. Living organisms were supposed to be made of 'protoplasm', which did not obey the laws of ordinary chemistry because it was living. Some contended that living issues were built of a peculiar 'biogen' molecule characteristic of life. Any chemical process in a living organism could be dismissed as a characteristic property of 'protoplasm' or 'biogen', and therefore unanalysable.

The organic chemists were absorbed in the development of their intricate dead geometry of molecular structures. The physical chemists were concerned with the transformations of energy in systems of uniform phases. Even the colloid chemists, who at first

[1] January 1941.

sight seemed much more promising, were conccrned mainly with the properties of featureless surfaces.

In England the divorce between biology and chemistry seemed profound. Nowhere were the two sciences being combined on equal terms. There were virtually no posts and no literature.

On the Continent the situation was better. In Germany the study of chemistry in every branch was more advanced. Great organic chemists like Emil Fischer had applied their technique to the elucidation of the constitution of living tissues, with magnificent success. Hofmeister, who had been appointed to his chair of experimental pharmacology in 1885, suggested in 1901 that the orderly sequences of chemical processes in the living cell were the result of the chemical organization of substances by individual chemical agents in the cell.

By 1900 thirty-one volumes of the German *Journal of Physiological Chemistry* had appeared, while the British *Biochemical Journal* was not founded until 1906.

Hopkins entered this scene in the 1890s with the conception of a chemical science, with its own principles, methods, and autonomy which would deal with organic chemistry, medicine, and biology on equal terms. It would explain the chemistry of living processes as well as the chemical constitution of living materials.

He developed his ideas for many years in his Cambridge lectures and discussions. He expounded them in a famous address to the Physiology Section of the British Association in 1913, on the 'Dynamic Side of Biochemistry', and on many occasions in his later years, especially in an address on 'Organicism' in 1927, and on the 'Influence of Chemical Thought on Biology' in 1936.

He described how Liebig, the founder both of organic chemistry and of modern animal chemistry at the beginning of the nineteenth century, had enthusiastically anticipated the great aid that organic chemistry would give to biology, but had been sorely disappointed. As a matter of fact, even in Germany the combination of chemistry with biology never reached, within the nineteenth century, the degree for which Liebig had hoped, and in England it happened hardly at all.

He thought that the fault lay partly in Liebig himself, who, in spite of his brilliance as a chemist and his intense desire to aid animal physiology and agriculture, did not succeed as he had wished because he lacked a biologist's training and a biologist's instincts. The other great mind of the time, Pasteur, was also not altogether fortunate in his influence, for when hc cntered biology from chemistry he 'became almost too much of a biologist' and tended to believe that nothing useful could be discovered about organisms except by studying them as wholes. This strengthened the biologist's belief that protoplasm contained elusive properties which were not susceptible of analysis, and that the chemist's investigations could not touch on the realities with which biology, as the science of life, was concerned.

Hopkins fought the notion of 'protoplasm' and 'proteid', which carried its mental associations, as Pasteur fought the notion of 'spontaneous generation'. He drove the word 'proteid' out of circulation, and when Michael Foster's picture, with this word 'proteid', had to be dated, somebody said at once that it must be 'pre-Hopkins'.

When the journal *Protoplasma* was founded in 1929, his colleagues waited with amusement to see whether he would allow it to be taken in his Institute's library.

Chemical investigation had revealed that the living tissues contained complicated substances such as proteins whose constituent molecules are large. It was difficult to elucidate the structure and properties of these big molecules, so the organic chemist left them severely alone.

The animal tissues yielded sticky, intractable substances, which prompted the organic chemist to the cynical comment that 'Thier-chemie ist Schmierchemie', that 'animal chemistry is the chemistry of mess'. So the organic chemist concentrated on his own nice, clean chemistry of the simpler molecules, on which he practised his 'magical synthetic art' that 'provided him with substances made not by nature but by himself', and which 'were in the highest degree suitable for the development of clear ideas concerning molecular structure.' Thus clear thinking on molecular structure became divorced from biological material.

The biologists seized on the big molecules and revelled in their vagueness and undefined qualities, attributing to them any properties that they pleased.

Hopkins' own patron and admirer, Michael Foster, had in 1895 supposed that when atoms of oxygen passed through the blood into the tissues, they disappeared 'into some protoplasmic complex'. The steps in which the oxygen passed through the processes in the cell before it reappeared in molecules of carbon dioxide had not as yet been traced. 'The whole mystery of life lies hidden in that process and for the present we must be content with knowing the beginning and the end.' With such ideas was associated the notion of a special chemistry of living matter. Hopkins said that in the 1880s a very distinguished organic chemist, on being questioned by him concerning the chemistry of living things, and his desire to pursue it, replied: 'The chemistry of the living? That is the chemistry of protoplasm; that is superchemistry; seek, my young friend, for other ambitions.' And so, as he tells us, his youth was fretted by the antagonism between organic chemistry and animal chemistry, and the belief of the organic chemists that the latter was an inferior discipline. Even in 1913 there was still a widespread distrust of methods of Biochemistry, which were supposed to be amateurish and inexact. He did not, by the way, find the new name 'biochemistry' very attractive. It seemed to contain a hint of biologism, with its theory of a special chemistry of life.

At first sight, physical chemistry seemed more sympathetic than organic chemistry to the elucidation of living processes. It at least was concerned with movement rather than the static geometrical shapes of molecules and structure, with function rather than morphology.

But in the 1890s a movement, led by Wilhelm Ostwald and Ernst Mach, arose out of the triumphs of the generalized theory of energy, which in turn had been born of the steam-engine and the universal impersonality of the industrial revolution, which set out to describe the world in mathematical formulae, without reference to its particular characteristics. They emphasized the current absence of rigid proofs of the existence of individual atoms and molecules, and therefore proposed that these hypothetical entities should be banished from science, together with the visualizations and pictures of them which scientists used to aid them in their researches.

This movement greatly hindered the application of physical chemistry to biology, because it denied the biochemist the use of visualizations of atomic processes which would assist him to imagine how the chemical processes of living matter might operate.

The proof by Jean Perrin of the reality of atoms, and the rise of atomic physics, with its vivid pictures of the atom, the revelation of the ordering of atoms in crystals by X-ray analysis, the extension of the discovery by Langmuir to the elucidation of the ordering of molecules on surfaces ended the non-atomistic, non-visual interlude of Ostwald and Mach.

In fact, the progress of physics has imposed on the physicist and on the organic chemist, at least for a period, 'a mode of thought which is essential for the biologist – the visualization of a mechanism.'

Hopkins told his audience in 1913 that 'in the study of the intermediate processes of metabolism we have to deal, not with complex substances which elude ordinary chemical methods, but with simple substances undergoing comprehensible reactions.' By simple substances he meant those with which organic chemists are accustomed to deal. The biochemist was not concerned only with the separation and identification of the products from the animal, 'but with their reactions in the body; with the dynamic side of biochemical phenomena'.

Thus Hopkins indicated that living processes did not consist of simple combinations of many different kinds of unanalysable, fundamental units, but of complicated combinations of simple units. Life was an enormously complicated variation on a few simple themes, so it might be successfully analysed, in its chemical aspects, by the unremitting pursuit of these few simple and therefore intelligible themes. On the other hand, if life consisted of comparatively simple variations on many unanalysable themes, there was little prospect of progress in its analysis.

The view of the living unit as 'very large and very labile' was 'as inhibitory to productive thought as it is lacking in basis'. There was an obsession that the molecules in tissues were necessarily of high molecular weight, and that simple molecules were of no importance. But, in fact, it was from the study of simple molecules that progress was to be expected. The extraordinarily wide influence of adrenaline in the body was a striking illustration of the importance of a simple molecule in its processes. The subtlety of living processes arises from the complication of their dynamics, and not from complexity of their units.

The life of the living cell 'is the expression of a particular dynamic equilibrium which obtains in a polyphasic system.' It 'is a property of the cell as a whole because it depends upon the organization of processes, upon the equilibrium displayed by the totality of the co-existing phases.'

Thus the presence of substances must not only be proved, and their nature identified. It was necessary to elucidate their role in the dynamics of the cell. Their quantity and prominence were not necessary signs of their importance – indeed, they might be there because they were not working in the cell; the important working molecules might, on the other hand, be too busy to be seen.

So he was certain that the search for simple tissue products would in the future receive important rewards, 'so long as physiologists are alive to the dynamical significance to all of them.'

One can deduce from the response of the living cell to foreign molecules of simple constitution results of profound biological significance. Before it had been observed that benzoic acid, when eaten by an animal, is converted in the body into hippuric acid, it had been believed that the animal body was incapable of performing chemical syntheses.

If one conceived the living cell as a system in dynamical equilibrium, one might be able to elucidate that equilibrium by studying the way in which it was disturbed by the intrusion of a new molecule.

The object of the process of digestion was to present the body with simple molecules. When one viewed the cell as a dynamical system, one gained the prospect of being able to follow these simple molecules through all the processes of metabolism, of their breakdown and oxidation.

Certain molecules may be of 'great importance, not to the structure, but to the dynamics of the body'. The hormones are instances. The structure of the living cell, which forms the most obvious parts, is made of highly complex, stable molecules. This character, and the colloidal condition in which they exist, give it its definite though mobile shape and its mechanical properties. The colloidal complexes form within the cell a milieu in which the dynamic chemical processes can proceed, providing, as it were, special apparatus and an organized laboratory. Within the cell, simple molecules undergo reactions promoted or catalysed by ferments or enzymes.

During the last fifteen years a doctrine had grown up which, while it had difficulties, had 'the supreme merit of possessing an experimental basis', and hence of providing a direction for further work. This was the conception that each chemical reaction within the cell was 'directed and controlled by a specific catalyst'. The evidence for the existence and activity of enzymes within the cell was extensive and increasing rapidly. One need not therefore pause at the 'myriad nature of the army of enzymes that seems called for' by the great multiplicity of chemical reactions which occur in the body and the narrow specificity with which each enzyme acts.

With the huge array of examples before us, we may logically conclude that 'all metabolic tissue reactions are catalysed by enzymes'.

The very complexity of the cell, with its myriad reaction enzymes and multitudes of exchanging atoms, all proceeding in a dynamic equilibrium, suggested that, underlying it, 'we may discover a simplicity which now escapes us'. Hopkins expressed the opinion that the investigation of the cell with the guidance of this dynamic conception was suitable to 'our natural tastes and talents'.

He invited organic chemists to join in the work, though he advised them to spend a year or two acquiring a biological discipline. The organic chemist could do useful work by determining the constitution of substances obtained from tissues, but he will not so become a biochemist. In order to do that, he must learn how the chemical reactions run in the organism. He must get in touch with the animal, which usually does the unexpected. The difficulties of his material and technique may cause the professor of organic chemistry to call his work amateurish, and he will find that, unlike the physical chemist, he can rarely put his results into mathematical form. He may occasionally meet the spectre of Vitalism and wonder, quite unjustifiably, whether his work may be to no purpose.

He should be content to describe the living animal in the sense that the morphologist has described the dead. If 'life' contains some residuum which will always remain elusive, then he can take comfort in the truth expressed by Robert Louis Stevenson:

> To travel hopefully is better than to arrive,
> And the true success is labour.

Sherrington, apparently, heard Hopkins deliver this address in 1913. He wrote in an obituary reminiscence in 1947 of how 'In an uninviting gaunt lecture room he descanted for nearly an hour on the cell as a theatre of chemical processes. Without any deliberate attempt at eloquence, and in a voice that, as he became more interested, fell to such a purely conversational level as to be a little difficult to hear, he conjured up for his audience – some thirty persons all told – a picture of the cell as a tiny sponge-work containing perhaps a thousand foci of different actions co-operatively confined within a unitary whole. An organized factory manifoldly hydrolizing, pulling to pieces, and contemporaneously constructing and reconstructing. And this unity bounded outwardly by a mosaic of countless poles and leaking like a sieve. As I listened I felt I was being privileged for the time to see something of the micro-cosmic world in which my friend's scientific thoughts took shape and did his bidding. One mental factor which, it seemed to me, such thinking must demand, was a peculiar intensity of visual imagination, continuously checked in factual knowledge.'

Through that address, Hopkins became the biochemical philosopher of his generation.

In addition to his onslaught on Vitalism, in the form of the theory of 'protoplasm' endowed with unanalysable qualities peculiar to life, he attacked the Organicism of J. S. Haldane and the Holism of Smuts. Haldane had asserted in 1885 that the problem of physiology is not to obtain piecemeal physico-chemical explanations of physiological processes, but to 'discover the details of structure and activity in each organism as expresssions of its nature as one organism.' Haldane held that a living organism was not a problem for analysis. Its essential attributes are axiomatic. Heredity, for example, is not a problem, but an axiom. He believed that biology and physical science would one day meet, but he confidently predicted that if 'one of the sciences is swallowed up, that one will not be biology.'

Latterly he had said that 'the attempt to analyse living organisms into physical and chemical mechanism is probably the most colossal failure in the whole history of modern science.'

Hopkins said that there was some subtle truth in what Haldane had been trying to say, but that unfortunately he never seemed to have succeeded in finding the right words in which to express his meaning. 'After some years of respectful effort' he could not pretend to have understood Haldane completely.

If the biochemist accepted and acted on Haldane's views, then he must retire gracefully from scientific endeavour. There is no doubt that, as a 'whole' the organism has properties not deducible from its parts, and that these properties are dominant. But the biochemist cannot proceed unless he believes that he is able to reconstruct from his data some significant aspect of the whole organism.

Biology started from the visual description of organisms, their forms, and evolution of their forms. It was concerned with 'what is visible in the phenomena of life', and it produced a kind of biological language which was 'apt to the description of the visible'. On the other hand, the chemist worked for a long time without concern for visible structure, and saw only in his mind arrangements of invisible atoms.

The biochemist must always picture in his mind the molecular events which underlie the changes of forms which interest the biologist. If he is to obtain a picture which will be of any use to him, 'it must be frankly mechanistic'. But he recognizes that the organism is not strictly a machine, for a machine loses its meaning unless it operates for some exterior purpose. An organism is 'adjusted as a mechanism in order that it may be itself'.

Haldane's teaching implies that if the results of experimental biology are to be of value, they must be made on the intact organism. This is contrary to experience. For instance, such very important agents as the hormones, which harmonize the actions of the body, could never have been discovered by experiments on the intact organism.

Where is the mistake of the organicist and the holist? It lies, in part at least, in exaggerating the difference between the whole and the sum of the parts; that the separation is so wide that the acquisition of knowledge of the parts goes only a little way towards the understanding of the whole.

No biochemist will forget that the materials with which he works are part of or have come from 'a sentient organism endowed with attributes which, having neither spatial extension nor numerical values, are beyond his scope.' The organicist, however, goes beyond this. He implies that the parts themselves are different when studied individually, than as in the whole. The biochemist does not desire to be assured of more than the fact that there is no such fundamental difference in category. He will not exaggerate the importance of his picture of living processes, but he claims for it a certain degree of validity.

He was far from maintaining that such fundamental properties as inheritance, growth, and differentiation might not depend upon organization above the chemical level. If there be a hierarchy of levels of organization, then the chemical level will have its place, which will be that of 'self-maintenance'.

He boldly claimed that the living cell is already able to maintain itself as an individual entity in equilibrium with its environment by its physical and chemical attributes, without calling on others at a possibly higher level. So, already, through these attributes, operating through the agency of highly specific catalysts, the cell is able to display one of the characteristic properties of a living organism. He believed that biochemists implicitly held this, or a similar, faith.

It would not be necessary to discover unknown principles before proceeding to describe in physical and chemical terms the cell's capacity for maintaining its identity.

Hopkins combated vitalism, biologist, organicism, and holism. He courteously but firmly fought the organic chemists who scorned the difficult molecules found in living tissues. He explained the error of a physical chemistry preoccupied with the thermodynamics of homogeneous systems, whose philosophers were trying to abolish the visualization of processes in their science, so that its power of dealing with complex living processes was 'strangely lessened'.

He described how, by dynamic equilibrium, the cell could preserve stability of form through incessant change. He marshalled the evidence which showed that the living cell is an interplay of simple molecules which in themselves might easily be known. This enormously increased the biochemist's confidence in the face of the prodigious complexity of living processes.

He produced conceptions which illuminated the biochemist's path, and revealed to the young the possibilities of research and discovery. He protected his colleagues from the miasmas of false philosophies. He provided the biochemist with a picture of the running internal mechanism of the cell, as a guide to thought and experiment. Sherrington compared his role in biochemistry with that of Kekulé in organic chemistry.

He objected to the impression sometimes given by mathematicians and cosmologists that biology and life were unimportant. Bertrand Russell had once ventured to suggest that in passing from physics to biology one is conscious of a transition from the cosmic to the parochial, because from the cosmic point of view life is a very unimportant affair. But 'those who know that supposed parish well are convinced that it is rather a metropolis entitled to much more attention than it sometimes obtains from authors of guidebooks to the universe. It may be small in extent, but it is the seat of all the most significant events.'

Hopkins had begun his scientific life as a naturalist. He was well acquainted with the 'inhabitants of the metropolis', and remained so always. His instinctive biological feeling and insight made his dynamic chemical picture of the living cell all the more fertile and remarkable. It was the product, not of an engineering, but of a biological mind.

Hopkins conducted all his teaching in the most engaging language and exquisite style. Marjory Stephenson rightly said that 'he alone amongst his contemporaries succeeded in formulating the subject.' This was his unique achievement, which, together with his experimental investigations, and his inspiration of others, made him, as E. Mellanby generously and justly said, 'in his sphere the greatest scientist of his time'.

Crowther's biochemical philosopher, Gowland Hopkins, was a forerunner of those who led us to biotechnology, a term which has become an everyday familiarity. It is concerned with the manipulation of the fundamental processes of life.

BIOTECHNOLOGY
from Shell Science & Technology

Products from life

Arriving at a satisfactory definition of what is meant by biotechnology is not easy. A simple statement along the lines of 'the commercial exploitation of living organisms and their chemical products' while accurate is imprecise. Agriculture could well be described in such a way but there are few who would nominate the traditional practices of farming as biotechnologies. Yet despite that, it *is* the combination of living organisms with some commercial end that is at the heart of biotechnology. Perhaps the easiest approach to deciding what can be reasonably included is to discuss some examples.

The best known and earliest biotechnologies are found in the food industry. For instance, in the baking of bread, yeast – a living organism – makes use of the sugar contained in flour to form bubbles of carbon dioxide. These bubbles are trapped in the bread dough and cause it to rise. Traditional cheese making is another biotechnology as it relies on a chemical found in the stomachs of calves to coagulate milk into the familiar dairy product. Of course it was thousands of years after man first began such activities before he had any real insight into what caused them. As understanding grew, though, so did the use of organisms and natural processes. One example has been the introduction of waste treatment plants that are based on microbes. In these plants waste, such as sewage and industrial chemicals which might otherwise pollute rivers and lakes, is converted into harmless carbon dioxide, oxygen and recoverable biomass.

In the types of biotechnology mentioned so far man has taken natural, living organisms or their products and used them to his advantage. The extra push of recent years has come from the discovery of ways to change and improve organisms so that they perform their activities more efficiently or, perhaps, carry out new combinations of tasks. In principle this is not a completely new development as plant and animal breeders have for many years been adapting living organisms. Breeders identify a particularly desirable trait such as speed in a racehorse or high yield in a rice plant and try to mate the parents that possess the best combination of those characteristics that will help to bring this trait about.

By contrast, instead of letting natural breeding cycles take their course, the biotechnologist seeks to alter directly the chemicals that control the workings of living organisms and then to arrive at a known product in a more effective way. For example, certain bacteria that can be grown easily have been altered so that they will make human insulin, a drug used in the treatment of diabetes that is found naturally only in man. Previously doctors used animal insulin, obtained in a very laborious manner from the slaughterhouse.

DNA – the code of life

One chemical lies at the heart of modern biotechnological techniques – deoxyribonucleic acid, better known by the initials DNA. Biotechnologists are interested in DNA because it contains the blueprints for the manufacture of proteins, living organisms' basic biochemicals (Figure 2). Proteins determine the development of an organism to maturity and all the processes of life.

Proteins perform a number of valuable services. Some facilitate the progress of chemical reactions by acting as biological catalysts, these are known as enzymes, while others, hormones, are responsible for the transport of internal messages that control the workings of organisms. Insulin is an example of a hormone, it controls the amount of sugar in the bloodstream and if the body cannot make enough insulin diabetes can follow. Proteins are also involved in the formation of hair, muscle and nails.

The structure of proteins is based on units called amino acids. Insulin consists of 51 amino acids, but proteins with chains over ten times longer are known. Only 20 different amino acids are normally found in the proteins of living organisms and the right

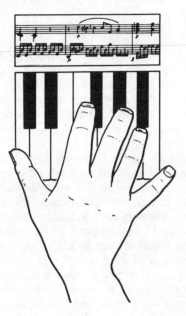

FIGURE 2. The relationship between DNA, amino acids and protein can be described by a musical analogy. If the music on a piano represents DNA then amino acids are the individual notes that the pianist can select. There is a limited number of notes available and when the right ones are played in the right order the result is a melody – the protein.

ones are joined together in the correct order through the information stored in DNA.

DNA's role in determining the sequence of amino acids in proteins stems from its own structure which contains molecules that are codes for individual amino acids. The codes run along the backbone of DNA in the same order as that required for the amino acids of the protein being produced. In this way the amino acids are assembled – and thus the protein is built up – correctly (see Appendix). The DNA molecule is itself extremely long and contains enough information to make thousands of separate proteins: special codes make it clear where the information for a particular protein starts and ends.

The molecular building site where the construction of proteins takes place lies within the cell, the smallest part of a living system capable of existing by itself. All creatures are made up of cells, each cell being potentially a microcosm of the entire animal or plant. Some creatures are unicellular, they consist of one cell only, whereas others, such as man are a highly complex array of specialised cells.

Genetic engineering

That part of a DNA molecule that contains the coded information for any one particular protein is known as a gene. (To be more exact, a gene specifies a physiological function or an anatomical feature. Often, but not always, such a feature or function depends on a single protein.) As understanding of how genes control the production of proteins grew, so did the possibilities for biotechnology.

As examples already quoted have indicated, changing or adding to an organism's DNA – often called genetic engineering – is *not* a prerequisite of biotechnology. If an organism can be found to carry out some task satisfactorily there is no merit in changing it for change's sake. However, it is undeniable that genetic engineering has broadened the horizons of biotechnology immensely. Where there is no suitable organism the biotechnologist can now consider the alteration of another.

All methods of altering genes can be considered to be genetic engineering but this term is most closely associated with one particular technique – recombinant-DNA. This technique became possible when it was discovered that certain enzymes can cut DNA at specific locations, enabling the removal of part of the DNA molecule. In this way the biotechnologist can isolate a gene or genes from a source organism and add them to another, known as the host. In theory it should be possible to transfer a gene – and with it the ability to make a certain protein – between any two organisms. A gene that has been successfully added to a host's DNA and brought about the production of its particular protein is said to have been expressed (see Appendix).

With recombinant-DNA a specific part of a DNA molecule is isolated and transferred, by contrast in 'shotgun cloning' DNA is cut into many pieces. Introduction of these pieces into host organisms leads to a variety of new DNAs and the desired one is detected and isolated.

Another way of altering the genetic composition of an organism is to strip away the outer walls from two cells and fuse the contents, a process called protoplast (a protoplast is the naked cell) fusion. Combinations of the DNA from the two original cells may then occur when the new cell is encouraged to grow and multiply.

At the time of the development of genetic engineering biotechnologists considered closely the risks involved in performing such research. As a result guidelines for safe laboratory practices were drawn up that have operated satisfactorily.

Before genetic engineering, scientists sought mutant organisms that had DNA that was out of the ordinary. X-rays, gamma-rays and some chemicals can increase the incidence of mutants, but they alter DNA in a random fashion; it is not, so far, possible to select a gene and bombard it with X-rays to get a specific end-result. Instead, following mutation, the organisms that have been altered have to be separated and desirable properties identified. Indeed, the ability to identify desirable properties in small numbers of organisms is of the essence in both random mutation and directed genetic engineering.

Starting with the simplest

Although one day it may prove possible for any living creatures to have biotechnological uses of commercial value, so far the greatest successes have been achieved with the simplest creatures – microorganisms (also known as microbes). And in practice the trend may be to transfer processes from the higher organisms to these smaller ones.

Microorganisms, as their name suggests, are the smallest living things – bacteria, algae, fungi, yeasts, etc. They can be classified into animal or plant types, but a more important difference to the biotechnologist is in the structure of their cells. If their cells have a nucleus most of the DNA will be found within it and will be less accessible than if they do not have a nucleus. Bacteria – often popularly known as bugs – are an example of creatures that do not have a nucleus in their cells and this factor has made them an especially fruitful area for biotechnology research.

Microorganisms have various attributes that make them attractive to the biotechnologist. Their metabolic work rates are very high which means that production of proteins proceeds at a faster rate than it would in, for example, a farm animal. Another great advantage is that they breed rapidly; cell division, the bacterial method of reproduction, can take as little as 20 minutes. This enables the success or otherwise of a particular experiment to be ascertained quickly and also means that production rates can be high. Some microorganisms have been studied very closely over the years and their inner, biochemical workings – their metabolisms – are well understood, which makes it easier to plan genetic manipulations within them. Bacteria, in particular, have very simple structures and are in fact unicellular, that is an entire bacterium is made up of one cell only.

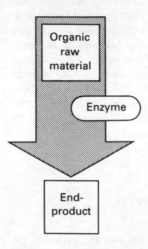

FIGURE 3. **Fermentation.** Proteins known as enzymes act as biological catalysts: they enable reactions based on organic raw materials, for example the conversion of sugar into alcohol, to take place.

The importance to the biotechnologist of understanding how an organism works cannot be stressed too highly. Because its biochemistry has been investigated so often in the past, one particular bacterium found in the intestine – *Escherichia Coli* – has become the most common agent for genetic engineering experiments. However, *E. Coli* suffers from the considerable disadvantage that most of the biochemicals it makes stay within the cell and therefore the cell wall has to be broken down and the desired product separated from the cell debris. There are other bacteria that will secrete proteins outside the cell, thus simplifying the biotechnologist's task. But *E. Coli* remains a laboratory favourite because it is so well understood.

Even with a microorganism that has a familiar biochemistry it is still a difficult exercise to complete genetic experiments successfully. As an illustration one might take the process of persuading a host cell to accept new DNA, which is known as transformation. The efficiency of transformation can be as low as one in 10,000. That means that for every 10,000 bacteria present in the test-tube, only one will incorporate and express the new genetic material and with it the ability to produce extra proteins. Low transformation efficiency leads to an obvious problem, how do you know which microbial cells have been successfully altered? One answer is to include in the genetic material being transferred a gene that provides protection against otherwise toxic conditions. Only those cells that contain the extra genes will survive such toxic conditions and they can thus be readily identified. Even then it is not certain that all the other genetic material transferred will be expressed and it can take a long time to identify the exact conditions in which the microorganism will thrive and produce the desired end product most efficiently.

From test tube to reactor

It is one thing to identify a microorganism in the laboratory that can be manipulated to provide a useful chemical, it is quite another to move to a large-scale manufacturing operation.

Biological factories
There are two main ways in which the cells of microorganisms are used industrially: one is as a direct source of protein: the other is to bring about specific chemical reactions. In the former case, the cells might be grown for use as food supplements –.the protein inside them has an intrinsic nutritive value – or the proteins might have properties that make them desirable in their own right, the example of insulin has already been quoted. The latter case – the industrial process by which microorganisms turn organic raw materials into useful end-products – is called fermentation (Figure 2). This is what is happening in wine making, for example, where enzymes manufactured by the microorganism yeast turn one chemical (sugar) into another (ethanol). The word 'enzyme' is, in fact, derived from the Greek for yeast.

To carry out their role as biological factories, microorganisms require energy and nutrients such as carbon and nitrogen. These enable the microorganisms to manufacture the enzymes that encourage chemical reactions to proceed. Biological processes are the only source of enzymes, no-one has managed to synthesise one chemically.

Because it is enzymes, with their ability to act as biological catalysts, that are at the heart of fermentation, it is logical to try to isolate particular examples and use them as biocatalysts on their own, much in the same way as conventional chemical catalysts. Before the mid-1970s enzymes were obtained by direct extraction from the organs of plants and animals, but now it is common to look to microorganisms, perhaps genetically altered, as the source.

Large-scale fermentation
In general, fermentation reactions have certain characteristics that make them quite different to conventional chemical industry processes. Their biological nature means that they proceed under relatively mild conditions – low temperature, pressure and acidity

– usually with water as the reaction medium. The extreme specificity of an enzyme – it will often catalyse one reaction only – can also be beneficial as this means it will tend to produce fewer unwanted byproducts. Biological catalysts can often achieve in a single reaction what would take a whole series of steps conventionally.

Fermentations, however, have the disadvantage of requiring exceedingly high standards of control. Temperature, acid level and oxygen supply must not be allowed to stray outside narrow bands. In addition, it is vital that sterility procedures are good enough to keep extraneous organisms out of the reaction vessel. As a result full-scale fermentation plants need a large investment in sophisticated sensors and control devices.

A common difficulty with fermentation is in the extraction of the end-product from the reaction mixture: enzymes are efficient, but their reactions often take place in an excess of water. Evaporation would be very energy intensive and so alternatives such as membrane filtration or extraction with pressurised carbon dioxide have to be considered. Another problem may occur in ensuring that reaction components are well mixed but that the microorganisms are not damaged by the mixing device. One way round has been to use air lift, which is more gentle than mechanical stirring. However, every technique has its own drawback, eg air lift can lead to foaming.

Plants and biotechnology

Many biotechnological projects use simple, unicellular microorganisms. Higher life forms are more complex and their biochemistries not yet sufficiently understood for them to be considered for more than a very few commercial purposes. However, plants are one, more complicated life form that offers opportunities for the biotechnologist.

Plant tissue culture
A basic advantage of plants is that many are totipotent, that is they have the ability to regenerate whole plants from mature single cells. The biotechnologist can thus work with single plant cells and then, because of totipotency, grow whole plants from them. Such a process can speed up a natural breeding cycle or enable unaffected cells from a diseased plant to grow into healthy mature plants. It can also be of immense benefit to plant breeders who use it to overcome natural obstacles to the crossing of different species.

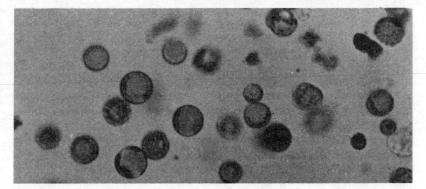

FIGURE 4. **Plant tissue culture.** Freshly isolated single cells of potato plants.

As well as the straightforward growth and harvesting of plants for their own qualities as foods, fuels and fibres, there is also the possibility, in some cases, of obtaining chemicals from them. Getting a chemical out of a plant can be laborious, requiring vast amounts of leaves or roots for relatively small yields. Plant cells, working in a similar way to unicellular bacteria, can provide an economic alternative. The chances of such a biotechnological approach to chemical production by plants are much enhanced if the plant cell secretes the particular chemical into its surrounding medium, rather than retaining it within the cell structure, as it is then far easier to isolate. Drugs, flavourings and insecticides are among the chemicals that are now being obtained in the laboratory from cultures of plant cells.

Genetic engineering of plants
As well as the added complexities compared to microbes, a significant obstacle to the genetic manipulation of plants was the lack of a suitable vehicle to transport genetic material from source to host. However, progress has been made in solving this problem and researchers are beginning to consider the many ways in which agricultural crops can benefit from genetic engineering techniques.

Some of the targets of genetic engineering research in this area are similar to those of traditional plant breeding: improved resistance to disease and drought, better crop yields, and easier harvesting. Knowledge of genetics, though, has opened up avenues inaccessible to breeders. For example, it could be possible to make crops more nutritious by supplying them with the genes for any amino acids – protein building blocks – that they either cannot make naturally or make in insufficient quantities for a balanced diet in animals. Already it has proved possible for genetically engineered plants to express bacterial genes, which leads to such prospects as

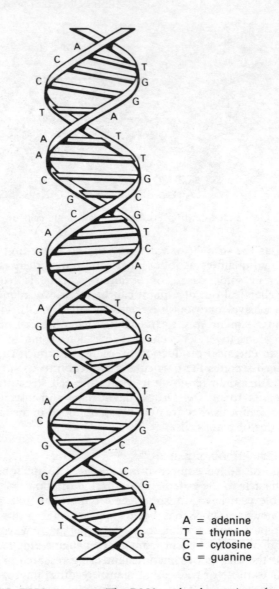

A = adenine
T = thymine
C = cytosine
G = guanine

FIGURE 5. **DNA structure.** The DNA molecule consists of two long threads of nucleotides that are formed together in specific pairs. Every 10 pairs the molecule completes a turn giving it its well known double helix shape. DNA was discovered in the nineteenth century but it was not until the 1950s that its structure was established.

crops that are independent of externally applied nitrogen fertilisers. This would arise if the genes that allow certain bacteria to absorb nitrogen from the air could be transferred to plants. The crops would then be able to reduce their need for nitrogen fertilisers. Alternatively, microorganisms that are already associated with the roots of particular crop plants could be genetically engineered to fix and transfer nitrogen into the plants. . . .

Appendix

DNA and protein synthesis

Deoxyribonucleic acid (DNA) is fundamental to life; it contains within its chemical structure all the information needed by an organism to create proteins, biochemicals that enable the organism to work.

DNA is an extremely large, thread-like molecule with a distinctive, helical shape (Figure 5). It has two interweaving chemical strands that are built up from smaller units known as nucleotides. There are four types of nucleotide, which differ only in one constituent part, an organic base. The four organic bases that differentiate nucleotides are often referred to by their initial letters: A – adenine, C – cytosine, G – guanine and T – thymine.

These organic bases play an important role in the structure of DNA as they form weak bonds between the two chemical strands and thus link the strands together. If one considers the whole DNA molecule to look something like a spiral staircase then the bonds between the nucleotide bases are the stairs. The bases form these bonds in specific pairs. A always pairs with T and G always pairs with C. As well as this structural role, nucleotides are vital to the transmission of information stored in the DNA, as they act as the code that makes sure proteins are built up successfully.

The whole process by which DNA in a cell of an organism causes particular amino acids to join up to form a protein is called expression (Figure 6). In essence expression has two stages: transcription – the creation of a nucleotide sequence in mRNA that follows that in DNA – and translation – the joining of amino acids in the correct order, made possible by the reading of amino acid codes contained within the nucleotide sequence.

The building blocks for proteins are chemical groups called amino acids. They have to be brought together in the right order to make the particular protein that the organism needs. While DNA contains the information for this exercise, it takes another complex molecule to relay the instructions. This chemical, which in composition has many similarities to DNA, is messenger ribonucleic acid or mRNA. mRNA can perform its task because during

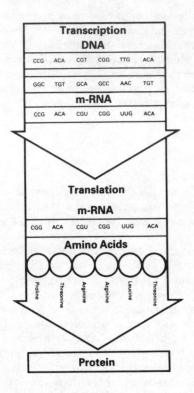

FIGURE 6. **Expression.** Expression is the synthesis of a protein in the cell based on information stored in DNA. It requires the creation of a vehicle – messenger RNA – to carry the information to where amino acids can respond to the codes and build up the protein.

its formation its own nucleotide structure follows that of DNA. Every A, C and G in the DNA is matched by an A, C and G in the mRNA, there is a slight complication with T which is replaced by a different base – uracil (U) – in mRNA. Such exact matching is again made possible by the complementarity of bases whereby they bond in specific pairs. This ensures that the sequence in the mRNA follows that of the DNA nucleotides. Creation of an mRNA molecule that mirrors a particular piece of DNA is called transcription.

Transcription is followed by the marshalling of amino acids to form a protein, a process known as translation. It is during translation that the nucleotides come into their own as codes. Every amino acid can be identified by a set of three adjacent nucleotide bases (a codon). For example, if the three bases GCA are found consecutively in the mRNA, the amino acid alanine will be iden-

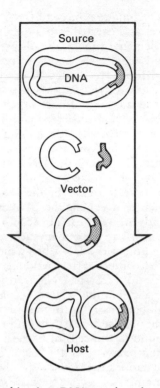

FIGURE 7. **DNA recombination.** DNA can be taken by a vector from one organism and inserted into another. In this way the ability to synthesise extra proteins can be imported.

tified while the base sequence GGA codes for glycine. Nearly all amino acids can, in fact, be coded by several base triplets, glycine, for instance, is also coded for by GGA, GGC and GGG. The sequence of perhaps several thousand bases that codes for a particular protein constitutes a gene and since DNA contains enough information for thousands of genes there are parts of the DNA molecule that indicate the starting and ending points for each one.

Recombination of DNA
With the discoveries during the 1970s of the principles which underly genetic engineering, biotechnology entered a new era. It became possible – in theory at least – to make up for any imperfections in the abilities of living organisms to carry out commercially useful tasks by altering their DNA. There are a number of ways in which this can be achieved (see page 166), but one method has attracted most attention: this is recombination of DNA (Figure 7).

Essentially in this technique a piece of DNA is snipped from one organism (the source) and carried into another (the host) where it is added to the complement of DNA already present. In this way the host organism should acquire the ability to produce new proteins.

Various biological tools are needed to carry out such an operation. Among them are restriction enzymes which cut the DNA molecules of both the donor and host organisms in such a way that the piece of DNA to be transferred will firstly be made available and secondly be accepted. A restriction enzyme 'reads' the bases contained in DNA looking for a particular sequence. If it finds such a sequence it will cut the DNA at a specific location. After making cuts and removing some DNA, particularly useful restriction enzymes leave the ends of the DNA strands unequal; short four base sequences overhang the DNA strands, one at either end. Such 'sticky' ends are very useful because if the DNA of both the donor and the host can be cut in such a way it is easier to persuade the host to accept the DNA being moved. Scientists now have several hundred different restriction enzymes available to them, each giving rise to different DNA transfer possibilities.

Another tool in recombinant – DNA technology is the carrier or vector that actually moves DNA from source to host. Two types of vectors are commonly used: plasmids and viruses. Plasmids are small DNA rings found in the cells of bacteria. They can replicate themselves and enter bacteria of the same or different species. Viruses are parasitic entities that invade specific hosts and multiply there. They have potential as vectors because some can inject DNA directly in the host's DNA in order to get it multiplied and expressed.

From biotechnology, we arrive naturally at the social impact of science on human institutions and individual behaviour. I recollect with great pleasure my occasional meetings with Waldemar Kaempffert. Although distinguished as science editor of *The New York Times*, he was for me a transplanted European. He loved 'the feel' of scientific behaviour and related it intimately to its European origins. When I left UNESCO, he wrote to me expressing his sadness at my going. He understood the significance of social responsibility.

CHANGE IN SCIENCE MEANS SOCIAL CHANGE
Waldemar Kaempffert

The authors of utopias and the creators of mechanised and electrified imaginary societies have usually been out-distanced by science. What Jules Verne did not predict and describe is far more important than what he did. Such inventions as the reaper, motion pictures, linotype and monotype machines, fast printing presses, and the developments in synthetic chemistry which enriched the world with new perfumes, flavours, drugs, and thousands of useful compounds made hardly an impression on him. Yet all these did more to change men's habits and lives than submarines that travelled 20,000 leagues under the sea.

Jules Verne did most of his imaginative writing in the second half of the last century. If he could be resuscitated and dropped in our midst he would be amazed, puzzled, and bewildered by what he saw, heard, and experienced. He would wonder what a reporter meant who, in summarising the proceedings of a medical meeting referred to 'vitamins', 'hormones', 'antibiotics'. There were no such words in Verne's time because the compounds for which they stood had not been discovered. Pasteur and Koch had not yet done their work when *A Journey to the Centre of the Earth* was written, so that very little was known about the mechanism of bacterial infections and still less about that of diseases caused by viruses. The physician of that day (1864) was no ignoramus, but he had only a smattering of what we would call 'scientific medicine'. There was no attempt to measure blood pressure in the routine practice of medicine, no electrocardiograph of the heart's action, no way of examining the chest and the stomach with X-rays because Roentgen, their discoverer, was only a boy. What progress in medicine has meant is shown by the life tables of insurance companies. The average baby born about 1850 had what statisticians call a 'mean expectancy of life' of 42 years, a girl baby one of 45 years. The corresponding figures for Britain in 1953 are 66 for men and 71 for women.

The lights of a provincial High Street would dazzle Jules Verne; the absence of horses in the streets would call for comment; the motor-cars that sweep silently past would astound and frighten him; the aeroplanes overhead would arouse wonder; the cinema, radio broadcasting, and television would entertain him but puzzle him, because he would not understand the underlying principle of operation. Skyscrapers would seem fantastic and not the economic necessities that they are.

The world of the Jules Verne who turned from the writing of plays to the writing of scientific fiction was much more self-satisfied than ours, so far as scientific beliefs was concerned. Newton's laws were considered inviolate. In accordance with the mathematical findings of James Clerk Maxwell light was conveyed to us from the sun or from a candle as a series of ripples in the ether, a medium that was supposed to pervade everything and that was more tenuous than any gas, yet more rigid than steel. Atoms were the smallest indivisible units of which matter was composed. The only disturber of mental peace was Charles Darwin whose *Origin of Species* (1859) and *Descent of Man* (1871) at first shocked the Western world but ultimately made it necessary to regard all living creatures as the warp and woof of a single fabric.

Nearly all the fundamental scientific conceptions in which Verne believed have been either shattered or modified, Darwin's included. Natural selection, a scythe, did not explain how new species originated, but De Vries and others who developed the mutation theory in the early part of this century did. The anthropologist of our day believes not that man is a descendant of an Old World anthropoid age, as Darwin concluded, but, as we have said, that apes and men are branches of the same limb of the family tree.

As for our conception of the world around us, Albert Einstein, a young examiner in the Patent Office of Switzerland, evolved a theory of relativity in 1905 and 1915, showed that Newton's laws did not conform with reality, and swept the ether away. Max Planck convinced physicists that light and radiation in general came not in ethereal waves but in packets called 'quanta'. Henri Becquerel in 1895 accidentally discovered that uranium was radioactive. Another accident led Wilhelm Konrad Roentgen in the same year to discover that from one electrode of an exhausted tube through which a current passed came invisible rays that would penetrate flesh and reveal the bones of the body. In 1897 J. J. Thomson of Cambridge found that the current in such a tube was composed of electrons, each of which was only $1/_{1840}$ as big as the hydrogen atom, so that the mid-century notion of an atom as the smallest possible chemical unit collapsed. In 1898 the Curies discovered in radium an element even more powerfully radioactive than uranium. Like uranium this radium shot out bits of itself that were smaller than atoms.

Because of these discoveries a new conception of the atom was necessary. It came from Ernest Rutherford, Niels Bohr, and others. Textbooks on physics had to be rewritten and physicists lost their old jaunty cocksureness. In Verne's prime everything was interpreted in terms of mechanism and laws of nature; today chance rules in the universe, and the laws of nature are recognized for what they are – mere statements of statistical averages. There are now doubt and uncertainty in physics. The Bohr-Dirac-Heisenberg school of scientists studied equations that were supposed to reveal the secrets of the atom and hence of reality, only to find that reality had vanished and that trees, houses, and stars were not what they seemed to be but only indications or 'pointer readings' of a deeper something that was real, something that science could never reach. Is it just a coincidence that the same uncertainty prevails in art, in economics, in international relations?

The changes in outlook brought about by research in fundamental science are of more practical importance than is at first apparent. Mendel's laws of heredity, when they were rediscovered in 1900 and later amplified by a score of geneticists, made it possible to breed animals and plants almost to commercial specifications. Out of all the studies of radioactivity and the formulation of new theories of atomic structure came a way of releasing energy from the nucleus of an atom – the supreme scientific achievement of man. Stars and atoms, relativity and nuclear physics, are linked together. Out of this linkage came electronics, a branch of engineering that has grown up within the memory of middle-aged men.

Like engineers of the late nineteenth century, Verne wrote about electrical machinery without ever knowing what electricity was. After Thomson, Becquerel, and the Curies, men knew. A current in a wire is a continuous flow of electrons, a lightning stroke is a gush of them. With this new knowledge came the electronic tube, or vacuum tube, which has ten thousand uses. One of the uses is the conversion of light into electric current or of electronic current into light, a conversion that made television possible – just the sort of invention that we associate with romancing of Verne's type. Now engineers are applying electronics in machines which solve in a few minutes problems that would ordinarily keep a mathematician busy for weeks and months. The creators of the electronic computers and other variations of 'thinking' machines are certain that in the factory of the future more machines with the skill of intelligent craftsmen will be seen than men. Holes punched in ribbons of paper will instruct and operate the machines. Anything from an electric toaster to a motor-car will be produced by the machines.

As we look back from the standpoint of recent electronic triumphs, three periods of invention can be distinguished. When Verne's first novels appeared, which was during the American Civil War, the world was still living in the first period, the period inaugurated by the introduction of the steam engine – period of steam-driven machinery on land and sea. The telegraph was the only electrical invention in wide use until the eighties, and the most potent source of electricity in the United States was a battery of 20,000 cells owned by the Western Union Company.

The second period begins with Thomas A. Edison – the period that exploited the discoveries of Michael Faraday, the period that the romanticists of the eighties and nineties called 'the electrical age'. In the history of invention no figure towers above Edison's. It was he who built the first steam-driven central station; he who invented an electric lamp that was equally useful in the home, the factory, or the street; he who devised electric motors that could drive factory machinery or trains. Fifteen hundred inventions are associated with his name.

Though he lived well into the twentieth century, Edison, like Verne, belonged to the nineteenth. There will always be lone, garret inventors, but Edison was probably the last colossus of invention. In his old age his place was already taken by the industrial laboratory in which trained physicists, chemists, and engineers worked in groups. No single genius could solve the problems presented by a modern telephone system, by an atomic bomb, or by the complicated electronic device of today.

With the development of the electron theory of matter we entered the third or electronic period of invention. Not only skill but something easily mistaken for intelligence has passed to the electronic machine.

With the evolution of the three periods of invention there has also been an evolution in our sense of social responsibility. Like most of the historians and philosophers of his time Verne did not see that a revolutionary process or a machine had social effects. The business man did not worry much over such matters as the displacement of labour that would follow the introduction of an invention, or the disfigurement of the countryside by heaps of slag, or the menace to health caused by chimneys that belched smoke all day long and sometimes far into the night. As for the deeper effects of progress in science and technology – the change in national habits and in modes of life – little attention was paid to them until the beginning of the present century.

There is not an industry, not a vocation, not a mode of life, but has felt the impact of science and technology. Old skills have been abandoned and new ones, highly paid, have been acquired. The enormous gap that separated the very rich from the very poor when Verne was a young boulevardier had been narrowed by science as much as by heavy taxation of the rich. This broader distribution of wealth has broken down social distinctions. Millionaires and machinists are indistinguishable in our cities. The mechanised society is not yet classless, but the bicycle mechanic who made a fortune out of motor-cars or the man who made his fortune out of a chain of cut-price stores is just as acceptable in café society or in Park Lane as a member of an aristocratic family, provided his manners are good and he is amusing. Family ties are looser than they were in the last century, and for that the cinema and the motor-car are partly responsible, also the migration of women from the home, first to the factory, later to the office when the typewriter and other office appliances were introduced. Many used to be their own masters in the sense that they were craftsmen who owned their own shops or business men who started with only a few pounds to make shirts or print books. Today most people are employees rather than self-employers, because it costs millions to build and equip a plant in which steel is produced, motor-cars are assembled, or textiles are woven.

Like his nineteenth-century contemporaries Verne assumed that institutions and customs would remain as they were regardless of the advance of scientific techniques. Yet Verne must have seen that Eli Whitney's gin was more than a device for picking seeds out of cotton. It revived the moribund institution of slavery and prompted the Republican party to formulate a high-tariff policy, which would force Southern planters to send their cotton to New England textile mills instead of to England. The motor-car proved to be more than a substitute for the horse and carriage, radio more than a means of dispensing with wires in electrical communication, the cinema more than an ingenious way of showing nature in action.

Why do scientific discoveries and inventions have this social effect? Because we cannot make use of a new discovery or invention unless we adapt ourselves to it. The more revolutionary the discovery or invention the more urgent is the need of adaptation. The introduction of the fountain pen did not demand any change in our writing habits. On the other hand the introduction of the motor-car made it necessary to spread a network of good roads over the entire country. Twenty miles was a long journey even in 1900. Today thousands live 20 miles from the nearest town and think nothing of travelling in their own cars across the continent. The cinema throttled the theatre and created Hollywood and with it an industry that is capitalised at thousands of millions and that entertains millions every day. Electric conveniences have virtually solved the servant problems just as tinned and frozen foods have solved the cooking problem. Back in 1851 and long thereafter a maid worked from six or seven in the morning until ten or eleven at night. For the last 10 years she has usually been hired by the hour at a wage that many a school teacher would be glad to earn. The permanently employed housemaid may be a passing phenomenon. Even today women whose husbands earn thousands a year do their own housework – but with the aid of electrically driven machinery. Changed economic conditions explain what has happened in the home and give inventors their chance, but the change in economic conditions and hence in industry is itself the result of the change that has occurred in science and technology.

In Verne's time the citizen was still much of an individualist. Science and technology have transformed his descendant into a standardised collectivist, in the sense that he is part of a social pattern which is as repetitive as the flowers on the wallpaper in his bedroom. The citizen of today is a consumer of mass-produced goods. Mass production is the distinguishing characteristic of our time. Without standardisation mass production as we know it is impossible. Hence the uniformity that prevails. One town looks very much like another through the windows of a railway train. This uniformity is less striking in Britain and on the European continent than it is in the United States. From New York to San Francisco grocers stock identical standardised packages of cereals, identical standardised tins of vegetables, identical standardised packages of frozen food. Very little latitude is allowed in the selection of apparel, especially men's apparel. It is harder than ever to be 'different,' and if we depart too far from standards of living or eating or dressing we may catch the disapproving eye of the police. Choice in recreation, choice in what we eat and wear, choice in everything, is restricted.

Even life in the rural districts has been standardised. The farmer of 1850 was a yokel who seemed out of place in a big city. His counterpart of our day rides to town in his own car, goes to a cinema, shops in a department store. So far as appearances go he hardly differs from the salesman who waits on him. When Verne was writing his novels in Amiens a farmer could still take his grain to a flour mill to be ground. Today's farmer may grow wheat but, like anyone who lives in a city, he buys flour in a bag, breakfast foods in cardboard containers, vegetables in tins. Gone is the isolation of old. A radio set tells the farmer what is going on in the world. The telephone enables him and his wife to gossip with friends who live 10 miles away, friends who are electrical neighbours.

Because we are so highly standardised by technology, artists have had to adapt themselves to new mass media of expression. When film plots or radio entertainment must appeal to millions it is hard to be an innovator. The standardisation of work, entertainment, and life which has occurred in a century is the price that we have to pay for more leisure and recreation and for the mass-produced necessities of today – necessities that would have been considered luxuries in Verne's youth, if they could have been conceived of at all.

Standardisation and the collective utilisation of energy have had the effect of knitting the population of a huge community together. A metropolis is a colossal organism, which is composed of such elements as skyscraper office buildings, blocks of flats and hotels, underground trains, telephones, water and gas mains. Millions in homes, factories, and offices are enmeshed in the hidden wires of a single central electric powerhouse. Let an accident occur like one of about 20 years ago which crippled a sub-station on New York's East Side, and lights go out, lifts stop running, ice cubes are no longer made in refrigerators, vacuum cleaners are impotent, candles have to be burned, homes and offices are cut off from one another, each fending for itself. Without the telephone the skyscraper would be impossible; with the telephone metropolitan cities like London,

Paris, New York, and Chicago develop into regions. Some remnants of sectionalism remain in the United States, but when ten million persons scattered between San Francisco and New York or St. Paul and New Orleans listen to the same radio programme or when one friend in Boston telephones another in New Orleans to wish him a Merry Christmas the effect cannot but be one of unification.

This mechanisation and electrification of life, this onward sweep of standardisation and the collective utilisation of energy, demands planning, organisation, and competent direction. Who are the planners, organisers, and directors? Scientists and engineers in industrial laboratories, corporation executives, bankers. A new class has arisen, a class of expert planners, designers and managers, that holds its place through sheer ability, unlike the old hereditary military caste of nobles, and that rules industrial empires unimagined by Verne – empires of oil, chemicals, synthetic fibres, electrical communication.

These experts at the top, on whom we are utterly dependent, are the moulders of our scientific and technologic culture. There are not many of them. If they were to disappear we should be helpless until we had trained their successors. They are the instruments of Progress – spelled with a capital – about which so much was written in the eighties and nineties. By Progress scientific and technologic advance was meant. Give us more machines to do the back-breaking, grimy work of the world, ran the formula, and there will be an end of misery and poverty; give us more international means of mass communication, like radio and films, and alien peoples will understand one another, with the result that there will be no more wars; give us more science and more international scientific congresses, and nations will learn to sink their differences in the common cause of enlightening one another.

None of the predictions has been fulfilled. The world has never been so restless, so uncertain of its future, so terrified at what may happen if another world war is fought with all the aid that science can lend. Every advance in science and technology has both improved man's material lot and heightened his military power. According to Pitirim A. Sorokin of Harvard, 957 important wars were fought between 500 B.C. and A.D. 1952, and we are now living in what he calls 'the bloodiest crisis of the bloodiest century'.

It is an old complaint that man's increasing technological ingenuity has not been accompanied by a corresponding sense of moral obligation, meaning that we have not yet succeeded in curbing abuses of scientific discoveries and inventions either in peace or in war. Dr. Norbert Wiener, distinguished mathematician of the Massachusetts Institute of Technology, is inclined to believe that we shall not do any better with electronic machines than we did when we permitted slums to grow around steam-driven factories. We still pollute the air with gases and smoke whenever we shovel coal into a furnace, though the chemists have been telling us for years that we are burning up the equivalent of chemists' shops in chemical values and that our fuel needs could be met by coke.

Science has so far been applied chiefly in fathoming the secrets of matter and motion, technology chiefly in winning wars and making profits, changing the environment, and improving man's material condition. This is perhaps inevitable.

Man has been on the earth about a million years, but it is only within the last 5,000 or so that his moral sense has been awakened. If the proper study of mankind is man, science is still in a primitive state. It knows less about man, the most interesting and important object in the universe, than it does about the atom. Science needs another Renaissance. It may dawn in another century. If it does, science will devote at least as much research to man's abilities and shortcomings as it has devoted to the synthesis of drugs and plastics. In the past science has taught man how to conquer his environment. It has still to tell him why he behaves as he does and to teach him how to conquer himself.

And then came the British scientist, Derek de Solla Price, who from his professorial chair in the history of science at Yale University founded 'scienomics', and gave the measurement of science development a quantitative basis. He made popular the Polish-born concept of 'the science of science', the name with which the International Science Policy Foundation was born.

BIG SCIENCE, BIGGER SCIENCE
H. J. Fyrth and Maurice Goldsmith

> Inside the ship he lay down like a hero
> The doors were sealed up and the count-down was near-o
> Ten, nine, eight, seven, six, five, four, three, two, one, zero
> And Yuri went up in the air.

> *(The Ballad of Yuri Gagarin*, a modern folk song)

> During a meeting at which a great number of physicists were to give firsthand accounts of their epoch-making discoveries, the chairman opened the proceedings with the remark: 'Today we are privileged to sit side-by-side with the giants on whose shoulders we stand.'

> (D. J. de Solla Price: *Little Science, Big Science*, 1963)

On April 12th, 1961, mankind took its first step into outer space when *Vostock I*, with a former ironmoulder Major Yuri Gagarin of the Soviet Air Force on board, orbited the earth.The exploits of Gagarin, Titov, the Americans Sheppard and Glenn, and other astronauts who followed them, had been preceded by the achievements of the remarkable International Geophysical Year of 1957–8, during which the Soviet Sputniks and the American Explorer satellites were launched. These first movements of men in space were made possible by vast expenditures of resources (men and money), and by the development of highly costly new devices and techniques in fundamental research – instruments costing tens of thousands of pounds such as linear accelerators and radio telescopes. Suddenly we became aware that the days of test tubes and Bunsen burners were gone: 'little science' had grown into 'big science'.

From 'little science' to 'big science'

Throughout what we, in *Book 2 Part 1*, have called the Age of Confidence scientific research was unco-ordinated and amateur: the tradition of the virtuoso still dominated, and facilities and funds for teaching and research were sparse. There were some with vision – Babbage, Pasteur, Playfair – who protested that it was not sufficient to rely upon the goodwill and ability of those engaged in scientific investigations, and that an appropriate administrative machinery should be set up to encourage and develop scientific research. Superficially their words have relevance today; but intrinsically they were involved in the early days of 'little science'. Those days have now gone, although there are still many who continue to project the past into the present, unable to discern that what they were taught to understand as being 'science' only a generation ago is no longer valid.

When we think of the 'great scientists', we picture Galileo and Boyle, Faraday and Darwin, but the first two would not have understood what the word 'scientist' meant and the others would have rejected it. The words 'scientist' and 'physicist' were coined in 1840 by Dr. William Whewell, the encyclopedic Master of Trinity College, Cambridge, in the introduction to his *Philosophy of the Inductive Sciences*, entitled 'The Language of Science'. He was concerned with preserving 'the purity and analogies of scientific language from wanton and needless violation' at a time when inadequate terminology and classification caused great confusion. Whewell wrote:

The terminations *ize* (rather than *ise*), *ism*, *ist* are applied to words of all origins; thus we have pulver*ize* . . . Heathen*ism*, Journa*list*, Tobaccon*ist* . . . As we cannot use physician for a cultivator of physics, I have called him a *physicist*. We need very much a name to describe a cultivator of Science in general. I should incline to call him a *Scientist*. Thus we might say, that an Artist is a Musician, Painter, or Poet, a Scientist is a Mathematician, Physicist or Naturalist.

However, Faraday would not use 'physicist' and T. H. Huxley rejected 'scientist', preferring to be known as a 'man of science' or 'natural philosopher'. Only during this century have the words come into common usage. Yet today's scientist is not the 'cultivator of science' that Whewell had in mind: his activities, his place of experimentation and his social status are very different from those of 120 years ago. He lives in the period 'big science'.

The term 'big science' was coined in 1961 by the American physicist, Professor Alvin M. Weinberg, to distinguish the modern scale of operations from the traditional 'string and sealing-wax' experimentation performed by gifted individuals under crude conditions. There were examples of 'big science' in the past, but they were exceptional. 'Little science' still exists, and the gifted amateur can still contribute to such fields of enquiry as the astronomy of the solar system, ecology and archaeology. But 'big science' is the typical form. In a sense the passing of the craftsman and the small shopkeeper is reflected in science. 'Big science' is not merely 'little science' grown bigger; it is as different from 'little science' as is the Ford Motor Company from a craftsman's workshop on the Left Bank of the Seine.

Derek de Solla Price, Professor of the History of Science at Yale University, has developed further the theory of the 'exponential growth' of science (that is, a mathematical consequence of having a quantity that increases so that the bigger it is the faster it grows). In the Western world science in all its aspects – numbers of people involved, number of periodicals, number of discoveries, etc. – has been doubling roughly every ten to fifteen years since the days of Galileo, and with each doubling it has formed a larger proportion of our cultural environment. At the time of Galileo scientific activity involved perhaps one in a million of the population; exponential growth means that the number of scientists at work today, or at any other time, is equal to 90% of all the scientists and research workers of the whole past and present combined. The same exponential growth is apparent in the number of scientific journals and periodicals: in 1850 there were about 1,000, in 1900 more than 10,000 and in 1960 about 100,000. Such an 'information explosion' makes it difficult to know what scientific work has been done, so that repetition of research is likely.

The same rate of growth is illustrated in the development of scientific equipment, for instance the energies of linear accelerators, popularly known as atom-smashing machines. In the late 1920s atomic particles could be accelerated to about 500,000 electron volts (½ MeV); in the 1930s this rose to 20 MeV; by 1950 it was 500 MeV; in the 1960s 30,000 MeV (or 30 GeV); and by 1970 it should be about 50,000 MeV (50 GeV). Thus there has been an increase by a factor of 10^5 in energy in thirty-five years, or multiplication by a factor of 10 every seven years. Similarly, the speed and capacity of computers in 20 years has gone up by a factor of 10.

The characteristics of 'big science', then, are that it has an enormously greater scale of operations than did science before the Second World War; it employs huge equipment compared, for example, with the apparatus with which Rutherford split the atom in 1919 and which he kept in a small brass box; it covers more fields of enquiry and produces much more information; the interconnections between the different branches of science are closer; there are many more research workers than heretofore and they can draw on much greater resources. All this means that science, from being one activity of Western civilization, has become the characteristic activity of our society. It permeates all creative fields; not only has technology become science-based, but the humanities employ scientific techniques, as when composers use electronic music or Biblical scholars consult computers to determine which Epistles St. Paul did not write. Modern archaeologists can use the proton magnetometer to map out a site before they dig; and the dating of finds by the rate of loss of radioactivity in the isotope carbon 14 has completely revised our time scales of the prehistoric past.

'Big science' also costs a great deal. Before 1918 the Cavendish Laboratory never spent more than £550 a year on new apparatus. Indeed, Rutherford was opposed to spending large sums, a process which he regarded as 'a substitute for thinking'. By the 1930s, however, more complicated equipment was needed, and Rutherford had built a machine for making powerful electric fields to be used by his Russian colleague Peter Kapitza, for which the Soviet Government paid £30,000 when it insisted on Kapitza's return home.

About the same time in the U.S.A. the 200-inch telescope was built for the Mount Palomar Observatory, at a cost of more than $1 million. Two approaches to scientific enquiry are illustrated by the story that Mrs. Einstein, on being told that the telescope was to help determine the shape of the universe, replied, 'But Albert did that on the back of an old envelope.' The cost of a million dollar telescope, which was exceptional in the 1930s, has been

dwarfed in the 1960s; the atomic particle accelerator at the Rutherford Laboratory in Berkshire cost nearly £10 million, and the
accelerator at Stanford University, U.S.A., cost more than $100
million. Such equipment can be almost as expensive to run each
year as it is to build, and a single experiment may cost £100,000.

Apparatus so costly may form a significant part of national
budgets for science, which in themselves consume a considerable
part of the resources of advanced nations – £750 million a year in
Britain in the mid-'sixties: that is, about 2% of the Gross National
Product (GNP). Therefore, important social and political problems
have to be answered: how much should a nation spend on science;
are particle accelerators and earth satellites justified while families
are homeless; is the science budget allocated where it will be most
fruitful or is research done to give prestige to certain institutions
or mollify military planners; does some expenditure do harm by
attracting scarce scientists from socially necessary research to where
there is expensive 'hardware'; how should the research budget be
divided between universities, institutes and industry, or between
science and technology; what proportion should go to 'pure'
research; is factual knowledge being accumulated faster than it can
be understood, judged and assimilated? These are some of the large
questions posed by 'big science', which scientists are not yet trained
to answer themselves, and on which non-scientists must be able to
judge if democracy is to have any meaning today. In these chapters
we examine some of the features of modern science which must
be taken into account if citizens are to make the best choice.

The science of science

If governments and electors are to use science wisely, they must
know far more than at present about the way in which science
develops and how it works. Today more is known about the moon
than about science itself; and the majority of people feel more and
more separated from science, which they tend to regard alternately
as a god and a devil. The needs of the age of 'big science' were
stated in 1949 by J. D. Bernal in *The Freedom of Necessity*:

> Already we require for the proper functioning of our society,
> a certain degree of knowledge of the facts of science and even
> more of its methods on the part of every citizen. The
> government cannot make decisions, the people cannot carry
> out the decisions reached, unless they have much fuller
> understanding than at present of what they are doing.

And a similar point was made by the American scientist and

educationalist James B. Conant in *On Understanding Science* (1947):

> Being well informed about science is not the same thing as understanding science, though the two propositions are not antithetical. What is needed are methods for imparting some knowledge of the Tactics and Strategy of Science to those who are not scientists.

The movement to study science in this new way began between the wars with conferences and publications on the history and philosophy of science (*Part II*, Chapter 8). It has now grown to the point where aspects of this are taught at many universities, and science and its social impact are permeating liberal and general studies courses in colleges and schools.

Now, too, the methods of science are being used to study science itself, a form of activity which was called 'the science of science' in 1936 by the Polish scholars Maria and Stanislas Ossowski. They wrote:

> The organization of institutions, the protection (patronage) of science by the State and by social organizations, the education of the scientists, all this – if it is to be truthful – cannot today do without studies just as specialized and complicated as those which are required for the construction of large industrial organizations.

The first of such studies was Bernal's *Social Function of Science* in 1939 (*Part II*, Chapter 8). By the 'science of science' is meant the study, under the conditions of pure science, of the tendencies in the development of science: internally, as a discipline with its own history and logic, and externally, in its relationships with society. This does not mean simply making generalizations about science: it means proceeding laboriously through the stages of observation, experiment and theory in detailed research into such matters as the evolution of science; the relations between science and technology; the best allocation of resources between teaching, research and production; the planning of research and development and so on. In this way the mechanisms by which science lives and grows may be revealed. In 1962 Derek de Solla Price said in his *Little Science, Big Science*:

> The disciplines which analyse science have been generated piecemeal, but show many signs of beginning to cohere into a whole which is greater than the sum of its parts. This new study might be called 'history, philosophy, sociology,

psychology, economics, political science and operations research (etc.) of science, technology, medicine (etc.)'. We prefer to dub it 'Science of Science', for then the repeated word serves as a constant reminder that science must run the entire gamut of its meanings in both contexts.

The new science became an organized study when in 1964 the Science of Science Foundation was set up in London, and research into the nature of scientific activity is going on in a number of British and American universities. In time the science of science should help to produce a new type of person, able to see the significance of science in all its aspects, and to act as the administrator, the critic or the communicator of science, so that, in the long run, these studies should contribute to a new climate of thought about, and understanding of, science among the entire population.

Scientific 'truth'

The new studies of science help us to realize that much of what we learnt about it at school and college is the mythology of science. We perhaps understood that science gives a wholly true picture of the workings of nature, that scientific laws have been, as it were, waiting since the beginning of time to be uncovered by the penetrating mind of a scientific genius. This attitude was derived from religion; it is not a scientific approach. When science, between 1500 and 1900, replaced that part of religion which 'explained' nature, it became for some (especially among non-practitioners of science) a substitute for religion, offering absolute certainty. This attitude drew strength from the more rigid concepts of science. Lord Kelvin expressed one of these vividly:

If you can measure that of which you speak, and can express it by a number, you can know something of your subject. If you cannot measure it your knowledge is meagre and unsatisfactory.

Since the universe was a machine, one should be able to measure it exactly and make models of nature. The term 'model' was first used in the age of Newton, and Kelvin said in 1844:

I am never content until I have constructed a mechanical model of the object I am studying. If I succeed in making one, I understand; otherwise, I do not.

Science does indeed subject its statements to tests which ensure that they are more certain than those of some other types of knowledge. Measurement and precision are still considered essential, but for the purpose of helping us to understand the relationships of fact and processes to one another. The Victorian interest in the 'arithmetic' of science has been replaced by interest in the 'geometry' of science (or perhaps more correctly the 'topology' of science). Very refined models are used today, but they are thought of as *analogies*; they do not reproduce nature but help us to understand something we do not know by something we do. Indeed, the modern scientist may be considered as one who creates images which help us to understand reality, although they are not pictures of reality.

A Victorian of vision was the mathematician William Kingdon Clifford, who wrote:

> Remember, then, that scientific thought is the guide of action; that the truth at which it arrives is not that which we can ideally contemplate without error, but that which we may act upon without fear . . .

This statement emphasizes that science does not simply record the world, it changes it. Science, we suggest, is a way of ordering our knowledge about the natural world (which includes ourselves), so that we can formulate 'laws', which we can use to make predictions. But these predictions do not tell us what *will* happen, only what is likely to happen if we choose one course of action rather than another. Moreover, our scientific laws are of varying degrees of 'goodness' or 'badness'. The Austrian-born philosopher Sir Karl Popper speaks of science not as 'a body of knowledge', but as

> a system of hypotheses; that is to say, as a sytem of guesses or anticipations which in principle cannot be justified, but about which we are never justified in saying that we know they are *true* or *more or less certain* or even *probable*.

The old tradition was that the scientist discovered truth as an onlooker of nature. But today we think of him observing nature actively as he helps to change it, and from the inside, for he himself is part of the nature he observes. We see him, too, as inside society, so that his ideas about nature, the questions he puts to it, the way he looks for answers and the use made of those answers, belong to a certain time, place and tradition. Newton's laws, for instance, could not have been discovered in the Middle Ages, nor in the China of his own time.

That scientific laws are not true always and everywhere does not mean that in science 'anything goes', that the difference between knowledge and opinion is only one of degree and that any one theory is as good as another. If this were so, science would be no better a guide than magic. Moreover, history suggests that in course of time scientists have constructed more and better hypotheses, consistent with more of the facts, explaining more fully the relations between the facts, and allowing us to make more complicated predictions. A more advanced theory embraces much of that which it replaces, and a core of knowledge of nature is preserved through all changes in theory. Thus the work of Ptolemaic astronomers was incorporated in Copernicus's system, and Newton's laws were embraced in, not overthrown by, Einstein's work. New hypotheses became necessary because increasing experience, through observation and experiment, shows up contradictions in the old ones; as when relativity theory was developed because of the difficulties of fitting new physical knowledge into Newton's system.

The idea that science tells us the whole truth and nothing but the truth belongs to the time when the universe was seen as a clockwork mechanism, not to an age which sees it as infinitely complex. The distinguished French theoretical physicist Jean-Pierre Vigier has said that since modern science suggests that matter may exist at an infinite number of levels of organization (sub-atomic, galactic, etc.),

> no single level of scientific knowledge can hope to cover the whole of reality. In fact, the progress of our vision of the world resembles that of a system periodically overthrown by revolutions succeeding each other with ever increasing rapidity . . . We are carried away by a movement whose end we cannot see, for the simple reason that there probably is no end. A physicist's view of the world is radically changed several times in his own lifetime.

Scientific 'method'

The superiority of science as a guide for choosing courses of action springs from its *approach* to problems. We were probably taught that a scientist begins by collecting many facts from which, by a process of induction, a hypothesis emerges; from this the scientist makes predictions and proceeds to test them by experiment. Today more regard is paid to the 'hypothetico-deductive' method. Here the scientist begins with a hypothesis, which may arise from his imagination, from his consideration of older ideas and so on. The essential business of science is testing such hypotheses, and therefore scientists must work out the appropriate tests and controlled experiments. The collection of facts is seen as important more for testing hypotheses than for making them, and more freedom is accorded to the creative imagination.

A modern school of philosophers, the Logical Positivists, hold that a good theory is one with which we can make predictions which can then be proved true, otherwise the theory is outside the business of science. Sir Karl Popper has argued that, on the contrary, science advances by refuting old theories, and that therefore only theories which are capable of being proved wrong belong to science. A test cannot prove a theory right, it can only support it by failing to prove it wrong. The best theories we have are those which have been tested the most often without being refuted; and if a theory predicts unlikely results and these are not refuted, it is a better theory than one which predicts results which we might expect.

We shall, however, learn much more about how scientific 'truth' is tested when we have more analyses of the work scientists have done.

Which are 'the sciences'?

The traditional view is also unhelpful today in deciding which 'the sciences' are. We have thought of chemistry, physics, biology and so on; but in the days of 'big science' these categories are dissolving. In 1961 a U.N.E.S.C.O. report on *Current Trends in Scientific Research* stated:

> A great deal of modern research is by its very nature inter-disciplinary, that is to say, it calls on all the resources of science, from mathematics to biology and the social sciences. It is, moreover, one of the most striking characteristics of present scientific work that it calls on these resources without taking account of the old divisions between the various disciplines.

Chemistry has been, for certain purposes, reduced to physics; chemistry, physics and biology merge and reappear in an inter-disciplinary variety of new sciences – such as biochemistry, astrophysics, bio-engineering and astro-geology; although the last two are as yet rather exotic. Biologists now classify themselves less by 'subjects' than by the level at which they work – molecular biologists or cellular, and so on. Indeed, one way of classifying the sciences is by the size of the objects they study (from the extra-galactic universe to 'fundamental' particles); another is by the level of organization in the objects of study (societies, man, organisms, cells, etc.); yet another by their service to society (food sciences, communication sciences etc.) Then again our era has its own fragmentations; for example, within organic chemistry there are the specialisms of polymers, stereochemistry, carbohydrates, steroids.

The important principle is that all these divisions and classifications are man-made; they do not arise from nature, but from the particular purpose we have in view in our attempt to understand nature.

Sources

'Foreword' by Sir Henry Tizard from *The Scientist and You*, edited by Maurice Goldsmith, London, Blackie & Son, 1959.

'The Story of the Periodic Table,' by Maurice Goldsmith from *The Young Scientist's Companion*, London, Souvenir Press, 1961.

'The Fantastic Nature of Reality', by Maurice Goldsmith from *The Young Physicist's Companion*, London, Souvenir Press, 1962.

'The Importance of Quanta' by Louis de Broglie from *The Revolution in Physics*, London, Routledge & Kegan Paul, 1954; New York, Noonday.

'Creation – and Warning,' by Maurice Goldsmith from *Frederic Joliot-Curie: a biography*, London, Lawrence Wishart, 1976.

'The Origin of the Universe,' by John Davy from *The Observer*, 12 February 1961.

'Statistics' and 'More about Statistics' by J.B.S. Haldane from *A Banned Broadcast*, London, Chatto & Windus, 1946; New York, Harper & Row.

'Biochemical Philosopher,' by J.G. Crowther from *British Scientists of the 20th Century*, London, Routledge & Kegan Paul, 1952; New York, Humanities Press.

'Biotechnology' from *Shell Science & Technology*, Issue 1, London, June 1984.

'Change in Science Means Social Change,' by Waldemar Kaempffert from *Explorations in Science*, by Waldemar Kaempffert. Copyright 1953 by Waldemar Kaempffert. Copyright 1946, 1947, 1948, 1949, 1950, 1951, 1952, by The New York Times Company; 1946 by Saturday Review Associates, Inc.; 1919, 1945, by The Curtis Publishing Company; 1943, 1948 by Tomorrow-Garrett Publications, Inc. Reprinted by permission of Viking Penguin, Inc.

'Big Science, Bigger Science,' by H.J. Fyrth and Maurice Goldsmith (*Science, History & Technology*, Part III, London, Cassell, 1969).

NOTES

CHAPTER 1 THE PROPER PUBLIC FOR SCIENCE

1 A distinction has been made between 'pop science' (that is, the portrayal of scientists and their work to the masses in comics, television cartoons, etc.) and 'popular science' (that is, science presented to an educated, limited segment of the population: in many cases, one group of scientists writing for another group) by George Basalla, associate professor of the history of S&T at the University of Delaware, in an essay, 'Pop Science: the Depiction of Science in Popular Culture' (*Science and Its Public: The Changing Relationship*, edited by Gerald Holton and William A. Blanpied, D. Reidel, 1976).

> Pop science is an aspect of the popularization of science that has never been systematically studied . . . . There exists a feedback loop between widely-held American ideas of science and their popular artistic representation in comic strips, television shows and feature films. Scientists are rarely the heroes in the current world of popular culture . . . one encounters the pop scientist as a villain (p. 261).

2 The French Christian mystic, Blaise Pascal in his *Pensées*, divides the mind between *l'esprit de géometrie* and *l'esprit de finesse*, the former being rational and analytical and the latter intuitive and sensitive.

3 I am indebted to Lynn White Jr for the detail. It appears in an interesting paper, 'Science and the Sense of Salt: the Medieval Background of a Modern Confrontation?', *Daedalus*, vol. 107, Spring no. 2, 1978, pp. 42f.

4 William Morris in 1878 in 'The Lesser Arts' commented:

> . . . Science – we have loved her well, and followed her diligently, what will she do? I fear she is so much in the pay of the counting-house, and the drill-serjent, that she is too busy, and will for the present do nothing. Yet there are matters which I should have thought easy for her; say, for example, teaching Manchester how to consume its town smoke, or Leeds how to get rid of its

superfluous black dye without turning it into the river, which
would be as much worth her attention as the production of the
heaviest black silks, or the biggest of useless guns.

5   The fruits of a three-year course are in: H. J. Fyrth and M.
    Goldsmith, *Science, History & Technology* – Book 1, *800–1840s*
    (London, 1965): Book 2– Part I, *The Age of Confidence
    1840s–1880s*; Part II, *The Age of Uncertainty 1880s–1940s*; Part III;
    The Age of Choice 1940s–1960s (London, 1969).

6   Quoted by Milan Kundera in his essay, 'Somewhere behind', in
    Granta, 11, p. 89, Penguin, 1984.

7   Compare: Michelangelo said to a friend admiring one of his statues,
    'The figure was always there inside the marble. I had merely removed
    the superfluous parts.'

CHAPTER 2 TECHNIQUES IN THE MEDIA CULTURE

1   His approach is expressed in 'How to write a popular scientific
    article' . . . in *A Banned Broadcast and Other Essays*, London, 1946.

2   In March 1947, at a meeting on 'The Outlook for Scientific
    Journalism,' held at the Society for Visiting Scientists in London.
    Calder recalled:

About 1930 I had a curious accident. I was in company with two
scientific people, and they talked about a story which, even to
me as an ex-crime reporter, sounded a good story. It was, I
remember, put out at a private dinner. Now, journalists do in
fact respect privacy, so I came to terms with them. If they would
give me a full account of this and explain it to me, I would
arrange to publish it on the same day as it was officially published.
The result was: on the same day of publication a story broke in
Holland, which was, by a most curious coincidence, tied up with
this one. I went to Holland and, during the weekend I was away,
what usually happens in Fleet Street happened: the editor was
sacked, a new editor took his place. He was an outsider, and he
decided, because I had hit the headlines with this scientific story
from Holland, that I must know something about science, so he
never discussed it with me at all. All he did was to push all
scientific stories at me, and being, I hope, a good conscientious
reporter, I went out and asked people who knew about it. He still
didn't ask any awkward questions, until one day he suggested I
should do a tour of laboratories and bring back a lot of scoops.
I couldn't just go to a laboratory and walk out with a newspaper
scoop for science does not work that way, but I was, at the same
time, looking for a descriptive series, and I took this opportunity
to suggest that I would explore this unknown hinterland of
science . . . to find out very simply what scientists were doing,
how they were doing it, and, later on, I added why they were
doing it, with no profound intention behind it. That was 1931
. . . I have gone on finding the things that interest people in
science, and also, more important than that, the things that interest
me about science . . . .

3 Crowther was 'drawn into writing for the press accidentally.' This story and of how he told the famous editor, C. P. Scott, of the *Manchester Guardian*, in 1928 that he proposed to invent the profession of science journalist is told in his *Fifty years with Science*, London, 1970, and in abbreviated form in 'Half a century of science writing', *New Scientist*, 30 July 1970, p.243f.

4 Edward Shils, 'The Religion of Science', unpublished MS, quoted by Perlman in *Daedalus*, 103, Summer 1974, p. 212.

5 For details and tables see 'Reshaping the BA' by Maurice Goldsmith, *Discovery*, November 1963, pp. 10f.

6 First, R. Davies and L. Sinclair in 'Science & technology in four British newspapers 1949–69', Mimeo LSE 1972; second, I. Connell, G. Jones, and J. Meadows. 'The presentation of Science by the media', PCRC. University of Leicester, British Library Report No. 5494, 1978.

7 Barry White, 'Three Mile Island and the British Press', A private communication in 1982.

8 During the accident at Three Mile Island, neither public information officials nor journalists served the public's right to know in a manner that must be achieved in the event of future accidents. Each side failed for different reasons, and to a different degree. The most common explanations – that the utility lied, that the NRC covered up to protect the nuclear industry, or that the media engaged in an orgy of sensationalism – do not hit the mark. Indeed, reporters often showed great skill in piecing together the story, and some NRC officials disclosed information that was truly alarming and damaging to the industry's image, because they thought the public had a right to know it. Even the utility's shortcomings in the public information area (and there were many) are attributable, in part, to self-deception, as well as to a lack of candor. Given the enormous investment at stake for Met Ed, the company's unwillingness to recognize the severity of damage to the reactor is not surprising. But such hesitancy presents serious problems for serving the public's right to know during the early stages of an accident.

Lack of Planning
The public information problems of Met Ed and the NRC were rooted in a lack of planning. Neither expected that an accident of this magnitude – one that went on for days, requiring evacuation planning – would ever happen. In a sense they were victims of their own reassurances about the safety of nuclear power. Coordination between the utility and the NRC was so weak that responsibility for informing the public in the first crucial hours of the accident was undefined. The NRC did not know when, or whether, to send its own public information people to the site, or when or where to set up an NRC press center. Met Ed's public information department was an operation with low status and no policy input.

Perhaps the most serious failure in the planning stage was that neither the utility nor the NRC made provision for getting information from the people who had it (in the control room and at the site) to the people who needed it. This group included other utility executives, the governor of Pennsylvania, the NRC's Region I headquarters in King of Prussia, Pa., the NRC's Incident Response Center in Bethesda, public information officials, journalists, and members of the public.

Given this confusion among sources, and given that reporters are almost entirely dependent on such sources for their information, it is not surprising that news media coverage of the accident in the first few days was also confused. For a number of issues during the accident – such as the danger posed by the hydrogen bubble, or the size of a radiation release that led to evacuation concerns on Friday – it is obvious that the only type of 'accurate' reporting possible, under the circumstances, was the presentation of contradictory and competing statements from a variety of officials.

Unprepared Media

The news media were also somewhat unprepared, and this added to the prevailing confusion. While it is a goal of many journalistic organizations to develop specialists who are expert in particular areas (such as science and medical reporting), few reporters who covered TMI had more than a rudimentary knowledge of nuclear power. Some, by their own admission, did not know how a pressurized, light water reactor worked, or what a meltdown was. Few knew what questions to ask about radiation release so that their reports could help the public evaluate health risks.

A number of reporters in the first group to arrive were assigned to the story because they were available, and because they could cover almost anything on short notice – not because they had nuclear power as regular beat. Good journalists can absorb vast amounts of unfamiliar material while on the job: that happened during TMI, but the effort required to make sense of the story was enormous. It was not like covering a political campaign or an airplane hijacking, where at least the vocabulary of the sources and the vocabulary of the reporters are the same.

The nuclear industry has developed its own language, and this was a handicap for the many journalists who did not speak it. A seemingly simple question of whether the core of the reactor had been uncovered and damaged elicited responses couched in terms of 'ruptured fuel pins,' 'pinholes in the cladding,' 'melted cladding,' 'cladding oxidation,' 'failed fuel,' 'fuel damage,' 'fuel oxidation,' 'structural fuel damage,' or 'core melt.' The distinctions are real, but for reporters not speaking the language, it was like suffering from color blindness at a water-color exhibition. Neither Met Ed nor the NRC provided enough technical briefers in the first five days of the accident to help journalists interpret what they were being told.

Reporters also arrived with different objectives. Some were science writers with an interest in the reactor. Some were medical writers with an interest in public health and safety. Others were sent to write 'color' stories and focus on reactions of citizens and evacuees. It would have been difficult under ideal circumstances for a public information program to serve the many needs of the reporters who covered the accident. Given the public information program in place when the reporters arrived, it proved to be an impossibility. In the important first few days of the accident, when evacuation decisions had to be made, the public's right to know was not served because the conditions under which all parties operated were such that the public's right to know *could not* be served.

Utility officials acknowledge they were unwilling to release 'pessimistic' information to the public until it was confirmed to their satisfaction. But, on the questions of core damage, operator error, and radiation release off-site, confirmation came well after some members of Congress and the NRC were aware of the problems. Utility officials were also more optimistic than others that the accident had run its course, and that the worst was over. Given the utility's stake in the resolution of the accident and in the public's perception of the accident, this optimistic approach is understandable, but not excusable. While there is not unambiguous evidence of cover-up, some utility officials showed a marked capacity for self-deception, and others hid behind technical jargon to obscure answers to troublesome questions.

The NRC's behavior in the area of public information is more difficult to categorize. From the first day the NRC distanced itself from the utility, refusing to participate in joint press conferences or issue joint press releases. In this regard the commission chose to adopt the role of regulator rather than the role of booster. With the exception of some of the NRC's Region I people on-site (who tended to side with the utility's view of the accident), NRC staff members spoke more bluntly about the accident. An analysis of 'alarming' versus 'reassuring' statements quoted by the press during the accident shows that the NRC in Washington/Bethesda was a chief source of 'alarming' statements about such subjects as the possibility of meltdown and the explosiveness of a hydrogen bubble in the reactor. Along with anti-nuclear groups, some NRC officials provided the news media with some of the most frightening information during the accident.

NRC Staff Undercut

On the other hand, the commissioners were frequently less candid than their own staff. On Friday afternoon, for example, the commission took the unusual step of writing a press release to undercut a UPI story on the chance of a meltdown at TMI. The story was based on information provided to the media by an NRC staffer – information which NRC Chairman Joseph Hendrie conceded was accurate. In another instance, a commissioner was reluctant to label what happened at TMI an 'accident' because of the damage this could do to the industry's image.

For the first two days of the accident reporters were given almost no help in reporting this extremely complex story. The catalyst for change came on Friday at about 9:00 a.m., because of the extraordinary confusion over a 1,200 millirem per hour release of radiation from the plant. Not only were the public and press unable to learn what Met Ed was doing that morning in releasing radioactive gas into the atmosphere, but the NRC in Bethesda could not either.

Alarm Sets In
Throughout Friday, Saturday and Sunday morning NRC experts in Bethesda viewed the reactor as less stable, and potentially more dangerous, than did Met Ed or NRC experts on-site. The alarm reached the public through the NRC's Bethesda press center. On Friday afternoon, a UPI story quoted an NRC official on the subject of meltdown as saying that although it was not likely, it was not impossible. And on Saturday night an AP story quoted an unnamed NRC offficial as saying that the hydrogen bubble in the reactor could explode in as little as two days. Both of these stories were considerably more pessimistic than the news coming from the site, and both alarmed the public in the area of the plant. The stories angered Gov. Thornburgh, the White House, and NRC officials on-site.

After the AP story, the White House requested that the press center in Bethesda be closed and that all NRC comment come from the site. By the time the President visited the site on Sunday, this policy had been carried out.

Given the prevailing confusion, the White House effort to centralize information was not an unreasonable response. Reporters did not make much of a fuss at the time. In retrospect, however, most journalists interviewed by this task force disapprove of this centralization. They point out that although Harold Denton of the NRC was reassuring, and an improvement over his predecessors, he had some faults. His language at press conferences was filled with jargon. He was usually unavailable to the press, and giving a single press conference each day, he could not possibly have discussed all the subjects of interest to the media. He did not really attempt to handle radiation information, and he had trouble explaining the term 'man-rem' to reporters. His first few press conferences at the Middletown press center were chaotic.

The most serious drawback to this approach, however, was that Denton could not speak for the many groups with a different perspective on the accident, such as the utility, and the various state and federal agencies monitoring radiation releases and reporting on public health consequences. Reporters believe that the public's right to know would be better served through some sort of structured contact with all these groups.

Was the public's right to know served by the NRC and Met Ed? Clearly not. It is not quite as clear how well the news media served the public's right to know, in large part because reporters do not act alone in covering such a story. They are dependent on the quality of their sources, and these sources, for the most part, were unavailable or misinformed at TMI. In addition, although the number of public information people at Met Ed, the NRC, and the state is relatively small, and their problems were similar, hundreds of reporters covered the accident for a wide variety of news organizations. Since no two performed the same way, it is more difficult to characterize the performance of the media.

In reporting events during the accident, confusion among sources was mirrored by confusion in the media. The media reported denials by Met Ed (now known to be misleading) that the core had been uncovered and that operator error had contributed significantly to the accident. The burst of radiation Friday morning found reporters quoting sources saying that the release was both planned and unplanned, controlled and uncontrolled, expected and unexpected. Directly conflicting statements on the possibility of a hydrogen explosion, the size of the bubble, and the status of the accident (growing more or less serious) also appeared with frequency.

Some stories, such as the core being uncovered and seriously damaged, and the relatively benign industrial waste water being dumped by the utility into the Susquehanna River, reached the media late. The media did not report at all during the week that Met Ed had to clear TMI's control room of unnecessary personnel because of high radiation levels; that coolant pumps had not been acting properly on Wednesday: and that NRC engineers had been wrong in their predictions that the hydrogen bubble would explode if left alone. None of this information was made available to reporters at the time, and they were unable to dig it out on their own.

Political Stories Missed

The media also missed a number of political stories during the week. They failed to report the dispute between the NRC and the state over who should take responsibility for the dumping of waste water into the Susquehanna on Thursday. They missed the confusion at the NRC Friday morning that led to the recommendation that the governor order an evacuation, a recommendation he did not accept.

The media were forced to report some stories secondhand because first-hand sources were not making the information available. These stories included the extent of fuel damage: the role of operator error in the accident: and the very fact that a general emergency had been declared.

Also significant for the public's right to know is that many 'facts' about the accident were not presented in a context that could be understood by the lay-person. For example, the media did not explain the significance of a general emergency; nor did they provide background or contextual information in their coverage of the health hazards posed by radiation releases at the plant. Here the reportage was almost always incomplete. Rarely was a radiation number accompanied by a unit of measurement (such as millirems), a rate (per hour), an explanation of the type of radiation being measured, an indication of where it was being measured, when it was measured, and the amount of time a person would have to be exposed to this source of radiation to incur some sort of health risk. Neither Denton nor Thornburgh was giving out such information with regularity or precision, and not many reporters had the background to know what questions to ask.

While the media can be criticized for missing some stories and failing to provide a context for others, they were generally not guilty of the most common criticism leveled at them: that they presented an overwhelmingly alarming view of the accident. To test this hypothesis the task force on public information noted each time the news media in the sample quoted a source on such topics as the likelihood of a meltdown or a hydrogen explosion, the status of the accident, the health threat posed by radiation releases, the general threat of danger, and the like. The numbers of alarming and reassuring statements were then compared.

Reassuring Statements

Overall, the media offered more reassuring statements than alarming statements about the accident. This was true for such topics as the status of the accident (seen as improving); the threat of danger (seen as none, or not immediate), the health effects of radiation exposure (seen as none); the necessity for evacuation of the area (seen as not necessary): and the management of the accident (seen as being handled by competent and knowledgable individuals.)

The media were somewhat more alarming than reassuring on meltdown (seen as a possibility); the quality of information being made available during the accident (seen as unreliable); the future of nuclear power (seen as not bright); and the state of evacuation preparedness (seen as inadequate). In general, no special effort by the media was detected to portray the accident as any worse than it was. This was true for both local and national media.

Nor was the hypothesis supported that other news media around the country were more sensational in tone than the main disseminators of news: the wire services, the broadcast networks, the *New York Times*, the *Washington Post*, the *Los Angeles Times*, the *Philadelphia Inquirer*, and the *Harrisburg News*. A non-quantitative examination of 43 newspapers from Alaska to Hawaii to Maine showed that, with just two exceptions (the *New York Post* and the *New York Daily News*) the papers did not sensationalize or overplay the story in headlines, captions, use of pictures, or choice of material. If anything, headlines throughout the first week were more sober than the *real* confusion at the site and in Bethesda warranted.

One positive result of the TMI experience is that reporters and nuclear engineers were introduced to each others' problems, needs, and peculiarities. Nuclear experts should now realize that technical language, which to them offers precision and economy of expression, can be maddeningly opaque when aimed at reporters under deadline pressure. Instead of leading to more accurate stories, a reliance on technical language can produce ambiguity, confusion and frustration.

Nuclear experts should also recognize that reporters are fond of 'What If?' questions that inevitably lead to discussions of worst possible outcomes. Unless such questions are handled candidly, but with sensitivity, the answers can produce an unintended sensational effect. The likelihood of a meltdown or hydrogen explosion occurring must be communicated in a concrete, specific manner. Journalists as a group view abstract constructs of uncertainty with distaste – particularly at a press conference called to discuss a potential public hazard about which there is already great concern. Over time, scientists and journalists may find it possible to develop jointly a commonly accepted set of definitions in this language of uncertainty.

To the extent that journalists learn about nuclear technology and develop a standard of comparison for future accidents, they can help reduce the confusion that surrounds such accidents. But the major and initial burden in the area of public information falls on those who operate and regulate the nuclear plants. They failed to serve the public's right to know at Three Mile Island.
NASW Newsletter, vol. 28, no. 1, January 1980.

9   From a paper presented at the annual conference of the Atomic Industrial Forum in Washington, DC, November 1976. Portions appeared in 'What makes a good science writer', *Nature*, vol. 250, pp. 747f, and in 'From our science correspondent,' *Science & Public Policy*, vol. 4, no. 3, June 1977, pp. 275f.

10   The editorial 'Keeping up with it,' *World Medicine*, August 1984.

11   I am indebted to J. G. Richardson, formerly of the Division of Scientific Research and Higher Education, UNESCO, for much of the information I have used from his paper, 'The Role of Periodicals in Scientific Communication as We approach the Year 2000: Trends in its Achievement,' September 1982.

12   You see, then, in the first instance, that a beautiful cup is formed. As the air comes to the candle it moves upward by the force of the current which the heat of the candle produces, and it so cools all the side of the wax, tallow, or fuel, as to keep the edge much cooler than the part within; the part within melts by the flame that runs down the wick as far as it can go before it is extinguished, but the part on the outside does not melt. If I made a current in one direction, my cup would be lop-sided, and the fluid would

consequently run over – for the same force of gravity which holds worlds together holds this fluid in a horizontal position, and if the cup be not horizontal, of course the fluid will run away in guttering . . . You see now why you would have had such a bad result if you were to burn these beautiful candles that I have shewn you, which are irregular, intermittent in their shape, and cannot therefore have that nicely-formed edge to the cup which is the great beauty in a candle. I hope you will now see that the perfection of a process – that is, its utility – is the better point of beauty about it. Michael Faraday, *The Chemical History of a Candle*

13 For a more detailed statement see 'Science Books for Children' by Eugene Garfield, *Current Comments*, December 24–31, 1984, vol. 6, no. 52.

14 Of course, this distortion still exists. A blatant present example appeared in the London *Daily Mirror* on 19 December 1984. The major front-page story by Roger Beam began, 'Five-week old Daniel Massey, who was born almost blind, can see again . . . thanks to new miracle of science.'

15 The TV critic in the London *Star* wrote (9 June 1956), 'TV Science marches on. Now the BBC, with an observant eye on the success of ITV's "Meet the Professor", have opened a new corner, "A Question of Science," to answer viewers' queries. The replies were given by experts who illustrated their words with film and discussion.' Of course, this is what we did on *Meet the Professor*.

16 How objective is a science programme, such as *Horizon*? An informed attempt to answer this question is given by Dr Roger Silverstone, a sociologist at Brunel University, England, who in Spring 1981 approached the BBC with an award for a Fellowship from the Joint Committee of the Science & Engineering Research Council (SERC) and the Economic & Social Research Council (ESRC). He was concerned with a study 'to look at science on television, to look at the structure and the content of the programmes, the way they were produced, and to make some limited sense of how they were received.' For one year his salary was to be paid, and he was to be free of his university teaching requirements.

In October 1981, he began his study in the *Horizon* offices in the BBC, where in 1971 he had been a documentary researcher. At the seventh request, he was fortunate enough to have producer Martin Freeth agree to allow him 'to hitch a ride' on a *Horizon* production, which he used as a basis for the study. The film, *A New Green Revolution*, took over two years to produce, cost over £80,000, and was transmitted on 23 January 1984. It was estimated to have been seen by an audience of 2.4 million, and by 0.5 million at the second transmission on 29 January.

Dr Silverstone was involved intimately from day-to-day in London, South America and Asia in the making of the film. His 'clear message' is 'that the presentation of science on television to a presumed non-specialized and non-student audience involves major transformations and translations from one discourse, that of science, to another, that of television.' For him, television's forms are both the product of, and the attempt to resolve, the real and the fantastic, which are two of the most fundamental forces in culture. He saw the producer seeking to secure a particular narrative balance: balancing realism and myth: 'our culture's perception of science lies at the heart of these dilemmas. Science holds the key to the meaning of the universe and its destruction.'

> (Roger Silverstone, 1985, *Framing Science: The Making of a BBC Documentary*, British Film Institute, London, pp. 179f.)

Another approach to the 'objective' nature of a TV programme is made by Virgilio Tosi, the Italian film maker. He wrote:

> . . . the problem of the relationship between information and culture; when this occurs training, education and the promotion of literacy are neglected or underestimated, or deformed, either for political motives or questions of economic profit. The problem is even more serious in industrialized societies where the quantity of messages . . . is so great that the ordinary man is submerged by them and tends to tolerate rather than use them. This means that in the end man places information, advertizing and entertainment on the same level . . We reach the high point when television brings us an event live, even a horrific one . . ., while it is happening and where the outcome is unknown. There results a climate of suspense and often morbid curiosity; reality has become entertainment.

> (Tosi, 1984)

17 André Leveillé, first director of the Palais de la Découverte in Paris, during a lecture in 1937, said: 'One rainy day about noon, a man, his wife and their two children entered the Palais. It's raining and the Palais seems a cozy place to have a quiet lunch. He has his provisions in a shopping bag. Having arrived at the Rotunda, he looks around for a long time, goes out, and says to the attendant in the cloakroom, "Look after my food: one does not eat in a cathedral." '

18 This is found in 'The nationwide provision and use of information', The Library Association, 1981.

19 Richard Gregory's Bristol Exploratory, which is modelled on the San Francisco Exploratorium, has as its essence Francis Bacon's 'House of Solomon', as defined in *The New Atlantis*, 1627 (see the *New Scientist*, 17 November 1985).

CHAPTER 3 SCIENCE FICTION: TRUTH AND REALITY

1 *A Sociological Study*, 1976, reprint, Kriger, P.O. Box 9542, Melbourne, Florida, USA.

2 The German-born now American rocket engineer, who joined the Nazi party in 1940.

CHAPTER 4 THE EDUCATIONAL SYSTEM NO-GO

1 Baez, 1969; Dainton, 1971; Thring, 1977; Rutherford, 1979. Science Council of Canada, Ottawa, *Science for every student:*

educating Canadians for tomorrow's world, Policy Report no. 36, 1984; *Science education in Canadian schools*, Background Study 52, in 3 volumes, 1984, provided a detailed analysis of the present situation in school science education in that country, and a rationale for basing future developments much more firmly on social issues.

2   Details of CASTME publications, including the quarterly CASTME journal, from the Education Division, Commonwealth Secretariat, Marlborough House, London SW1Y 5HX.

CHAPTER 5 THE THIRD WORLD (TW): ILLITERACY AND COMMUNITY

1   The term TW originated in postwar France. It appeared first in print in the radical Socialist, non-Communist newspaper, *l'Observateur*, on 14 August 1952, in an article entitled 'Trois Mondes, Une Planete,' by the demographer, Alfred Sauvy. He wrote, the TW 'ignoré, exploité, méprisé, comme le tiers état, veut lui aussi être quelque chose.' (The TW 'ignored, exploited, scorned, like the third estate, also wishes to be something.') As Peter Worsley points out, 'The analogy was a peculiarly French one: the *tiers monde* was the analogue of the *tiers état* of pre-revolutionary France: the estate of the bourgeois, the petty bourgeois, the artisans, the peasants and workers who lacked the privileges of the first and second estates, the clergy and the nobility.' (Worsley, Peter, *The Three Worlds*, London, 1984, p. 307). The First World broadly is the free enterprise, capitalist countries; the Second World the closed economy, socialist countries.

2   Laedza Batanani is the name given to the initial experiment begun in 1974, for which the University of Botswana's institute of adult education was largely responsible, to use popular theatre as a tool for programme development and community education. 'Laedza' means 'wake up – it's time to get moving,' and 'Batanani' means 'let's unite and work together'. See Kydd, Ross & Byram, 'Laedza Batanani – popular theatre and development: a Botswana case study', *Convergence*, vol. 10, no. 2, October 1977 (available from the Institute of Adult Education, University of Botswana, Private Bag 0022, Gaborone).

3   A statement. Popular Theatre Dialogue, International Workshop. Dhaka, Bangladesh, Feb. 1983.

4   ISWA Newsletter, August 1984. p. 6.

CHAPTER 6 THE SCIENCE CRITIC: A CASE FOR NEW CREATION

1   Earl Ubell, a leading American science communicator and a former president of the NASW, whom I asked to comment on this chapter, wrote (September 6, 1984):

I applaud your call for a science critic, who can certainly function to feed back to the science community the public's response to scientific enterprise. Back in 1952, when I won the Lasker prize, I asked for such science critics in my acceptance speech. Albert Deutsch was sitting in the audience and I pointed to him as one

of the first critics in medicine and social science. His critical
abilities flowed from his wide ranging knowledge of psychiatric
practice . . . and from his social conscience . . .

I think your criteria for the training of such critics are
insufficient. Unlike a theater critic who can become such by
sitting in on a variety of plays and by having an innate ability to
write (cuttingly is preferred), a science critic must have a hands-
on knowledge. (I might add that the best American theater critic
in the generation, Walter Kerr, had just such knowledge, having
been a producer, director and playwright.) Therefore, I believe
that our future science critics will mostly come from science –
members of the community who for some reason or another see
themselves in that role and put themselves at risk by entering
the critical lists.

Ubell went to cite as examples, Carl Sagan 'by his examination
of the nuclear winter,' and Hans Bethe 'has done it by his expose
of the star wars fallacy.'

2 In October 1985, the Ciba Foundation, a London-based
independent scientific and educational charity, set up the Media
Resource Service (MRS) to improve links between journalists and
scientists in order to help them in communicating to the public.
The MRS provides access to 1000 leading scientists without charge.
All that is required is a phone call to MRS on either 01–631 1634
or 01–580 0100. The steering committee is made up of scientists
and practising journalists.

BIBLIOGRAPHY

Ansah, P., Fall, C., Kouleu, B. C. and Muraura, P. (1981), 'Rural Journalism in Africa', *Reports and Papers on Mass Communication*, no. 88, UNESCO, Paris.

Ashton, B. G. and Meredith, H. M. (1969), 'The Attitude of sixth formers', in *School Science Review* vol. 51, no. 174, pp. 15–19.

Baes, A. V. (1969), 'The spirit of science world-wide', in *The Science Teacher*, vol. 36, no. 4, pp. 15–18.

Barrileaux, B. E. (1967), 'An Experiment on the Effects of Multiple Library Resources as Compared to the Use of a Basic Textbook in Junior High School Science', *J.Exp. Educ.*, vol. 35, no. 3, pp. 27–35.

BBC *Handbook* (1960), London.

Bernal, J. D. (1939), *The Social Function of Science*, London.

British Association for the Advancement of Science (1918), 'Science in secondary schools': report of the committee, etc., in *BAAS Annual Report*, London, pp. 123–207.

Bronowski, J. (1973), *The Ascent of Man*, London.

Bullock, Lord (1976), Presidential address: 'Science – a tarnished image', in *School Science Review*, vol. 57, pp. 621–7.

Canada, Science Council of (1984), 'Science for every student: educating Canadians for tomorrow's world', *Policy Report* no. 36.

Cantril, H. (1940), *The Invasion from Mars*, Princeton University Press.

Caska, F. (1978), 'Peasant Commercialization in the Serranos of Cochabamba, Bolivia', Ph.D. Thesis, University of Michigan.

Chapdelaine, P. A. Sr, Chapdelaine, Tony and Hay. G. (1985), *The John W. Campbell Letters*, vol. 1, Franklin, USA.

Clarke, G. C. S. (1981), 'Evolving an exhibit on evolution', *Museums Journal*, vol. 81, no. 3, pp. 147–51.

Cornell, J. (1977), 'Ibero-American Science Writers Meeting', *NASW Newsletter*, vol. 25, nos 1, 2, 3.

Crowther, J. G. (1970), *Fifty Years with Science*, London.

Dainton, F. S. (chairman) (1968), *Enquiry into the Flow of Candidates in Science and Technology in Higher Education*, London, HMSO.

Dainton, Sir F. S. (1971), *Science: Salvation or Damnation*, University of Southampton.

Debray, R. (1969), *A Criticism of Arms*, vol. 1, Harmondsworth.

Devonshire Reports, The (1872), *Reports of the Royal Commission on*

Scientific Instruction and the Advancement of Science: 1st and 2nd reports, HMSO, London.

Europe, Council of (1972), European Curriculum Studies No. 6: *Physics*.

Firth, J. D. E. (1949), *Winchester College*, Winchester, P&G Wells.

Fisher, B. (1980), 'Using Literature to Teach Science', *J. Res. Sci. Tech*, vol. 17, pp. 173–7.

Fletcher, T. J. (1983), 'Microcomputers and Mathematics in Schools: a discussion paper', Department of Education and Science, London.

Flower, Sir W. H. (1893), *Modern Museums. Report of Proceedings, 4th Annual General Meeting of the Museums Association*, pp. 21–48.

Fontenelle, B. le Bovier de (1686), *Entretiens sur la Pluralité de Mondes*.

Gardener, P. L. (1974), 'Mathematical Learning and the Sexes: a review', in *Journal for Research in Mathematical Education*, vol. 5, pp. 126–39.

Gardener, P. L. (1975), 'Attitudes to Science: a review', in *Studies in Science Education* 2.1.41.

George, J. T. G. (1974), 'An Investigation of the Attitudes of Secondary School Pupils to the Social and Humanitarian Values of Science Undertaken in a Representative Sample of Schools', Unpublished M.Sc. thesis, University of Sheffield.

Glass, B. (1970), *The Timely and the Timeless: the interrelations of science, education and society*. New York & London.

Griggs, S. A. (1983), 'Orienting visitors within a thematic display', *International Journal of Museum Management and Curatorship*, vol. 2, pp. 119–34.

Haggis, S. M. and Adey, P. (1979), 'A review of integrated science education world-wide', in *Studies in Science Education*, vol. 6, pp. 69–89.

Haldane, J. B. S. (1939), *Science and Everyday Life*, London.

Halstead, L. B. H. (1978), 'Whither the Natural History Museum?', *Nature*, vol. 275, p. 683.

Herschel, J. (1830), *Discourse on the Study of Natural Philosophy*.

Hurd, P. D. (1969), *New Directions in Secondary School Science*, Rand McNally, Chicago.

Jennett, B. (1984), 'High Technology medicines: its benefits and burdens', *World Medicine*.

Jones, G., Connell, I., Meadows, J. (1978), 'The Presentation of Science by the Media', PCRC, University of Leicester, *Brit. Lib. Report* no. 5494, March.

Kidd, Ross and Byram (1978), 'Popular theatre as a tool for community education in Botswana', *Assignment*, no. 44, Unicef, Geneva.

Leder, G. (1979), 'Fear of Success, Mathematics Performance, and Career Choice', a paper presented at the Third International Congress on Gifted and Talented Children, Jerusalem.

Leprince-Ringuet, L. (1983), 'Le scientifique et la vulgarisation', *Bulletin d'Information. AESF* no. 32.

Levine, A. G. (1982), *Love Canal-science, politics and people*, Lexington.

Lewis, B. N. (1980), 'The museum as an educational facility', *Museums Journal*, vol. 80, no. 3, pp. 151–5.

Low, C. (1974), 'Media as a mirror. A Resource for active community', *Canadian Radio-Television Commission*, Ottawa.

MacBride, Sean et al. (1980), *Many Voices, One World*, UNESCO, Paris.

McElheny, V. K. (1985), 'On Impacts of presentday popularization', in *Sociology of the Sciences Yearbook*, vol. ix, Reidel, Dordrecht.

Melton, A. W. (1972), 'Visitor behaviour in museums: some early research in environmental design', *Human Factors*, vol. 14, no. 5, pp. 393–403.

Miles, R. S. (1981), 'Information as an experience; exploitation of a museum's resources', in *The Nationwide Provision and Use of Information*, The Library Association, London, pp. 112–18.

Musgrave, P. W. (1967), 'Towards a sociological theory of occupational choice', in *Sociology Review*, pp. 33–46.

Newton, G. (1962), 'The Financial Times', in *World Press News Supplement*, 30 March.

Nicholls, P. (1982), *The Science in Science Fiction*, London.

Perlman, D. (1974), 'Science and the mass media', *Daedalus*, vol. 10, no. 3, p. 212.

Pluche, l'Abbé (1732), *Spectacle de la Nature*.

Prigogine and Stengers (1979), *La Nouvelle Alliance, Metamorphose de la Science*, Gallimard, Paris.

Rutherford, F. J. (1979), 'The Implementation of socially relevant science education', in P. Tamir et al. (eds), *Curriculum Implementation and its Relationship to Curriculum Development in Science*, Israel Science Teaching Centre, Hebrew University, Jerusalem.

Shaulle, R. (1983), in Foreword to *Pedagogy of the Oppressed* by Paulo Freire, New York.

Shinn, Terry and Whitley, R. (eds), 'Expository science forms and functions of popularization', *Sociology of the Sciences Yearbook*, vol. ix, Reidel, Dordrecht.

Silver, H. (1974), 'Extending science to the social context', in *Times Educational Supplement*, 11 October.

Singer, A. (1966), 'Science Broadcasting', *BBC Lunchtime Lectures*, 4th Series, no. 5.

Snow, C. P. (1968), *The Two Cultures and a Second Look*, CUP.

Solana, F. (1980), *Inaugural Address*, 12th International Conference of the International Council of Museums, Mexico City.

Spottiswoode, R. (1935), *A Grammar of the Film*, London.

Steiner, G. (1984), *A Reader*, London.

Third World Popular Theatre Newsletter, no. 1, p. 6, January 1982.

Thring, M. W. (1977), 'Can science save society?' in *School Science Review*, vol. 59, pp. 210–222.

Tosi, V. (1984), *How to make Scientific Audio-Visuals*, Paris.

Walden, S. M. (1973), 'Some of the Reasons why Relatively Few Girls Choose Science', Unpublished M.Ed. dissertation, Chelsea College, University of London.

Weholt, R. (1984), 'The hard, the soft, and the squishy', *Science and Public Policy*, vol. 11, no. 5.

Weiskopf, V. (1968), *Scientific World*, no. 1.

William, R. (1965), *The Long Revolution*, London.

Williams, W. F., 'The social function of science education', in *Trends in Education*, vol. 1, pp. 32–5.

Wolfe, R. L. and Tymitz, B. L. (1981), 'The Evolution of a Teaching Hall:

"You can Lead a Horse to Water and You can help it Drink", A Study of the dynamics of Evolution exhibit', National Museum of Natural History, Smithsonian Institute, Washington, DC (Office of Museum Programs, Smithsonian Institute).

Worsley, P. (1984), *The Three Worlds*. London.

Zuckerman, Lord (1976), 'Advice and Responsibility', a Romanes Lecture given at Oxford.

INDEX

THE SCIENCE CRITIC

We are at the beginning of the age of public
participation in science. The once widespread
feeling that scientists alone should rule over
scientific enterprise is being challenged, and
participation will increasingly take place as it
becomes clear that science is not a private game
for scientists but is a vital part of society, involving
us all. It is not separate from, but is part of, our 'one
culture'.
In this thought-provoking book Maurice Goldsmith
raises many vital questions – both current and
emerging – on the relationship between science
and society. Drawing on forty years' experience in
science communication as writer and educator, he
deals with such issues as: What forms can public
participation in science take? Who will take part in
establishing the control of science? How will such
controls be organized? How far will they influence
detailed decisions concerning the nature and
procedures of research? What are the implications
not only for science but also for the spread of
scientific information?
Goldsmith looks at the way science is presented to
us through the various media: newspapers,
magazines, books; television, film, radio; and
museums. He examines the role of science fiction
in advancing scientific concepts, looks at the
special problem of 'popularizing' science in the
Third World, and proposes the creation of 'the
Science Critic' as an effective way of improving the
dissemination of scientific information.
The book includes a list of national and international
associations of science journalists and
communicators, and a list of national and
international scientific awards.

Cellular
Interactions in
Animal Development

Cellular
Interactions in
Animal Development

ELIZABETH M. DEUCHAR
Lecturer in Anatomy
University of Bristol

CHAPMAN AND HALL
London

A HALSTED PRESS BOOK
JOHN WILEY & SONS
New York

First published 1975
by Chapman and Hall Ltd
11 New Fetter Lane, London EC4P 4EE
© *1975 Elizabeth M. Deuchar*

Printed in Great Britain

Library of Congress Cataloging in Publication Data

Deuchar, Elizabeth Marion.
　　Cellular interactions in animal development.

　"A Halsted Press book."
　References: p. 256
　　　1. Embryology. 2. Developmental cytology.
3. Cellular control mechanisms. I. Title.
[DNLM: 1. Cell differentiation. 2. Cells—Physiology.
QH631 D485c]
QL955.D38　1974　　　　591.8'761　　　74-960
ISBN 0-470-20961-5

Contents

Preface

During the past ten to fifteen years, Cell Biology has come to be recognized in many Universities as a new branch of biology. It could be argued, however, that nearly all of biology is concerned with the study of cells, either individually or in complex groupings which we call organisms. One can re-frame many courses in biology for undergraduates, and one could rewrite many textbooks, under titles which emphasize the 'cellular' aspects of the subject. But besides this, we now have far more information to add which is based on experimental studies of cells in tissue culture, where observations can often be made on individual cells for a long period of time. As a result of advances in techniques for observing living cells, we now know far more than we did fifteen years ago about how cells move, how they make contact with a surface and with each other, how they interact in small groups and what metabolic processes they undergo. In addition, studies with the electron microscope have enabled us to see much more detail of the changes in structure that take place as cells move, interact and differentiate. Nevertheless, having seen these extra details in individual cells, we still have to look back at the larger groups and whole organisms in which they normally function, in order to see the full significance of all these cellular events. One cannot get far in ones studies, if one remains a '*single*-cell biologist'.

I have never regarded myself as a cell biologist. Hence it was with some reluctance that I eventually agreed to write a book under the present title. For a long time I hoped to find a co-author (or, better still, a substitute author!) who would be able to provide the necessary emphasis on cells as biological entities. Failing this, I still hoped to find an expert on invertebrates, to contribute more examples of cellular interaction in this vast majority of the animal kingdom. However, neither hope was realized and I alone, a mere vertebrate embryologist, have had to try to discuss the whole range of cellular interactions in animal development. As a result, I have followed my natural 'homing' instincts and have orientated most of the discussions towards groups of embryonic cells, whole tissues and whole embryos. This has left many gaps, I know, and several readers may feel that the mechanisms of cell-interaction, as studied in tissue culture, are not adequately discussed. However, first- and second-year science and medical students whose courses include either 'cell biology' or 'developmental biology' may find in these pages an apéritif to their future interests, and the extensive bibliography should help to satisfy their further appetites.

This book is going to press at a time of industrial setbacks which may cause a long delay in its publication: hence there will inevitably be some out-of-date statements in it, if research in cell and developmental biology continues at its present rate. However, the students of next year or the year after will always, I hope, be alive to the possibility of new findings in these subjects, and if any of my over-pedestrian remarks stimulate them to look up the latest research papers, I shall have achieved one of my main objects, which is to provoke thought, criticism and a thirst for more knowledge in this fascinating field of developmental biology.

I owe a great deal of thanks to colleagues in Britain and abroad who have taken an interest in this book. Several authors have kindly provided me with original illustrations from their own work: these are acknowledged below. Many friends have helped me by discussing certain points and by reading parts of the manuscript. I owe particular thanks to Dr Ruth Bellairs and Dr Jim Dodson, who each nobly read through the whole book in its first draft, and offered a number of helpful criticisms. I should also like to thank Dr Brian Pickering and Dr Roger Taylor for their comments on Chapters 10 and 11.

I am indebted to the following authors and publishers for several of the illustrations used in this book.

Professor W. Beermann and Springer-Verlag, Berlin-Heidelberg-New York (Plate 1.1); Professor R. A. Flickinger (original of Plate 1.2); Dr S. Pegrem (original of Plate 1.3a); Dr V. Gabie and the Editor of *Acta Sbryologiae Experimentalis* (Plate 1.3b); Professor A. S. Curtis (original of Plate 1.4); Dr B. Mintz and the Company of Biologists Ltd. (Plate 2.1); Dr R. Bellairs (original of Plate 2.2); Dr R. Presley (original of Plate 3.2); Professor M. Mitchison and the Company of Biologists Ltd. (Plate 3.3); Dr M. Tegner and the American Association for the Advancement of Science (Plate 3.4); Dr D. Szolossi and the Rockefeller University Press (Plate 4.1); Dr J. van den Biggelaar (original of Plate 4.2); Dr M. Kalt and the Company of Biologists Ltd. (Plate 4.3); Mrs P. Hyatt (original of Plate 5.1); Dr P. Flood (original of Plate 6.1); Dr T. E. Schroeder and the American Society of Zoologists (Plate 6.2); Dr R. DeLong and Academic Press Inc., New York (Plate 9.1); Professors G. Veneroni and M. Murray and the Company of Biologists Ltd. (Plate 9.2); Dr S. Bradley and Cambridge University Press (Plate 9.3b); Dr R. Hauser (original of Plate 9.4) and Professor L. B. Arey and W. B. Saunders Co., Philadelphia (Plate 11.1).

Finally, I should like to thank Messrs Chapman and Hall Ltd. for providing some professional help with the line drawings.

Elizabeth M. Deuchar

Bristol, January 1974

Introduction

The scope of this book

This book attempts to provide students of animal biology with some insight into the variety of mechanisms by which the cells within an animal's body interact with one another and exert controls over each other's development and differentiation. Far more is known about cellular interactions and development in vertebrates than in invertebrates. This is mainly because vertebrate embryos are on the whole larger and easier to study under laboratory conditions than are those of invertebrates. Hence, all through this book, most has been said about vertebrates. Reference has been made to invertebrates when there are specially interesting features to note.

This book is written from the point of view of the embryologist, or 'developmental biologist' as he may now prefer to be called. Students who have not done any animal embryology will find it necessary to read some introductory book on this subject, in order to understand all the terms used in the following pages. For briefer, lighter reading, a book such as *Principles of Development and Differentiation* by C. H. Waddington, or *Living Embryos* by Jack Cohen, might be suitable. Those who have time for more comprehensive reading will find B. I. Balinsky's *Introduction to Embryology* very useful. Fuller details of these books are given in the Bibliography on pp. 256–283.

1

1

Cells in the developing organism: some general considerations

Cells are the units which make up the body of an animal or plant. Under special conditions, individual cells may be capable of life on their own, but normally in the intact organism they are mutually dependent for their survival and maintenance. In this book we shall deal with animal and not plant cells, and shall be considering only the Metazoa (multicelluar animals), not the Protozoa (in which each organism consists of only one cell), for it is only in the Metazoa that one can discuss cellular interactions within a single developing organism.

The cells of metazoan animals have many different shapes, sizes and functions: the more complex the structure of the animal, the greater is the variety of cells which compose it. The cells are grouped into larger functional units, the tissues and organs, and within each tissue or organ, particularly close cellular interactions take place. Tissues and organs also interact with each other however, thus ensuring that all the functions essential to the life of the organism are carried out efficiently. This continual co-ordination of functions in all its cells is essential to the life and integrity of every animal, from the moment that it starts to become multicellular when the fertilized egg divides into two cells, right through development, adulthood and ageing. At all stages in its life history an animal's cells are interacting, normally in such a way as to promote resistance to any adverse environmental conditions that may arise, and to make good any damage or loss of cells that may occur through external agencies. In later life, some of the cellular co-ordination processes

3

may begin to fail, so that death from injuries or disease is more likely to occur. But in the early embryo there are remarkable powers of recovery by means of cellular interactions, even after the most drastic experimental procedures.

In this book we shall concentrate on those cellular interactions that occur during embryonic and juvenile stages of life. Before we define the kinds of interactions that may occur, however, we must first give critical consideration to some of the basic 'facts' that are usually accepted as true of the cells in an animal at early embryonic stages.

1.1 The problem of identity of genotype in different cells

It is a commonly accepted generalization that all the cells in any one embryo contain identical sets of *genes*, the factors inherited from the parent organisms via the chromosomes in the nuclei of the egg and the spermatozoön. It is assumed that from the zygote (fertilized egg) stage onwards, at each cell division the daughter cells receive identical sets of genes, since they normally receive identical sets of chromosomes. What evidence have we for this assumption, however? None of it is very direct or conclusive. To start with, there are a few known exceptions to the rule that identical sets of chromosomes pass into all cells during early embryonic cell divisions. In the nematode worm *Ascaris,* and in Chironomid flies, certain chromosomes are lost from all except the future germ cells. In other species of the insect order Diptera, cells of certain tissues such as the salivary glands acquire giant, 'polytene' chromosomes in which all the gene loci have been replicated several times. In all these cases, cells from different tissues clearly do *not* contain identical sets of chromosomes and genes. Even in those animals where there are no gross chromosomal differences between the different cells, we are not in a position to be sure from direct observations that all the genes on corresponding chromosomes in all cells remain identical throughout embryonic development. Even the highest power of an electron microscope does not show enough detail of chromosome structure for every 'cistron' (the active unit of the gene that codes for a polypeptide) to be identified and compared with those of other cells. We have to rely, then, on indirect evidence to support the assumption that there are normally identical sets of genes in all the cells of any individual embryo.

An important line of indirect evidence comes from nuclear trans-

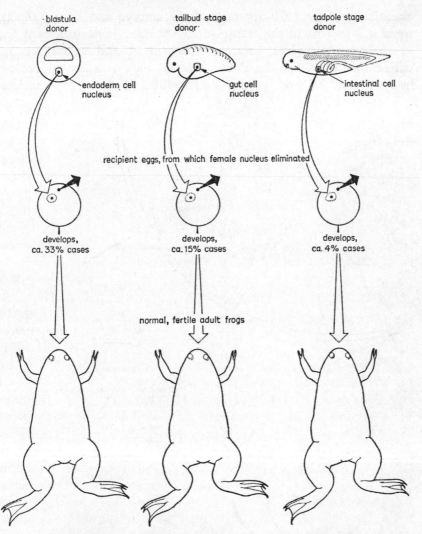

blastula
donor

tailbud stage
donor

tadpole stage
donor

endoderm cell
nucleus

gut cell
nucleus

intestinal cell
nucleus

recipient eggs, from which female nucleus eliminated

develops,
ca. 33% cases

develops,
ca. 15% cases

develops,
ca. 4% cases

normal, fertile adult frogs

Fig. 1.1 Nuclear transplantation into amphibian eggs.

plantation experiments. Illmensee's work (1972) on *Drosophila*, added
to the work of Briggs and King in the 1950s on the American frog
Rana pipiens, and that of Gurdon and his collaborators in the 1960s
on the South African frog *Xenopus laevis*, have shown that nuclei
from the cells of quite advanced embryos or even from tadpoles can,
when transplanted back into an egg whose own nucleus has been
removed, set off normal development and give rise to normal adult
individuals in several cases (Fig. 1.1). This shows that all the genes

5

essential for normal development of the embryo and larva (tadpole) were still present in the transplanted nucleus: none had been lost as the various tissues became differentiated and specialized for different functions in the donor animal. The transplanted nucleus had retained the same genetic potentialities as the zygote nucleus.

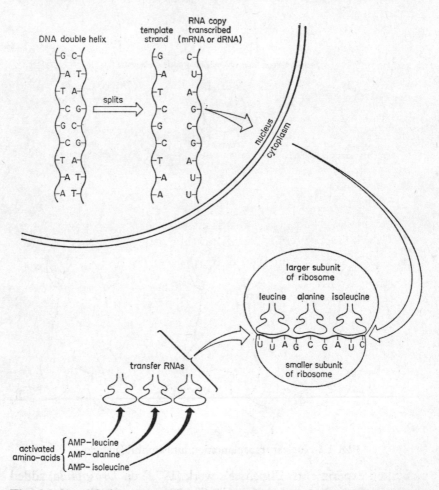

Fig. 1.2 Simplified representation of the genetic control of protein synthesis. A single-stranded DNA template is transcribed as dRNA, which enters the cytoplasm and becomes attached to the smaller subunit of a ribosome. Cytoplasmic amino acids, after activation by attachment to AMP, become attached to small RNA molecules called 'transfer RNA', which carry them to the larger subunit of the ribosome and attach them to suitable triplets of bases (codons) on the dRNA. Adjacent amino acids become bonded together and a polypeptide is thus formed. Several ribosomes may combine as 'polysomes' and thus facilitate the synthesis of larger protein molecules.

If we accept, however, that all cell nuclei of an embryo or larva have identical genetic potentialities, it is not easy to account for the fact that cells differ in form and function, except by assuming that different genes are in some way inactivated in different cells. If, for instance, the genes for haemoglobin production and for myosin synthesis are present in all cells, it is hard to explain why all cells do not become red blood cells or muscle cells, unless one assumes that the haemoglobin and myosin genes are inactivated in all except a very limited number of embryonic cells. But again, the actual evidence for this assumption is not yet very substantial. It all hangs, too, on acceptance of the current hypothesis for gene action. According to this, DNA of active genes is transcribed as 'messenger' RNA (now usually called dRNA), and the sequence of bases in this dRNA determines the order in which amino acids are assembled into a polypeptide on the ribosomes of the cytoplasm. (see Fig. 1.2). This hypothesis has stood up to several experimental tests, but cannot yet be said to be proved beyond all doubt.

On the hypothesis just stated, one could argue that the mere presence of different proteins in different cells must imply that different genes are being transcribed – i.e. that other different genes are inactivated too. A more practical line of approach, however, is to look for differences in the base sequences of dRNA extracted from different types of embryonic cell. If there are differences, this will imply that different genes have been transcribed in the various cells. This experiment was in fact tried by Paul and Gilmour (1968), using molecular hybridization as a test of the degree of similarity of dRNAs from different tissues of the rabbit embryo. The principles of this method are explained in Fig 1.3. Many people now doubt the reliability of some hybridization techniques, however, so although Paul and Gilmour did find differences in the hybridization properties of dRNAs from different tissues, their results are not considered to be conclusive.

Another type of evidence that different genes may be active in different tissues of an individual organism is the presence of 'puffs' at various points on the polytene chromosomes of insects. Plate 1.1 shows such a puff described by Beermann (1973). The puffs lie at different points on the chromosomes in cells of different tissues, and at times such as moulting and metamorphosis, new puffs appear. It has been shown by labelling the cells with radioactive uridine, which is taken up into newly-synthesized RNA, that there is par-

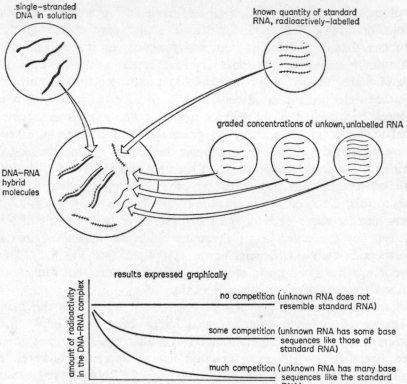

Fig. 1.3 Principles of competitive hybridization experiments. A known quantity of standard, radioactively-labelled RNA is added to single-stranded DNA in solution. Sequences of bases which are complementary to those on the DNA will hybridize with it, forming radioactive DNA-RNA hybrid molecules which may be collected and their radioactivity measured. The experiment is then repeated with the addition of increasing concentrations of the unknown RNA which is to be compared with the standard, and which is not radioactive. The more base sequences it has in common with the standard RNA, the more it will compete with it for hybridization, resulting in less radioactivity in the RNA-DNA hybrid molecules (see graphs.)

ticularly active RNA synthesis in the puff regions. This suggests that they are sites where genes are being transcribed most actively. However, the fact that puffs have been seen in these giant chromosomes of insects cannot be taken as evidence for what happens in all other animals. Labelling work has been carried out also on the chromosomes of amphibian embryos, using radioactive thymidine which is taken up during DNA synthesis, and this has shown that some regions of certain chromosomes replicate later than the rest. Stambrook and Flickinger (1970) compared the locations of late-

8

replicating regions in some easily-recognizable chromosomes of the frog *Rana pipiens,* and they found that in different tissues of the embryo, these regions differed in number and in distribution. Some of their findings are shown in Plate 1.2. It is rather a remote argument, however, to suggest that the late-replication has much bearing on which genes are transcribed as RNA in the different tissues.

The argument that some genes are inactivated (though still present) in certain embryonic cells is one way of explaining why cells produce the different proteins which are the basis of their differences in structure and function. Some have argued that there could be a process of selection of dRNA types in each cell, so that although all genes were transcribed, only some dRNAs were able to function. This argument requires that there is some different cytoplasmic material in different cells, to do the selecting, however: and to explain the origin of the cytoplasmic differences one has again to revert to possible differences in gene activity: so it becomes a circular argument.

If we are now prepared to accept, in the absence of any better hypothesis, that different genes are inactivated in different embryonic cells, the next question that has to be answered, is what agents inactivate the genes? Jacob and Monod (1961) envisaged the existence of 'regulator' genes, producing substances that inhibited the action of neighbouring genes. Others have considered the possibility that chromosomal proteins 'wrap round' the DNA molecule at certain points. It has been suggested that histones, which form a high proportion of chromosomal protein, do this. But no one has yet been able to explain convincingly how *different* parts of the DNA could be so obscured, in different cells. An extension of Jacob and Monod's idea is to suggest that agents from outside the nucleus, or even outside the cell, may cause inactivation of some of its genes. Gurdon and his collaborators argued from their nuclear transplantation work that the egg cytoplasm was responsible for *reactivating* the genes in the transplanted nucleus. Substances which pass into the cytoplasm of a cell from outside, perhaps coming from an adjacent cell, may therefore be thought of as possible agents for the activation or inactivation of certain genes in the cell. This is thought to be the basis of many influences on cell differentiation that are exerted by neighbouring cells.

We are now in a position to try and define some of the ways in which cells of an embryo may interact and may influence each other's course of differentiation.

1.2 The nature of cellular interactions

The term 'cellular interactions' covers any situation in which one cell or a group of cells is affected in any way by the presence of other cells nearby. In practice it includes *all* situations so far studied in animals, since cells *always do* interact and we know of no case where they are absolutely unaffected by the presence of other cells. In embryonic cells the response to the presence of other cells may be visible immediately, when they are viewed alive in culture under the microscope. Cells of early amphibian embryos, for example, throw out pseudopodia round their surface and attempt to adhere to the other cells. Fibroblasts, the long spindle-shaped cells found in connective tissue and mesenchyme, send out a thin sheet of cytoplasm known as the 'ruffled membrane' (Plate 1.3a) and when this makes contact with another fibroblast, the two cells adhere to each other and become immobile: a phenomenon known as 'contact inhibition' of movement (Abercrombie and Heaysman, 1954). In contrast to this, cells of the embryonic neural crest (Plate 1.3b) which form nerve ganglia and melanophores, repel each other, and whereas one crest cell isolated in culture is immobile, as soon as another cell is placed near it, the two cells start to move away from each other (cf. Fig. 8.3). It is thought that neural crest cells must exude some chemical substance, to which other crest cells respond by negative chemotaxis (Twitty and Niu, 1948). But in the kind of interaction seen in fibroblasts, contact between the cells is essential and it is probable that the agent responsible is not any chemical exudate. Some mechanical stimulus may be responsible, or it is possible that some electrical or chemical signal is passed between the surfaces of the fibroblasts.

It may be helpful at this point to list some of the possible mechanisms by which cells may interact at different stages in the life history of an animal. We shall see examples of these kinds of interaction in later chapters.

1.3 Types of interaction in animal cells

1.3.1 *Aggregation and adhesion*

This is the interaction shown by cells of early embryos of blastula

10

and gastrula stages, before any distinct tissues are present. These cells are mobile and throw out pseudopodia, seen most easily in the large cells of amphibian embryos (Plate 1.4). When the pseudopodia make contact with adjacent cells, adhesions will occur between cells of the same 'germ layer' (ectoderm, mesoderm or endoderm). This implies that cells of the same germ layer are able to 'recognize' each other by some property of their surfaces which come into contact. It has been suggested, as we shall see later, that the cells carry surface molecules which are specific to the germ layer. The actual adhesion may take place by electrical bonding, or by some more complex fitting together of stereochemically complementary molecules (see Fig. 1.4).

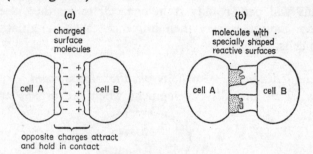
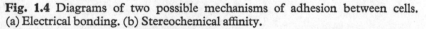

Fig. 1.4 Diagrams of two possible mechanisms of adhesion between cells. (a) Electrical bonding. (b) Stereochemical affinity.

1.3.2 De-adhesion, following on aggregation and adhesion
This occurs when cells of blastula and gastrula stages make contact but are *not* of the same germ layer, so their surface properties are not sufficiently similar or complementary for adhesions to be maintained.

1.3.3 Contact inhibition of movement
This is the interaction characteristic of fibroblasts. For it to take place, the fibroblasts must first be mobile so that they can make contact with others of their kind, then a mutual confrontation of ruffled membranes is necessary before the movement stops and the cells adhere. Another component of this behaviour is that the cells adhere to a substratum too, if this is present, and form a sheet of cells with none overlapping, known as a monolayer.

1.3.4 Passage of an electrical 'signal' from one cell to another
This involves an electrical charge being emitted from the surface of one cell and impinging on the surface of a neighbouring cell,

then perhaps entering this cell. Electrical discharges have been shown to pass between early epidermal cells (Roberts, 1969; Roberts and Stirling, 1971) and even between the undifferentiated cells of a blastula (Palmer and Slack, 1969).

1.3.5 *Passage of molecules or ions between cells*
We know relatively little about the extent to which free ions pass between cells, but many embryonic cells bind ions from the medium, particularly calcium and magnesium ions, to their surfaces. Calcium and magnesium play major roles in governing the adhesiveness of these cells too. There is increasing evidence from labelling with radioactive and fluorescent tracers that even such large molecules as steroids will pass readily from one cell to another (see Chapter 10). These molecules may then influence the differentiation of the recipient cells.

1.3.6 *Long distance interactions via the bloodstream*
In stages of an animal's development when the organs are formed

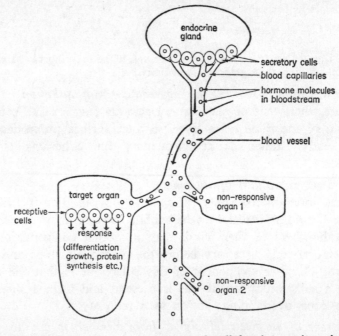

Fig. 1.5 Diagrammatic representation of cellular interactions involving a hormone, secreted by a specialized endocrine gland and received through the bloodstream by a target organ, which responds to it in specific ways.

and the blood circulation is complete, a special kind of long-distance cellular interaction becomes possible by means of hormones carried in the blood stream. Cells of an endocrine gland can act on cells of a target organ at some distance from it, by releasing a blood-borne agent known as a hormone (usually a steroid or a polypeptide). This hormone has a selective action, and only certain cells in certain target organs respond to it, though the hormone is potentially able to reach all cells in the body as it travels through the blood system (Fig. 1.5).

All the types of interaction that we have just listed may either be thought of as *unidirectional* – one cell being the initiator of the process and the other the recipient of the contact, signal or substance – or they may be thought of as *reciprocal*, two-way interactions. Fig. 1.6 gives a scheme of the various possibilities. In early embryos it is quite common for cells to interact reciprocally, but later when certain tissues or organs become specialized, their most characteristic interactions with other organs are in one direction only.

Another general and important point about all the types of cellular interaction listed above is that they are all likely to have an effect on the future differentiation of the recipient cell. The aggregation, adhesion and de-adhesion events (Sections 1.3.1 and 1.3.2) are responsible for determining the relative positions of germ layers in the embryo, and this enables the germ layers to interact, perhaps by passages of signals or molecules between them (see Sections 1.3.4 and 1.3.5). The contact inhibition reaction (Section 1.3.3) has the result that groups of fibroblasts tend to flatten and form monolayers of connective tissue. When passage of signals or molecules occurs, as in Sections 1.3.4, 1.3.5 and 1.3.6, these may either interact with the surface of the recipient cell or may pass into its cytoplasm, and as we have seen earlier, cytoplasmic changes may affect gene activities, so that the course of differentiation in the recipient cell may be altered. We shall be saying a great deal more about cell and tissue interactions which alter the course of differentiation in animal embryos, in subsequent chapters.

It should be emphasized, finally, that in all those cellular interactions that have been recognized, some definite *response* is seen in the recipient cell (Fig. 1.7). It may change its mobility or shape, as in Sections 1.3.1 and 1.3.2 above, or it may undergo special morphogenetic movements in conjunction with other cells responding to the same stimulus, electrical or chemical (Sections 1.3.4 and 1.3.5). Another kind of response is for the recipient cells to proliferate,

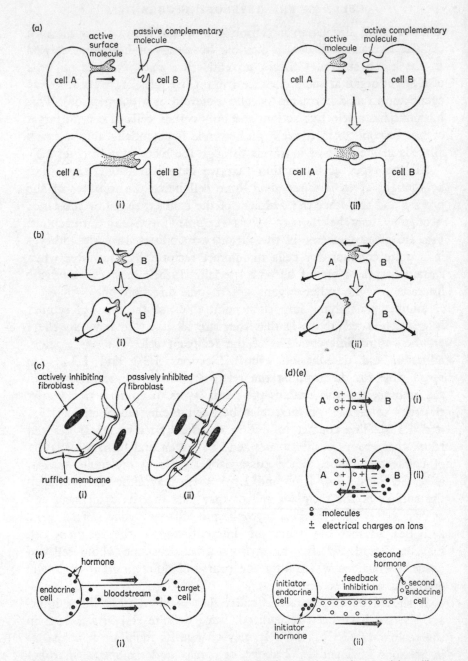

Fig. 1.6 Diagrams of some types of cellular interactions, shown either (i) as one-way processes, or (ii) as reciprocal interactions. (a) Aggregation and adhesion; (b) De-adhesion; (c) Contact inhibition of movement; (d), (e) Passage of molecules or ions between cells; (f) Passage of molecules via the bloodstream.

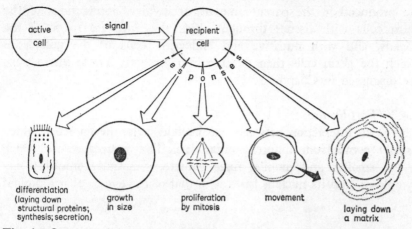

differentiation (laying down structural proteins; synthesis; secretion)

growth in size

proliferation by mitosis

movement

laying down a matrix

Fig. 1.7 Some possible types of response by a cell to signals received during interactions with other cells.

and/or to synthesize special proteins such as enzymes, haemoglobin, myosin or collagen. Sometimes these proteins and other substances are laid down either as stores or structures within the cell: at other times they may be laid down as an intercellular matrix. The responses of tissues and organs to hormones (type (f) in Fig. 1.6) are even more varied and specialized, and individual examples will be considered in later chapters. Different target organs may respond in very different ways to the same hormone: the best example of this is the varied effects of thyroxine on tissues of amphibian larvae at metamorphosis. The tail tissues break down, while limb tissues start to develop, in response to thyroxine. There are many other points of contrast, which will be reviewed in Chapter 10.

Having discussed in a preliminary way the interactions that can occur between individual cells, we now need to see these in the context of the development of the animal organism. A few introductory remarks about the sequence of events in animal development are necessary here, before we go on in the following chapters to consider each phase of development and its cellular interactions in more detail. Fig. 1.8 summarizes the developmental phases.

1.4 Phases of animal development and the cellular interactions involved in each

1.4.1 Development of the germ cells

This is the process by which ova (eggs) and spermatozoa (sperm), the *gamete* cells which fuse to initiate a new developing organism,

15

are produced in the parent animals. It involves interactions of the germ cells with tissues through which they migrate to reach the gonads, and with nutritive and endocrine cells of the gonads in which the germ cells then develop into gametes. These interactions are discussed in Chapter 2.

1.4.2 *Fertilization*

This is an interaction between two haploid cells, the gametes, which usually come from different organisms. The spermatozoön, which is the smaller and mobile, male gamete, penetrates into the egg cytoplasm, and its nucleus fuses with that of the egg so that a diploid

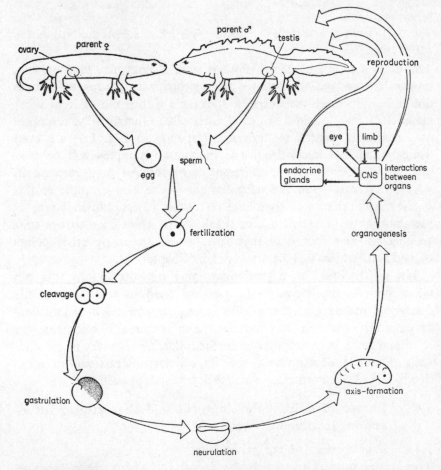

Fig. 1.8 Diagrammatic representation of the phases of animal development, as seen in amphibians.

number of chromosomes is restored. In many animals there is a complex sequence of interactions to ensure that the sperm reaches the egg, that it remains in contact with it and is able to penetrate it, and that after this no additional sperm can penetrate the same egg. In some animals, notably the echinoderms, the entry of the sperm activates metabolic processes in the egg cytoplasm so that there is an immediate increase in respiration rate and in the uptake of raw materials into proteins. In the Amphibia, a special interaction occurring at fertilization is a movement of cortical material in a region diametrically opposite to the point at which the sperm enters. This movement forms a grey crescent in the cortex, which determines the dorsal side of the embryo. Fertilization and the reactions which follow it are discussed in Chapter 3.

1.4.3 *Cleavage*

This is the phase when the fertilized egg (referred to as the *zygote*) divides to form a number of cells. These cells may have characteristic differences in size, shape and cytoplasmic contents, depending on the original distribution of cytoplasmic constituents in the egg and the quantity of yolk that it contained. In some animals, for instance the mammals, the cells do not adhere to each other very closely at first (cf. Fig. 1.9a). In eggs where there is much yolk present, the cleavage divisions may be incomplete at first so that the cells have some continuity with their neighbours and can easily exchange cytoplasmic materials. This situation is seen in amphibians, for instance (Fig. 1.9b). Even more possibility of cytoplasmic communication during cleavage exists in the insects, for in them no cell membranes are formed until the zygote nucleus has divided several times and a layer of nuclei surrounds the periphery of the embryo.

We shall see in Chapter 4 that experimental work has shown that

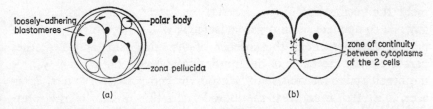

loosely-adhering blastomeres

polar body

zona pellucida

zone of continuity between cytoplasms of the 2 cells

(a) (b)

Fig. 1.9 Different degrees of adhesion and continuity between cleavage cells. (a) four-cell stage in mammals, external view; (b) two-cell stage in amphibians, seen in section.

cleavage cells do interact and influence each other's differentiation in a variety of ways.

1.4.4 *Gastrula and neurula stages*

The stage defined as 'gastrulation' in animals is characterized by movements of cells which bring them into different positions relative to each other, and create in most cases three distinct layers: *ectoderm* outside, which will form the skin, nervous system and associated structures; *mesoderm* in the middle which gives rise to muscle, skeleton, connective tissue, vascular system and excretory system; and *endoderm* innermost, forming the lining of the gut and the germ cells for the next generation of gametes.

The process of gastrulation gives opportunities for a major phase of interactions between cells of the three layers, ectoderm, mesoderm and endoderm. The most important of these for the future of the embryo is the action of dorsal mesoderm on ectoderm overlying it which causes the ectoderm to form a *neural plate,* the beginnings of the nervous system. *Neurulation,* the rolling up of this plate into a tube which becomes the brain and spinal cord, then ensues. At the same time many other groups of cells interact to form the rudiments of the embryonic organs and tissues. We shall see in more detail some of the cellular interactions that occur during gastrulation and neurulation, in Chapters 5 and 6.

1.4.5 *Organogenesis*

Each individual organ of an animal can be very complex and is often made up of a number of distinct types of cells, arranged in elaborate groupings and layers. This orderly complexity is produced by a whole series of cellular interactions, and it is only in a few organs of vertebrates that the sequence of these interactions has been worked out (see Chapters 7 and 8). At this phase of development the reciprocity of the interactions becomes most marked: if any one component of an organ is isolated, it is rarely able to continue normal development in the absence of the other part, and this other part is equally incapable of differentiating normally on its own. Where the organ rudiment consists of mesoderm and ectoderm layers, however, it is often found that the mesoderm is the initiator of the interactions and has more independence than the ectoderm. This is the same situation as in the gastrula, where mesoderm acts on ectoderm first.

1.4.6 *Interactions between different organs and tissues*
Besides being dependent on intrinsic cellular interactions for their normal development, some organs also depend on interactions with other organ systems. The nervous system, for instance, undergoes a number of interactions with the end-organs that it innervates. These are reciprocal interactions, and not only does the end-organ require innervation in order to develop properly, but the part of the central nervous system which supplies it will not develop to full size unless the end-organ is present. For instance, fewer neurons develop at limb levels in the spinal cord if the limb buds are removed, and limb buds deprived of their innervation become stunted in growth. Further examples of interrelationships between the nervous system and end-organs are discussed in Chapter 9.

1.4.7 *Interactions with specialized cells*
A special case of interaction betwen different organs is the action of endocrine glands on a number of different processes of development in the animal body. The organ which co-ordinates many of these endocrine controls is the pituitary gland: if this is removed early in development, the growth and differentiation of many organs in the body will suffer. More will be said about hormonal cellular interactions in Chapter 10.

Finally, there are several cellular interactions which protect developing and adult animals from the deleterious effects of abnormal and foreign cells, including infective and toxic agents. Among the most important protective interactions are those of the immune system, in which a specialized line of cells, the lymphocytes, produce specific molecules called 'antibodies' which react with complementary 'antigens' of the foreign cells and thereby eventually destroy them. We shall refer to the immune system and other protective interactions, in Chapter 11.

From this introductory outline it can be seen that we have a very wide range of cellular interactions to consider, in the development of animal organisms. Some may seem to be only minute events, such as the passage of a few ions between two adjacent cells, but because they set off a long chain of responses in the recipient cell, including possibly an alteration in the activities of its genes, and with the further possibility that it may in turn carry out new interactions with other cells, even such a very minute initial event can have far-reaching effects.

2 Cellular interactions during the development of germ cells

Any account of the cellular interactions that control animal development should start with a consideration of the development of the germ cells, since these give rise to the gametes, or sex cells, which fuse to form the single zygote cell from which the new individual originates. The male and female germ cells develop within 'parent' organisms of the preceding generation, and it is perhaps surprising that in many animals they appear while that 'parent' is still itself an embryo. In some species their origin has been traced from very early stages indeed, when the embryo consists of only a few, scarcely differentiated cells. In these species the germ cells have usually been recognized either because they have a distinct morphology, or because their cytoplasm has special staining properties. So long as they retain these properties, they can be traced throughout embryonic development until they become incorporated into the gonads (Fig. 2.1). In vertebrates, the germ cells have been found to migrate considerable distances in order to reach the developing gonads, entering blood vessels in their course in some cases, so that several interactions with other cells in the body may occur at these stages. Prior to this, however, there must be interactions that cause certain cells to differentiate into germ cells. These we must now consider.

2.1 Factors influencing the initial differentiation of germ cells

It is only in a very few species of animals that we know some of the processes that determine which cells are to become the *germ*

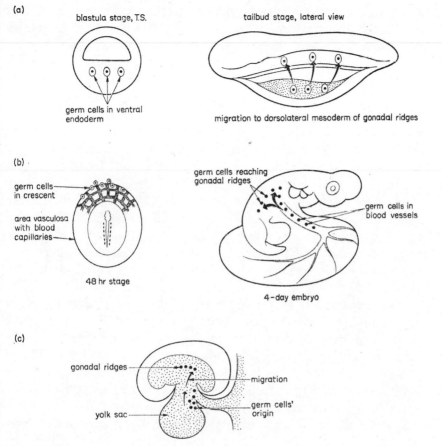

(a)

blastula stage, T.S.

tailbud stage, lateral view

germ cells in ventral endoderm

migration to dorsolateral mesoderm of gonadal ridges

(b)

germ cells reaching gonadal ridges

germ cells in crescent

germ cells in blood vessels

area vasculosa with blood capillaries

48 hr stage

4-day embryo

(c)

gonadal ridges

migration

germ cells' origin

yolk sac

Fig. 2.1 Diagrams to show the origin and migration of the germ cells in (a) amphibians, (b) birds and (c) mammals.

cells. These cells are distinct from the *somatic cells* which form all other tissues of the body. One highly specialized mechanism by which germ cells become determined has already been mentioned in Chapter 1; we saw there that in certain insects and nematodes, during the mitoses of early cleavage stages, chromosomes are lost from the majority of daughter cells. In fact only one or two cells retain the same chromosome complement as the original zygote. It has been found by tracing them to later stages of development that these cells with complete chromosome sets are the stem line for the germ cells, while all the others which have lost chromosomal material become somatic cells. These findings have been reviewed by Tyler (1955). One can say in general terms that it must be the genes on those chromosomes that are lost from somatic cells, that control the differentiation of the germ cells. But we still do not know what controlling factor causes the chromosome losses. Since they are very regular and constant for each species, they appear to be under genetic control in the first instance. It is also noticeable that the cell divisions (cleavages) in these species have a very constant pattern, with no individual variations. So far it has not been possible to alter the pattern of cleavages experimentally, or to rearrange the cells, in order to find out what sequence of interactions during cleavage may influence whether or not chromosomes are lost from each cell. Centrifugation experiments on eggs of Ascaris, however (Boveri, 1910) were found in some cases to result in *two* stem cells for the germ line instead of the usual *one*. In other cases, centrifugation caused chromosome losses from *all* cells, so that *no* stem cell for the germ line was left. This suggests that some cytoplasmic component, displaced by the centrifugation, is involved in the control of chromosome losses.

In some animals, as already mentioned, there is a recognizable cytoplasm in the germ cells. This 'germ plasm' is a part of the contents of the egg cytoplasm which becomes segregated into the germ cells during cleavage. The sequence of events has been followed in detail in the Amphibia (see Bounoure, 1934; Blackler 1958). The cytoplasm of the egg owes its origin to raw materials carried to the ovary in the bloodstream of the female animal, and this involves a series of cellular interactions which we shall be considering later (see p. 32). It has been shown by experiments that both in the amphibian *Xenopus* and in the fruit-fly *Drosophila* the germ plasm is essential for the differentiation of the germ cells. In *Xenopus*, Buehr and Blackler (1970) identified the germ plasm at the vegetal ends of the cells at 2-4 cell stages, using histochemical methods which

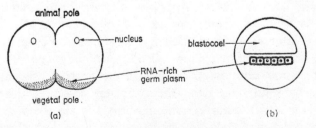

Fig. 2.2 Position of the germ plasm and germ cells in early embryos of *Xenopus*. (a) Section of embryo at two-cell stage; (b) section of blastula showing germ plasm at the periphery of cells in floor of blastocoel.

22

showed that it was rich in RNA (see Fig. 2.2). If some or all of this plasm was removed, by pricking the vegetal ends of the blastomeres, few or no germ cells developed and the resulting animals were partially or totally sterile. In *Drosophila*, if the cytoplasm of the pole cells (which become germ cells) is exposed to ultraviolet light (Geigy 1931), this results in animals with sterile gonads, containing no germ cells but only interstitial tissue (Fig. 2.3).

Special cytoplasm is present in the germ cells of many other invertebrates; for instance in scyphozoans, chaetognaths, rotifers and crustaceans (see review by Tyler 1955) and also in many vertebrates, and this cytoplasm is probably essential for the differentiation of the germ cells, as it is in *Drosophila* and *Xenopus*. In these two species, germ cells have also been transplanted into other individuals at early

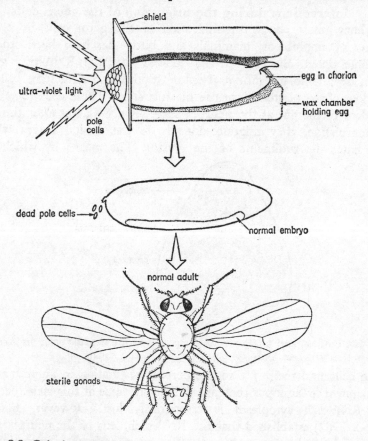

Fig. 2.3 Geigy's experiment in which the germ cells of *Drosophila* were irradiated.

embryonic stages (Blackler, 1960; Illmensee 1973) and it has been found that their determination as germ cells persists, even when they are placed in different tissue environments. So this is further proof that the initial differentiation of the germ cells is controlled internally – presumably by factors carried in the germ plasm – and is not affected by neighbouring cells. When the germ cells begin to migrate to their definitive position in the gonads, however, there are a number of essential interactions with other cells *en route*. Subsequently when they undergo maturation processes to form the gametes, there are interactions with the surrounding cells of the gonad, and more remotely with endocrine cells that control gonadal activity. We shall now go on to consider some of these later events.

2.2 Interactions during the migration of the germ cells
We know most about the migration of the germ cells in a few species of amphibians, mammals and birds that have been studied in some detail. Blackler (1958) showed that the RNA-rich germ plasm of amphibian embryos became localized in endoderm cells of the floor of the blastocoel, at the blastula stage. These cells remain in the floor of the gut at the neurula stage (Fig. 2.4), but after the end of neurulation, they migrate dorsally to the genital ridges where they enter the rudiments of the gonads. The means by which the

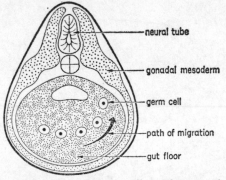

Fig. 2.4 Position and migration of germ cells at the neurula stage in *Xenopus* seen in transverse section.

germ cells find their way to the correct destination in amphibians is not known; in Xenopus they are difficult to trace at this stage because the RNA-rich cytoplasm is temporarily lost. However, Blackler (1960, 1962) established that the RNA-rich cells of the neurula stage were indeed the future germ cells, by experiments in which he transferred them between different sub-species of *Xenopus laevis*:

X.l. laevis and X.l. victorianus. In order to be able to distinguish the transferred cells from those of the host, he used a mutant of *X.l. laevis* which had only one nucleolus instead of the normal two, and he was thus able to establish that the grafted cells did in each case reach the gonads. Female recipients of grafts which developed into oocytes, moreover, laid eggs which had the colouring characteristics of the donor sub-species. So these endoderm cells with the RNA-rich cytoplasm were without doubt the germ cells. In further corroboration of this result, the donor embryos from which the RNA-rich endoderm cells had been removed were found later to have no germ cells and they developed into sterile animals.

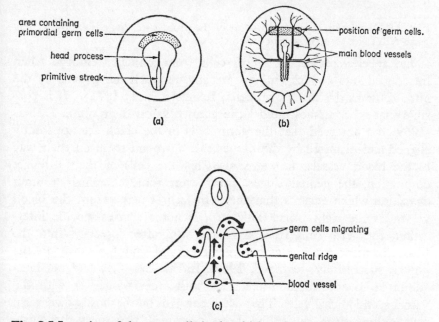

Fig. 2.5 Location of the germ cells in the chick embryo at successive stages of development. (a) Primitive streak stage; (b) 2–3 days' incubation; (c) 4–6 days' incubation. Whole blastoderm in surface view in (a) and (b); transverse section of dorsal axis in (c).

In birds, the germ cells originate from a crescentic area of endoderm at the anterior border of the embryonic disc (Fig. 2.5a). From here they migrate towards the midline after the folding of the embryonic disc has brought this area of endoderm to lie ventrally, in the yolk sac region of the gut floor (cf. Fig. 2.5c). The germ cells enter blood cells in order to carry out their migration, pushing their way between the cells of the capillary endothelium by a process

25

known as diapedesis (Fig. 2.6). As far as is known the germ cells do not penetrate any other epithelia in the embryo, so there must be some specific cellular interaction by which they recognize the blood vessels, and the endothelial cells allow them to penetrate. We do not yet know the mechanism of this interaction, but we do know that it

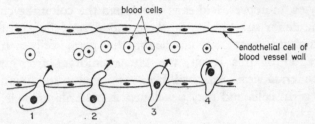

Fig. 2.6 Diagrammatic representation of the process of 'diapedesis' by which germ cells enter blood vessels in the chick.

is not species-specific, for germ cells from a mouse embryo when introduced into a chick embryo are able to reach the gonads by the same route as the host's own cells (Rogulska *et al.*, 1971). It is possible that other tissues 'guide' the germ cells on their route. Dubois (1969) has observed that the germ cells in the chick are apparently aligned and distorted by the tissues that surround them on their way to the blood vessels. It seems also that the cells of their ultimate destination, the genital ridges, must exert some chemical or other attraction which ensures that the germ cells emerge from the blood vessels at the right place. This, too, is not a species-specific interaction, for germ cells from a turkey will, after injection into the bloodstream of a chick embryo, emerge and find their way to the gonads successfully (Reynaud, 1969). In this case they had not been subjected to any initial interactions with other tissues or with the vascular endothelial cells. They were not 'led' by the host's own germ cells either, since these had been destroyed previously by ultra-violet irradiation.

In mammals, the germ cells originate from endoderm at the posterior region of the yolk sac (Plate 2.1) and they migrate dorsally from here to the genital ridges, but not by an intravascular route as in birds. Their migration was traced in the mouse by Mintz (1957), who recognized them by their strong alkaline phosphatase reaction. More recently, Jeon and Kennedy (1973) have followed them by electron microscopy too. Since it is not so far possible to carry out germ cell transfers in mammals, the interactions by which mammalian germ cells reach the genital ridges are almost unknown: it is a wide-open field for research, which could perhaps be tackled by tissue-culture methods.

We know very little about the interactions between germ cells and other tissues in invertebrates. In Crustacea, the germ cells are already included in the rudiment of the gonad at a very early stage of embryonic development when the ectoderm, mesoderm and endoderm cells have not yet begun to differentiate (Manton, 1928). The large gonadal cells lie mid-ventrally and are flanked by endoderm and mesoderm (Fig. 2.7). Later the whole genital rudiments migrate to the thoracic region, carrying the germ cells with them, and take up a ventrolateral position there.

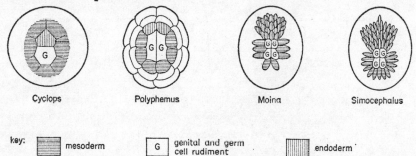

Cyclops   Polyphemus   Moina   Simocephalus

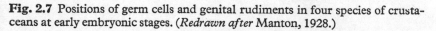

key: ▦ mesoderm   G genital and germ cell rudiment   ▥ endoderm

Fig. 2.7 Positions of germ cells and genital rudiments in four species of crustaceans at early embryonic stages. (*Redrawn after* Manton, 1928.)

In insects, which have been studied more extensively than the Crustacea, the germ cells are separate from the gonadal rudiments and lie at the extreme posterior end of the embryo (Fig. 2.3), for this reason they are often referred to as the 'pole' cells. They are large and spherical, and as we saw above contain a special cytoplasm. This is called the 'oösome' and may be distinguished in the uncleaved egg before the pole cells form. Irradiation of the oösome (cf. Geigy, 1931) does not affect the gonads since these develop from mesoderm in a more anterior position. They are, however, smaller and less differentiated than would be expected simply from the absence of germ cells, and it seems likely that they lack some essential interaction with the germ cells to promote their growth. In normal development the pole cells migrate inwards as the embryo develops, to reach a position in contact with the gonadal rudiments.

In vertebrates, a number of interactions are known to take place between the germ cells and gonadal tissue, after the phase of migration is over. Having arrived at the genital ridges, the germ cells become surrounded by mesoderm cells which will form the primitive sex cords and the epithelium of the gonad. Whether these cords develop fully or not, and the course of differentiation they follow, towards either male or female gonadal tissue, depends primarily on what sex chromosomes the cells contain, but it is also influenced by the germ cells which reach them. In the absence of germ cells, the gonadal tissue does not differentiate fully. The gonad also influences the differentiation of the germ cells, for if germ cells of the opposite sex are substituted for those of a host embryo, they may then

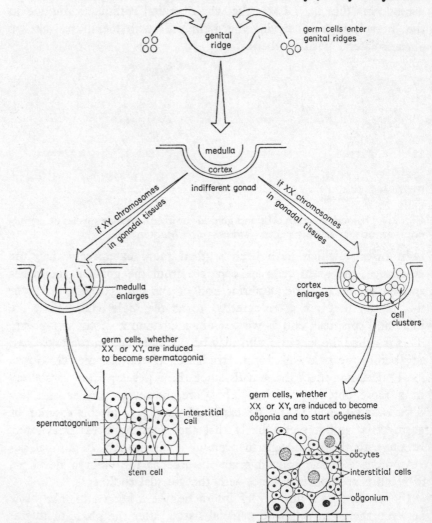

Fig. 2.8 Diagrammatic representation of the interactions between gonads and germ cells in vertebrates.

develop into gametes appropriate to the chromosomal sex of the host. The possibilities are illustrated in Fig. 2.8. The relative influences of gonadal and germ cells on each other's development varies in different groups of vertebrates. In mammals, for instance, the gonadal tissue is more likely to alter the course of differentiation of opposite-sexed germ cells, than are germ cells to influence an opposite-sexed gonad. Also in mammals, the male gonad has a more active influence on differentiation of associated tissues than does the ovary. But the same is not true of birds and amphibians. We shall see more examples of this when considering the influence of the gonads on the development of their ducts (see Chapters 7 and 10). We will now consider some of the special exchanges that take place between gonads and germ cells during oogenesis and spermatogenesis.

2.3 Nutrition and protection of the developing female germ cells

In the gonads of female animals there are special nutritive relations between the developing germ cells and other cells adjacent to them. As the oocytes (which are derived from the primordial germ cells by mitosis, see Fig. 3.1) develop, they increase enormously in size and lay down vast food reserves in the cytoplasm. In certain invertebrates that have been studied, these food reserves are seen to be derived from specialized 'nurse' cells, or 'trophocytes' which are associated with each developing oocyte. Some of the different types

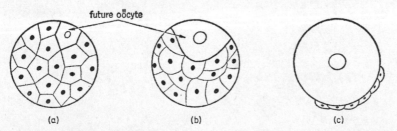

Fig. 2.9 Trophocytes in the leech, *Pisciola* at successive stages of development of the oocyte. (*Modified from* Jörgensen 1913.)

of arrangement of trophocytes are shown in Figs. 2.9 and 2.10. The first example (Fig. 2.9) is seen in the leech *Pisciola* (Jörgensen, 1913). In these animals, each oogonium, the progenitor cell of the oocytes (cf. Fig. 3.1) divides into about ten cells, then all the cells derived from a group of five oogonia associate together. Only one of these

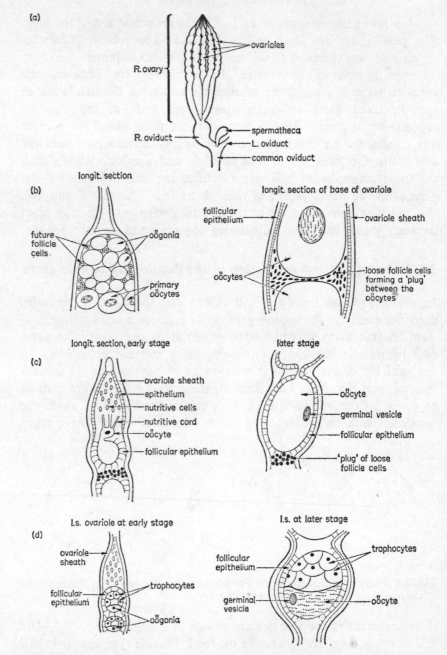

Fig. 2.10 Ovarioles in insects. (a) R. ovary of the bug *Rhodnius*, seen in ventral view (b) Detail of oocytes and follicle cells, at two stages, in an ovariole of the 'panoistic' type (c) Similar detail of 'telotrophic' ovariole. (d) 'Polytrophic' ovariole. (Drawings diagrammatic and based on Davey 1965.)

cells continues to grow as the oocyte, while all the other cells in the group (about fifty) now become trophocytes. These dwindle in size gradually as the oocyte grows (Fig. 2.9b and c) and they evidently contribute materials to it from their cytoplasms and perhaps also from their nuclei.

The relationships between trophocytes and oocytes have received much study in the insects (see Johanssen and Butt, 1941; Davey, 1965). The developing oocytes lie in long branches of the ovary called ovarioles (Fig. 2.10a) which vary in number from four in butterflies and moths to two thousand or more in termites. The more primitive groups of insects have 'panoistic' ovaries in which there are no trophocytes but only small, epithelial follicle cells which cannot provide much in the way of nutrition, though they may protect the oocyte (Fig. 2.10b). In insects which have 'telotrophic' ovaries, the trophocytes form a cluster alongside each developing oocyte, and surround a core of cytosplasm, which sends extensions into the follicle, attaching to the surface of the oocyte (Fig. 2.10c). Finally, in 'polytrophic' ovaries there are fifteen intact trophocytes round each oocyte, and these shrink as the oocyte grows (Fig. 2.10d). In all these types, the trophocytes are derived from earlier divisions of the germ cells, and we do not know what it is that determines which cell becomes an oocyte while the rest become its nurse cells.

The passage of materials into the oocyte from the nurse cells has been followed by electron microscopical work (e.g. King and Devine, 1958, in *Drosophila*), by autoradiography (Ullmann, 1973) and by tracing the movement of labelled antigenic proteins (e.g. Telfer, 1961, in Saturniid moths). In *Drosophila* the nuclei of the trophocytes extrude material which passes into the oocytes. Certain antigens have been traced from the haemolymph (blood) of female Saturniid moths into their oocytes, but at least one of these, called 'antigen 7' by Telfer, passes via the follicle cells into the oocyte, and not via the trophocytes.

It has been shown recently that the synthesis of yolk protein in insects is under hormonal control from the brain and from the ovary. Hagedorn and Fallon (1973) report a series of experiments on the mosquito, *Aedes aegypti*, which show that after a blood meal, a hormone which stimulates enlargement of the oocytes is released from neurosecretory cells of the brain. A few hours after this, the yolk protein 'vitellogenin' begins to be synthesized in the fat body. It is released into the haemolymph and from there reaches the ovaries.

31

B*

Ligaturing the neck to cut off communication from the brain to the ovaries, or ovarectomy, prevents synthesis of vitellogenin. When the fat body is incubated *in vitro,* it will continue synthesizing vitellogenin only if the ovary is also present in the medium.

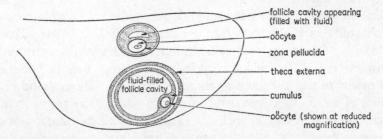

follicle cavity appearing (filled with fluid)

oöcyte

zona pellucida

theca externa

fluid-filled follicle cavity

cumulus

oöcyte (shown at reduced magnification)

Fig. 2.11 Diagram of two stages in maturation of the mammalian oocyte and its follicle, in the ovary.

Specialized trophocytes are not seen in the ovaries of vertebrates that have been studied. The follicle cells are more numerous than in invertebrates, however, and function to varying degrees in the nutrition of the oocyte; they are particularly prominent in mammals, where they also have important secretory activities (Fig. 2.11). In birds and amphibians, as in insects, one of the main features of the developing oocyte is its accumulation of large stores of yolk, and it has been shown that the raw materials for yolk synthesis pass through the follicle cells, though they mostly originate from further away. Some whole proteins have been traced into the developing oocyte from the maternal bloodstream, by immunofluorescent labelling methods. It has been shown that these proteins are synthesized in the liver, in response to gonadotrophic hormones secreted by the pituitary gland. A particular class of protein, called 'serum lipophosphoprotein' (SLPP) by some authors and 'vitellogenin' by others, has been traced clearly from its synthesis in the liver, into the bloodstream, and then via ovarian tissue to the follicles and oocytes. This route involves passing through several layers of cells, and it is not known how this occurs while the molecule still retains its structure intact, so that it is still recognizable by immunological methods. It is possible that some of its transfer is by the engulfing process known as pinocytosis (Fig. 2.12). In birds, Bellairs (1965) has observed long processes from the follicle cells which indent the surface of the oocyte and bear special structures on them which she has named 'lining bodies' (Plate 2.2); these appear to be engulfed by the oocyte. She suggests that this may be a mechanism for transfer of high molecular weight materials into the oocyte. In both amphibians and mammals, long processes penetrating from the follicle cells into the oocyte have also been described, but in no case has there been experimental proof that materials pass inwards via these processes.

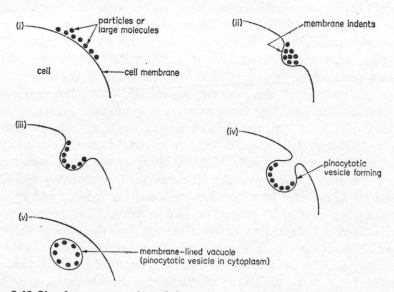

Fig. 2.12 Simple representation of the process of pinocytosis. Stages (i) to (v) show the gradual engulfment of particles from the surface of the cell by invagination of part of its membrane, forming a pinocytotic vesicle in the cell cytoplasm.

The main work on SLPP was carried out by Wallace and Jared (1969) on *Xenopus*. They showed that slices of liver tissue would respond to pituitary gonadotrophins *in vitro* by synthesizing this protein, and that the liver of the male responded in just the same way as that of the female. So they brought to light a chain of cellular interactions involved in this aspect of the development of germ cells in vertebrates, which is comparable with the events in insects. We shall see in the next chapter that hormones also play essential roles in ensuring that the germ cells are shed at appropriate times and become fertilized. The cells of the anterior lobe of the pituitary gland interact, at a distance, directly with the germ cells, using hormones as intermediaries.

An essential end-event in the development of the oocyte is the laying down of protective coverings round it: membranes, albumen and/or shell, according to the species, so that it can survive when

33

it is expelled from the ovary and then, except in viviparous animals, into the external world. Usually it is the follicle cells that are responsible for laying down the innermost of these membranes, in both vertebrates and invertebrates. Electron microscope work on *Drosophila* (King and Devine, 1958), has shown that in this species even the egg membrane immediately surrounding the cytoplasm is a secretion of the follicle cells. In mammals, the follicle cells secrete an additional, jelly-like layer called the 'zona pellucida', and some of these cells remain attached to it, forming the 'corona' (Fig. 2.13) which gives extra protection as the egg is expelled from the ovary into the opening of the oviducts. In animals where the egg is extruded to the exterior, further, more impermeable coverings are added. The tough, impermeable chorion of insect eggs is a secretion of the follicle cells (Davey, 1965), but in vertebrates the outer layers are added by secretory cells of the oviduct, as the egg passes down it. In the hen, for instance, the albumen, shell membranes and shell itself are all secreted by successive regions of the oviduct and applied to each egg during its passage to the exterior. This necessitates a perfectly-timed interaction between the egg and the cells of the oviduct. The oviduct cells are in their turn brought into a state of secretory activity by oestrogenic hormones secreted by the interstitial tissue of the ovary, which are circulating in the bloodstream; so the ovarian cells are involved in the chain of interactions too.

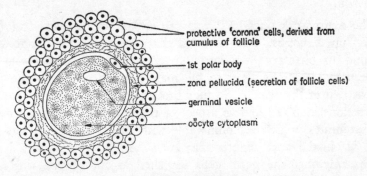

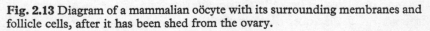

protective 'corona' cells, derived from cumulus of follicle

1st polar body

zona pellucida (secretion of follicle cells)

germinal vesicle

oöcyte cytoplasm

Fig. 2.13 Diagram of a mammalian oöcyte with its surrounding membranes and follicle cells, after it has been shed from the ovary.

In mammals, there are special interactions between the embryo and the uterus which facilitate the implantation of the embryo. These have been summarized by McLaren (1969).

2.4   Maturation of male germ cells, and the control of the rate of production of gametes in both sexes

We know much less about cellular interactions that take place in the male gonad as the spermatozoa develop than we do about equivalent processes in the female. Spermatogenesis occurs by a series of steps essentially similar to those of oögenesis (cf. Fig. 3.1). An important difference is that, since large food reserves are not carried in the male gamete, large numbers of spermatozoa can be produced continuously without exhausting the resources of the parent organism. It is therefore common to find that the males of a species are capable of being fertile all the year round, whereas the production of ova by the female is limited to certain times or seasons.

The activity of the testis in vertebrates is, like that of the ovary, controlled by gonadotrophic hormones from the pituitary gland. These hormones appear to be produced steadily and continuously in the male in many species, whereas in the female they are released only periodically. Studies on mammals have shown that the inherent rhythms of female sexual cycles are controlled by rhythmic activity of the hypothalamus, a region of the forebrain which exerts several controls over the release of pituitary hormones. Further references to these findings can be found in Austin and Short (1972). Thus in vertebrates there are several steps in the chain of cellular interactions that controls the rate and timing of production of the gametes, after the germ cells have reached the gonads and have begun to mature.

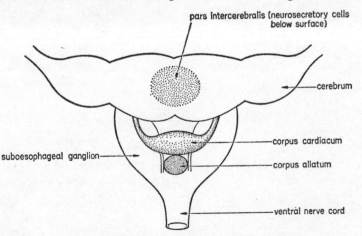

Fig. 2.14 The cerebral ganglia and the endocrine tissue associated with them in the bug **Rhodnius**. (*Modified from* Wigglesworth, 1954.) *Note:* many insect species have separate, paired corpora cardiaca and corpora allata (cf. Fig. 10.4).

In invertebrates, the 'brain' usually consists only of a few large nerve ganglia, so that there is of course no region comparable to the hypothalamus, and no endocrine gland of anything like the complexity of the pituitary gland of vertebrates. In insects, however, it has been shown that gonadotrophins exist (Davey, 1965). A gland known as the corpus allatum (Fig. 2.14) near the brain, is essential for the maturation of ova; if it is removed, no yolk is deposited and the oocytes remain small and immature. Several species of insects studied also have some neurosecretory cells in the median area of the brain (Fig. 2.14) (the pars intercerebralis) and extirpation of this area in the blowfly *Calliphora* causes regression of the ovaries. The neurosecretory products appear to be stored in another pair of organs associated with the brain, called the corpora cardiaca, whose removal also inhibits the growth of the ovaries. In other insects such as the milkweed bug *Oncopeltus fasciatus*, however, the maturation of the germ cells is unaffected by removal of these various glands. So it is impossible to generalize about this extremely varied and specialized group, the insects. What needs to be stressed is that as we learn more about control processes in the development of invertebrates, we may expect to find in them, just as has been found in the better-known vertebrates, that a whole series of cellular interactions, including hormonal ones, is involved in the development of the germ cells.

3

Cellular interactions concerned in fertilization

There are at least six phases of cellular interaction during the process of fertilization in animals. First, there is interaction during the maturation and shedding of the gametes, then during their passage to the exterior; next there are the interactions between eggs and spermatozoa, the penetration of the egg by the sperm, the several events that take place in the egg after the sperm has penetrated, and finally the fusion of the male and female pronuclei. In the account that follows, we shall consider each phase in turn. These are not strictly separate sets of events, however, and the phases overlap in a continuous process leading to the production of the zygote cell and the initiation of embryonic development.

3.1 Gonadotrophins and the shedding of the sex cells (gametes)

It is vital for the successful sexual reproduction of any organism, that the male and female gametes – spermatozoa and eggs – should mature and be shed in both sexes at the same time and in the same locality, so that there is a good chance of their meeting and of the eggs becoming fertilized. In bisexual animals, one way of ensuring this synchrony of gamete-production is for each individual to be able to respond to the presence of an individual of the opposite sex by setting in train mechanisms which cause its own germ cells to mature and to be shed reasonably quickly. The complete process of maturation of the germ cells, which is illustrated diagrammatically in Fig. 3.1, may take days, months or years according to the species, but it is usual for a proportion of the cells to remain 'dormant' at an advanced stage, ready for rapid maturation on receiving a suitable stimulus. Thus in some mammals, including man, primary oöcytes

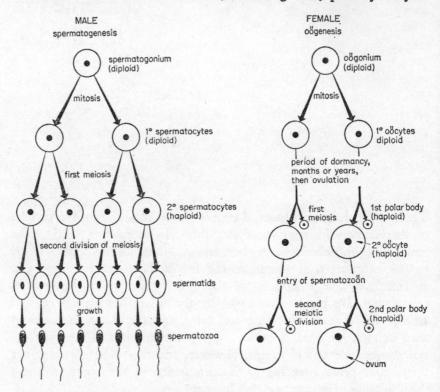

Fig. 3.1 Gametogenesis: conventional diagrams of the series of maturation divisions of the germ cells in both sexes.

already reach the prophase of the first meiotic division (see Fig. 3.1) during fetal life, and they remain dormant at this stage until the female is sexually mature and hormonal stimuli (see below) cause ovulation to occur.

Responses of a whole animal to the presence of a member of the opposite sex can be translated into responses of its germ cells by means of a series of controls by hormones, which operate long-distance cellular interactions. The hormonal control of reproduction has received much study in vertebrates, and we know a few examples of similar hormonal controls in invertebrates too. In vertebrates, one well-known class of hormones concerned with reproduction are the *gonadotrophins*. These are secreted by cells of the pituitary gland, a very important endocrine gland which lies at the base of the brain, and they act on the cells of the gonads. Two main groups of gonado-trophins are recognized: follicle-stimulating hormones which stimu-late the fiinal stages of gametogenesis and the growth of the ovarian follicles (Fig. 2.11) and so-called 'luteinising' hormones which cause the ripe gametes to be shed from the gonads into the gonoducts which lead to the exterior. The term 'luteinising' arose because in mammals these hormones cause the follicles from which eggs have been shed to develop into 'corpora lutea' (i.e. 'yellow bodies'). These become secretory endocrine organs and produce the hormone pro-gesterone which helps to maintain pregnancy. Luteinising hormones of a wide range of different vertebrates have similar chemical struc-tures and produce similar effects – i.e. shedding of the gametes. It is generally true of vertebrate hormones, in fact, that they are not species-specific. Their effects are tissue-specific, however, as we shall see in Chapter 10 when we consider the events of metamorphosis in amphibians, in which different larval tissues respond in very different ways to the same hormone, thyroxine.

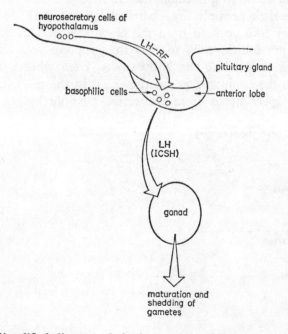

Fig. 3.2 Simplified diagram of the hormonal pathways that control shedding of the gametes in vertebrates. LH = luteinising hormone; LH–RF=LH-releasing factor; ICSH = interstitial cell-stimulating hormone.

The precise mechanism by which gonadotrophic hormones secreted by the cells of the pituitary gland act on the cells in the testis and ovary is not fully understood. The release of the hormones from the pituitary cells into the bloodstream is controlled by the *hypothalamus,* a region in the floor of the forebrain. In addition, the amounts of these hormones released are partly regulated by feedback controls from the gonads themselves. Fig. 3.2 shows some of these control pathways diagrammatically but does not show the two-way interaction between gonad and pituitary cells. Follicle-stimulating hormone (known as FSH) on reaching either the testis or the ovary, has the effect of stimulating growth of those germ cells that are nearly mature. In the testis, it also causes divisions of further stem cells to take place, so that new generations of spermatozoa begin to develop, ready to replace those that are ready to be shed. In both males and females of Amphibia, FSH also stimulates the production of a phospholipoprotein by the liver cells. This protein is liberated into the blood serum (cf. page 32), and in the female becomes incorporated into the storage proteins of the yolk, in the enlarging oöcytes. The effects of luteinising hormone (or LH) on the cells of the ovary is to increase their permeability – particularly that of the follicle cells – so that they take up fluid from the blood serum. This results in a swelling of the follicle cells so that they rupture and liberate the large, mature oocyte into the body cavity, from whence it will be carried down the oviducts (see Section 3.2). In the male, LH causes shedding of spermatozoa from the testis, possibly by some similar

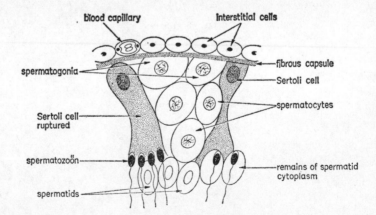

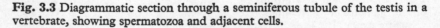

Fig. 3.3 Diagrammatic section through a seminiferous tubule of the testis in a vertebrate, showing spermatozoa and adjacent cells.

process of rupture of the cells to which they are attached (Fig. 3.3). It also stimulates the interstinal cells to secrete testosterone, and hence is also called 'interstitial cell-stimulating hormone', or ICSH.

We know very little about possible gonadotrophic hormones in invertebrates, but in groups such as the arthropods there are identifiable endocrine glands lying near the large head ganglia (cf. Chapter 2) and these may have among their functions the production of gonadotrophins. They are also responsible for controlling the metamorphosis of the larva into the adult, a process which includes the development of the gonads and germ cells. It is probable that the release of gametes in many invertebrates is under less precise hormonal control than in vertebrates, and is more susceptible to environmental factors. For instance, echinoderms may be made to shed their gametes by mild electric shock treatments (Metz and Tyler, 1959). There is also a famous Palolo worm which swarms and spawns in the Pacific at full moon. Many other, less spectacular responses to environmental conditions are known in the reproductive behaviour of invertebrates.

3.2 The passage of the gametes down the gonoducts

Once shed from the gonads, the germ cells are dependent on a new series of cellular interactions to ensure their progress along the gonoducts. The eggs, after ovulation, are propelled along the oviducts by ciliary action and also by muscular contractions of the duct wall. In female mammals, these contractions are enhanced just after ovulation, by the action of oestrogenic hormones which are secreted by ovarian cells. So there is in this case an interaction between cells of the ovary, cells of the oviduct and the ova themselves. In the male genital tract, the propulsion of sperm from the testis to the exterior appears to be due mainly to muscular contractions of the ducts. The seminal fluid has important energizing properties too, enabling the sperm to become motile after they have been ejaculated. The sperm tail, which contains contractile fibres arranged in characteristic patterns (cf. Fig. 3.4), is used for its swimming towards the egg. This motility is essential, for internally fertilizing sperms have to penetrate up the female genital tract, sometimes moving through a highly viscous medium which may impede their progress. In animals where fertilization is external, the sperm may have to swim through a foot or two of open water.

The motility of spermatozoa may be affected by many environmental factors before they reach the eggs. This applies, whether fertilization is external or internal. For instance, spermatozoa of sea-urchins, which fertilize externally, become more active when placed

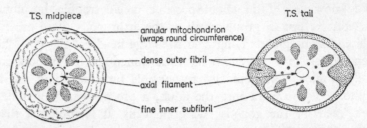

Fig. 3.4 Patterns of fibres in the midpiece and tail of the mammalian spermatozoön, as seen by electron microscopy. Diagrams of transverse sections.

in 'egg-water' – i.e. sea water which has previously contained sea-urchin eggs. Internally-fertilizing mammalian spermatozoa may be either facilitated or inhibited in their passage through the female genital tract, according to the physiological condition of the female, which affects the viscosity of the fluids in the tract. Just after ovulation, when oestrogens are in circulation, the mucus of the cervical passage which leads into the uterus has a consistency and structure which favours the penetration of sperm; at other phases in the female cycle, this mucus impedes sperm passage. Higher up, in the oviduct, antibodies are secreted which can interact with foreign proteins carried by genetically incompatible sperm, and so can immobilize these sperm. It is now well established that spermatozoa carry a number of distinct antigens. Injection of mammalian spermatozoa into another species initiates antibody reactions (Metz, 1967). In the spermatozoa of guinea-pigs, as many as eleven antigenic factors have been distinguished: six in the sperm tails, four in the head and one common to head and tail. Some antigens are an integral part of the membrane of the sperm head. In amphibians, it has been shown that addition to the medium of an antiserum to sperm of another species prevents sperm of this species from fertilizing the eggs in this medium. Besides species-specific antigens, individual strain differences in sperm antigens exist within species, this has been shown in crosses between inbred strains of mice.

Not only the sperm, but also the egg may carry antigens on its surface which affect its ability to be fertilized. Tests by Shaver *et al.* (1970) in various species of the frog *Rana* and the toad *Bufo* indicate that antigens are secreted by the middle region of the oviduct and are deposited from here into the jelly envelopes that surround the eggs. Treatment of the eggs with antisera to oviducal antigens from the same species, prevents the eggs from being fertilized.

In mammals, there is an important interaction betwen secretions of the oviduct and the sperm, which causes 'capacitation' of the sperm. Mammalian sperm which has not passed through the female tract is found not to be capable of fertilizing eggs *in vitro*, unless it is first treated with other 'capacitating' agents. This process of capacitation is not fully understood, but it is thought to depend on the rupture of interlocking molecules at the surface of the sperm membrane. This causes the membrane to break down and expose the enzymes of the acrosomal cap region (see Fig. 3.5). This event is also called the 'acrosomal reaction.' Ericsson (1969) observed that the type of white blood cells called eosinophils caused capacitation of rabbit spermatozoa, apparently because they were able to split the

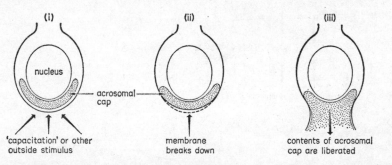

Fig. 3.5 Diagrams to show the sequence of events in the 'acrosomal reaction which forms part of the process of sperm 'capacitation'.

molecular complex at the acrosomal membrane. The reaction was not species-specific, for eosinophils from mules were used successfully on rabbit spermatozoa.

Something similar to 'capacitation' seems to be necessary for eggs too, for Kirby (1962) showed that if mouse or rat eggs were retrieved from the vicinity of the ovary immediately after ovulation, and then either transplanted to the uterus, or placed *in vitro*, without first having passed down the oviducts, they were not able to be fertilized. In amphibians too (Metz, 1967), eggs taken from the coelom before they have passed down the oviducts cannot be fertilized. It is interesting to speculate how these essential interactions between gametes and secretory cells of the female tract may have evolved gradually towards a mechanism whereby, in viviparous animals, the embryo eventually attaches to the uterine wall and completes its development there. Fig. 3.6 compares the events diagrammatically.

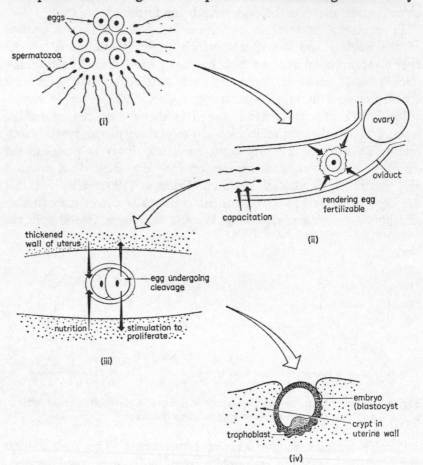

Fig. 3.6 Hypothetical steps by which viviparity may be thought to have evolved, in a series of increasing interactions between the female genital tract and, first, the gametes, then later the embryo. (i) External fertilization, in which gametes are extruded by both sexes. (ii) Internal fertilization, possibly due to ovulation occurring too late for the eggs to be extruded before the male approaches the female. Ovulation may then in some species become a *response* to mating. In addition, the female tract may acquire a *capacitating* influence on the sperm and may render the egg fertilizable. (iii) The fertilized egg is retained in the female tract: here it may derive nourishment from the oviduct or uterus. In turn, it may acquire the faculty of *stimulating* the proliferation of nutritive tissue. (iv) The embryo attaches to and *implants* in the now specialized uterine wall. Specialized *trophoblast* cells of the embryo both digest and invade the uterine tissue in some species. Nutrition maintains the embryo until it has developed sufficiently for independent life and is *born*.

44

3.3 Interactions between spermatozoa and eggs before fertilization

These interactions can only be studied outside the female tract; hence observations on animals that normally have external fertilization are more reliable than those made *in vitro* on species whose fertilization is normally internal, in which there is some doubt as to whether the events seen are the same as those occurring *in vivo*. Most of our more certain knowledge comes from studies on aquatic invertebrates whose fertilization is external.

In echinoderms and in the mollusc *Mytilus,* which have been studied in some detail, the first evidence of chemical interaction between eggs and sperm is that an active substance is secreted at the surface of the egg. This substance has been called 'fertilizin' by Lillie (1919) and identified as a glycoprotein (Runnström, 1952). It has three main actions on the sperm. First, it increases their motility (as we saw above) when the sperm are put into 'egg water', so that they have a better chance of reaching the eggs. Secondly, fertilizin causes the spermatozoa to clump, or 'agglutinate' in the vicinity of the eggs, which increases their chances of making firm contact with the eggs' surfaces. Finally, fertilizin in some way increases the fertilizing capacity of the sperm, since in its absence they are unable to fertilize succesfully.

It is thought that fertilizin is secreted at the surfaces of numerous microvilli which penetrate from the egg cytoplasm into the jelly which surrounds it (see Hörstadius, 1973). In sea urchins there may also be other active substances which originate in the jelly and are exuded into the surrounding medium. At least nine active substances have been identified in eggs of the echinoderm *Arbacia,* and some of these tend to promote fertilization while others appear to prevent it, when they are tested experimentally. The preventive substances may come into play after one sperm has entered an egg, to prevent polyspermy occurring which would lead to abnormal development. In many animals the egg also undergoes a cortical reaction after it has been fertilized by one sperm, and this makes it impervious to further sperm (see Section 3.5).

We do not know if any equivalent of fertilizin exists in vertebrates. The jelly of amphibian eggs has been found to carry antigens however, at least one of which appears to activate the sperm. There are also antigens in the egg cortex of amphibians, and one of these, called 'antigen F', is thought to facilitate fertilization.

So far, the evidence suggests that spermatozoa of most animals have to undergo a rigorous selective process which ensures that each egg is fertilized by only one, genetically compatible spermatozoön. However, the sperm have in their turn a number of mechanisms by which they interact with the egg and its surrounding cells and membranes and thus facilitate their entry into the egg. At the time that fertilizin was discovered in echinoderms, it was also found that the sperm produced an antibody-like substance, called 'antifertilizin', which seemed to react specifically with fertilizin and also to cause an 'acrosomal reaction' in the sperm. In this acrosomal reaction (already mentioned above in connection with sperm capacitation in mammals) there is a dehiscence of the outer membrane of the sperm head, exposing the acrosome and the fluid surrounding it (see Fig. 3.5c). The acrosomal fluid in a number of different animal species has now been found to contain enzymes which attack the protective coverings of the egg. One of these enzymes, hyaluronidase, disperses any follicle cells which may adhere round the egg membranes. This is particularly important in mammals, where the ovulated egg is surrounded by a thick 'corona' of follicle cells. There are also other lytic enzymes in the acrosomal fluid, which partly dissolve the jelly

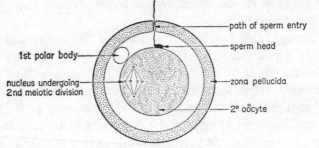

1st polar body

nucleus undergoing
2nd meiotic division

path of sperm entry

sperm head

zona pellucida

2° oöcyte

Fig. 3.7 Diagram to represent fertilization in the rat's egg. (*Simplified from* Austin, 1965.)

or other mucoprotein coating of the egg. (In mammals, this coating is called the 'zona' and is a secretion of the follicle cells, as we saw in the last chapter.) Thus a pathway is made through which the sperm can penetrate (cf. Fig. 3.7). Mammalian sperm of several species – e.g. ram, bull and rabbit – have been shown to produce special zona lysins, these have been isolated and are found to be lipoglycoprotein complexes, capable of attacking the follicle cells too. In rabbits, according to Piko (1969) the dispersal of corona cells may also be facilitated by contractions of the uterine tubes.

As soon as the sperm has passed through the protective coverings and has made contact with the plasma membrane of the egg, it is able to penetrate the egg cytoplasm by special processes which vary in different animal species.

3.4 The penetration of the egg by the spermatozoön

In the majority of animals, the spermatozoön can penetrate the egg at any point on its surface, but there are some exceptions to this. In mammals, Piko (1969) has observed that one small area, just above the meiotic spindle of the female nucleus, is not receptive to the sperm. This region also lacks cortical granules, which are present in all the rest of the surface cytoplasm and which play a role in the post-fertilization reactions of the egg (see Section 3.5). Other exceptions are the eggs of teleost fish and of some invertebrates (e.g. the insects) which have a tough chorion impervious to enzymes from the sperm. These eggs have a small channel at one end, the micropyle (Fig. 3.8), through which entry of the sperm is possible. Hörstadius (1973) has described a micropyle in eggs of echinoderms too. In insects where fertilization is internal, sperm entry may be facilitated

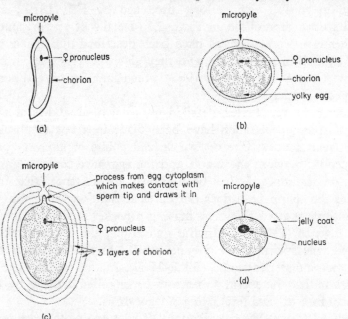

Fig. 3.8 Examples of animal eggs which have micropyles. (a) Egg of the fly, *Musca*; (b) the herring, *Clupea*; (c) the sturgeon, *Amia*; (d) an echinoderm, *Paracentrotus lividus*.

by the fact that the egg is orientated with its micropyle opposite the sperm receptacle in the female (Davey, 1965). The events at fertilization in fish have been described by Nakano (1969). He observed that sperm-activating substances are extractable from the micropylar region, but that there is no visible acrosomal reaction in the sperm. After one sperm has penetrated, the chorion and egg membranes harden and the micropyle is closed.

There is not space here to consider the variety of mechanisms of sperm penetration in the many species of animals that have been studied. The general point should be made, however, that it is not always purely by its own efforts that the sperm enters the egg; in many instances there is active participation in this process by the egg also. For instance, in echinoderms, molluscs, ascidians and arthropods – the majority of the invertebrates studied – a filament or tube is seen connecting the sperm head to the surface of the egg. In at least some species this originates as a cone-shaped outgrowth from the surface of the egg, in response to the contact of the sperm. This 'fertilization cone', as it is sometimes called, then contracts gradually, drawing the sperm into the egg as it does so. The process was described in starfish eggs a long time ago by Fol (1879), and his drawings are reproduced in Plate 3.1. Details of sperm penetration in other invertebrate species have been described during many years' work by Colwin and Colwin, and they give full references in a recent review (Colwin and Colwin, 1967) which includes observations with the electron microscope.

Among the vertebrates, detailed ultrastructural studies of the process of sperm penetration have been made in a few amphibian and mammalian species. One or two special points of interest about the interactions between the sperm and the egg have come to light from work on mammals. It has been shown, for instance, that in some species the sperm head does not enter at right-angles to the surface, but insinuates itself sideways, making a pocket in the egg membrane as it does so (Plate 3.2). In the guinea-pig, the rat, the rabbit and the pig, the sperm is found lying parallel to the surface of the egg just after it has penetrated (Piko, 1969). In the rat, another special feature is that the sperm appears to be engulfed actively by the egg surface, in a process resembling phagocytosis.

Bedford (1972) has described in detail some of the events at fertilization in the rabbit, from carefully-timed studies *in vitro*. He notes that by 11-15 hours after mating, many of the spermatozoa

48

have undergone an acrosomal reaction and are penetrating the corona cells. These cells are mostly dispersed by the enzyme hyaluronidase which is liberated from the acrosomal fluid at this stage; however, a surprising observation is that some corona cells persist and are later seen actively ingesting spermatozoa by means of pseudopodia. This suggests a new and possibly important role for the corona cells, in removing excess sperm. By the time that certain sperm have made contact with the zona of the egg, they have entirely lost the contents of their acrosomal sacs, and the inner membrane of the acrosome is exposed, this inner membrane makes contact with the surface of the zona. After this, the sideways penetration of the sperm head begins (Fig. 3.9i) and is brought about by the fact that the mid to posterior region of the sperm head, and not its anterior tip, is the first region to fuse with the egg membrane. Subsequently the rostral end of the sperm head is drawn into the egg surface, clothed with a vesicle formed by the inner acrosomal membrane and the vitelline membrane jointly (Fig. 3.9iii). Finally the whole head is drawn in, and meanwhile its nucleus seems to de-condense into a web of strands. Vesicles then enclose it, forming the membrane of what is now the male pronucleus. The mid-piece and tail of the sperm also enter the egg in many mammalian species studied, and become incorporated into the egg cytoplasm. Other descriptions of sperm penetration in mammals mention essentially similar processes to those described by Bedford (e.g. Yanagimachi and Noda, 1970, in hamsters; Thibault, 1969, in rabbits).

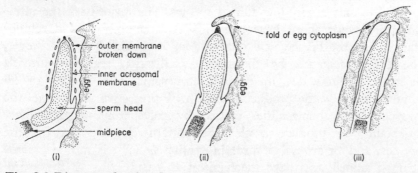
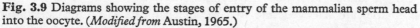

Fig. 3.9 Diagrams showing the stages of entry of the mammalian sperm head into the oocyte. (*Modified from* Austin, 1965.)

It can be seen from this brief survey of methods of sperm penetration that the main activity at first, in reaching the egg surface, is that of the sperm, but that after this the egg often plays an active part in drawing the sperm into its cytoplasm. To do this, it must be receptive to the sperm, and in some cases a local specialization of its surface, the micropyle, is also essential. Once one spermatozoön has penetrated an egg, however, in many animals changes take place which make it impossible for further sperm to enter that egg. These events will be described in the next section. It is interesting that in birds, these reactions do not take place, and polyspermy often occurs.

3.5 Events in the egg after it has been penetrated by the sperm

As soon as one spermatozoön has entered the egg, there are, in the eggs of many animals, several rapid and immediate reactions. These comprise: (a) breakdown of cortical vesicles or 'granules' which lie near the surface of the egg, and discharge of these into the space underlying the vitelline membrane; as a result, the membrane lifts off the egg and is now called the fertilization membrane; (b) a change in the physical properties of the egg surface, which prevents the entry of further sperm; (c) release of antigens, possibly also blocking further sperm entry, and also a release of proteases; (d) further changes in the egg cortex leading to a redistribution of pigment and to orientation of the future embryo; (e) activation of synthetic processes in the egg cytoplasm.

Not all of the above reactions take place in the eggs of all species of animals. We have already noted the absence of reaction (b) in birds. It should also be stressed that each type of reaction has been studied in only a very few animal species. The majority of the ultrastructural observations have been made on amphibians and echinoderms, and the biochemical observations on echinoderms. However, the effects achieved by these post-fertilization reactions provide essential preliminaries to normal embryonic development in all animals, namely, the production of a new protective membrane, the prevention of abnormalities caused by polyspermy, a preliminary orientation to produce a cleavage pattern characteristic for the species, and an arousal of protein synthesis. So it is fair to assume that all animal eggs have mechanisms to ensure that all these preliminaries are achieved, and that the mechanisms will often be similar to those that have already been discovered in the few species studied so far. We will now consider observations on these post-fertilization reactions in a few individual species where they have been described in some detail.

3.5.1 *Breakdown of cortical granules*

This was described in the amphibians, *Bufo* and *Xenopus* by Balinsky (1966). His were some of the earliest studies of this process with the electron microscope, and it was then realized that these so-called 'granules' at the periphery of the egg cytoplasm were in fact hollow vesicles. Immediately after fertilization, these cortical vesicles come to the surface and discharge their contents into the perivitelline space, under the former vitelline membrane. This perivitelline space increases, as the vitelline membrane lifts away from the egg surface and becomes the fertilization membrane. The surface cytoplasm of the egg then becomes reorganized, by the breakdown of stacks of membranes to give a new, vesiculated cortical zone. Some of Balinsky's descriptions of these events are sketched diagrammatically in Fig. 3.10.

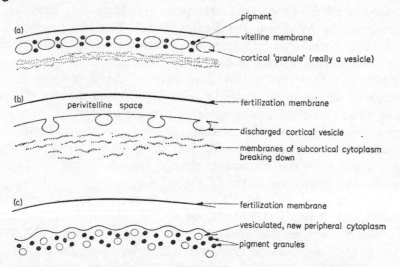

Fig. 3.10 Diagrammatic representation of the discharge of cortical granules at fertilization, described in amphibians by Balinsky (1966). (a) before fertilization; (b) immediately after entry of sperm; (c) ½ h later.

Nakano (1969) has described the process of breakdown of cortical vesicles in teleost fish. In them and in amphibians, it looks as if material from the cortical vesicles may contribute to the structure of the fertilization membrane. The surface ultrastructure of this membrane in the sea-urchin *Strongylocentrotus purpuratus* has been studied by Inoue *et al.* (1967), using a surface replica method with the electron microscope. They found that the membrane had a crystalline structure reminiscent of tropomyosin, and also that there were a number of protuberances from its surface, like microvilli. It is possible that these microvilli could be sites for the secretion of antigens and proteases (process (c) listed above).

3.5.2 *The block to polyspermy*

It has been known for a very long time from observations on amphibian and echinoderm eggs that as soon as the fertilization membrane has formed, it is impossible for further sperm to enter the egg. This phenomenon is known as the block to polyspermy. In living sea-urchin eggs viewed by polarized light or by phase contrast, it is possible to see the changes in surface structure which cause this block, passing rapidly over the egg surface from the point of sperm entry. A zone of birefringence is seen to radiate over the surface in all directions from this point (cf. Plate 3.3, based on a description by Mitchison and Swann, 1952). The birefringence has been taken to indicate an alignment of molecules tangentially to the egg surface, thus making it more impermeable. Recently, however, Tegner and Epel (1973) have been able to obtain pictures of the surface of the sea-urchin's egg by scanning electron microscopy, and these show that there is also apparently an active rejection of supernumerary sperm from the surface. At first, numerous sperm are attached all over the membrane, then these appear to be sloughed off gradually, as the membrane change spreads. Some of Tegner and Epel's pictures are shown in Plate 3.4.

In amphibians, the post-fertilization reactions were studied in *Xenopus* by Balinsky (1966). As we saw in Section 3.5.1, he noted that new cortial cytoplasm was laid down: Recently Wyrich *et al.* (1974) have re-described this new layer, which is derived from the discharged cortical granules, and which constitutes the block to polyspermy. In some urodeles, on the other hand, polyspermy occurs regularly, as it does in birds (Romanoff, 1960). As soon as one sperm pronucleus has fused with that of the egg, however, the other spermatozoa begin to disintegrate in the egg cytoplasm.

In mammals, there is recent evidence that the cortical granules of the egg may themselves mediate the block to polyspermy. Barros and Yanagimachi (1971) found that treating eggs of the golden hamster with a suspension of cortical granules blocked the entry of sperm into them. In this case the granules were outside the egg membranes, however, and it cannot be certain that their action within the membrane would be the same. It is significant, though, that the same authors observed polyspermy in some untreated hamster eggs, and noted that these were always ones in which the cortical granules had not yet been discharged into the perivitelline space.

3.5.3 *Release of antibody-like substances and proteases*

Fertilized eggs of sea urchins emit a sperm agglutinin into the sea water round them (Gregg, 1969). In other species this same phenomenon has been observed, and the antibody-like material has been named 'cytofertilizin' to distinguish it from the fertilizin which is released before fertilization. Gregg showed later by immunological methods that cytofertilizin and fertilizin are chemically identical. Besides the difference in timing of their release, they come from different sources however: 'cytofertilizin', as its name suggests, comes from the cytoplasm of the egg. It is not released until the sperm has entered and the cortical granules have broken down.

A release of proteases from fertilized echinoderm eggs has also been observed (Vacquier, Epel and Douglas, 1971). Protease activity was assayed by measuring the time course of release of TCA-soluble, ^{14}C-labelled material from protein or esters added to the medium. The same protease activity was observed in parthenogenetically activated eggs, showing that the sperm were not the source of the proteases. Incidentally, it should be remarked that all of the processes (a) to (e) which we are now considering also take place in eggs which have been activated to develop parthenogenetically, in the absence of live sperm. So these processes are evidently capable of being triggered off by quite non-specific stimuli.

The cortical granules seem to be implicated again in the release of proteases, for treating fertilized echinoderm eggs with a trypsin inhibitor blocks the breakdown of the granules, and also blocks the release of protease activity. It is not clear in what way these two processes are dependent: whether, for instance, the same proteases that are liberated, also attack and break down the granules, or whether the granules actually contain the proteases and liberate them as they break down.

3.5.4 *Changes in the polarity of the egg cortex*

The eggs of many invertebrates – for example those of marine annelids and of ascidians – have special, coloured regions of cortical cytoplasm which become distributed into certain regions of the egg after fertilization. The egg of the ascidian Styela has been studied in some detail (Reverberi, 1961), and it shows a streaming of yellow cytoplasm followed by greyer material down from the upper, animal pole towards the point where the sperm has entered. As the sperm nucleus travels up towards the animal pole to fuse with the egg nucleus, some of the coloured cytoplasms follow it, and eventually they form two crescents, the grey and yellow crescents, which mark the plane of future bilateral symmetry of the embryo (Fig. 3.11). The broadest zone of the crescents is on the future ventral side of the embryo, and the yellow crescent material goes into mesoderm cells. Inasmuch as the position of these plasms is influenced by the point of entry of the sperm and its subsequent path of movement, it can be said that the orientation of the future animal's body is governed by this early interaction between the sperm and the egg cytoplasm.

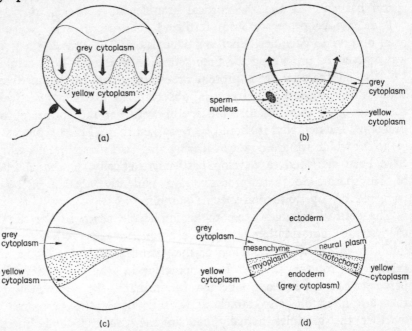

Fig. 3.11 Diagrams to show the movements of the coloured cortical plasms in the egg of the ascidian *Styela* after fertilization, and the final localization. (a) At the moment of sperm entry, both grey and yellow plasms stream down towards it. (b) They subsequently move upwards again, as the sperm nucleus migrates up to fuse with the female nucleus at the animal pole. (c) Final positions of grey and yellow plasms. (d) Map of future destinies of these areas of the egg. ((a)–(c) *modified from* Waddington, 1956; (d) *adapted from* Reverberi, 1972.)

Among the vertebrates, the Amphibia provide an example of similar redistributions of cortical cytoplasm according to the point of sperm entry. The upper half of the amphibian egg is brown or black because of melanin granules in the cortex, while the lower half is white. After fertilization however, some of the dark cortical cytoplasm contracts upwards towards the animal pole, starting in a region diametrically opposite to the point of sperm entry (Fig. 3.12). This movement leaves a pale crescent in the equatorial region, known as the 'grey crescent'. The broadest region of this grey crescent marks the future dorsal side of the embryo. This grey crescent material goes into dorsal mesoderm cells, and we shall see in the next chapter that it influences the pattern of cleavage of the embryo. The dorsal mesoderm cells also have far-reaching effects, for they induce the formation of the nervous system (cf. Chapter 5). So, from this initial interaction between the sperm and the egg cortex, two major features of the orientation and differentiation of the embryo are decided.

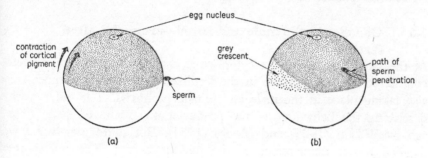

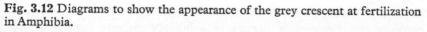

Fig. 3.12 Diagrams to show the appearance of the grey crescent at fertilization in Amphibia.

3.5.5 *Activation of synthetic processes in the egg*

A long series of observations on echinoderms by Monroy and his collaborators (see Monroy, 1967) have shown that immediately after fertilization in these animals there is an awakening of metabolic activity in the egg. Zygotes (i.e. fertilized eggs) take up tracers and synthesize protein far more rapidly than do unfertilized eggs. A number of suggestions have been put forward to explain this phenomenon. For instance, it could be that the raw materials are in some way unavailable until after fertilization, perhaps because they are contained within yolk or other cytoplasmic stores which are not broken down until the sperm enters. Or it could be that the

C

ribosomes are inactive till after fertilization, so that polypeptides cannot be assembled on them. Alternatively, messenger RNAs may be masked and unable to provide templates for protein synthesis on the ribosomes. Earlier work on the activity of ribosomes isolated from unfertilized and fertilized eggs suggested that these did differ in their protein-synthesizing ability. But more recent tests (Stavy and Gross, 1969) indicate that ribosomes from unfertilized eggs are capable of slow amino acid incorporation, and that if given an artificial nucleotide template such as polyuridine, they will synthesize polyphenylalanine just as rapidly as do ribosomes from fertilized eggs. So it has been concluded that it is the nucleotide templates that are unavailable in unfertilized eggs. It has also been noted that the ribosomes undergo morphological changes after fertilization, but this may be secondary to their increased rate of protein synthesis. Thus, Denny and Reback (1970) noted an increased number of polysomes in fertilized as compared with unfertilized eggs of *Lytechinus pictus*.

3.6 Changes in the male and female pronuclei after fertilization

At the same time as all the cytoplasmic changes that have just been considered are taking place in the fertilized egg, certain changes are also taking place in the male and female pronuclei. The older work describing the behaviour of the pronuclei was reviewed some time ago by Wilson (1925) and Moore (1937). But there are very few detailed recent descriptions of the nuclear events. The nuclei are of course more difficult to observe than any of the cytoplasmic components, since they occupy a relatively minute region of the egg, which might easily be missed in electron microscope sections or other preparative procedures. However, Gondos and Bhiraleus (1970) succeeded in following the ultrastructural changes in the pronuclei of the zygote of the rabbit. They observed that there is a complex interdigitation of the two nuclear membranes, with channels communicating between them, for a long time before they eventually fuse and the intervening membranes break down.

Other observations, which throw some light on normal events, have been made on hybrids. The male pronucleus in amphibian hybrids may enter the egg normally, but may then not be able to survive and form chromosomes. Ferrier (1967 a, b) showed, for instance, that when eggs of the salamander *Pleurodeles* were fertilized with sperm from *Ambystoma,* the two pronuclei managed to fuse, but then did not form a mitotic spindle because no male chromosomes appeared. In other intergeneric hybrids of amphibians, male chromosomes and a mitotic spindle may appear, but the movements at anaphase may then be abnormal, and non-disjunction (failure of chromatids to separate) may occur, suggesting that some of the DNA may not have been able to replicate normally. Interspecific hybrids are not quite so abnormal, however, and in them development may proceed as far as the beginning of gastrulation (cf. Chapter 5). It is thought that this is the first stage at which paternal genes from the foreign sperm come into play, and that up to that time the maternal mRNAs and synthetic processes initiated in the egg cytoplasm at fertilization provide all materials necessary for early embryonic development. These findings in hybrids serve to emphasize that it is not only a physical union of the nuclear contents of the male and female gametes that occurs, but also a balancing of gene activities so that these take place in the correct sequence. Many genes must at first be repressed, to become active later only in certain kinds of cells. The programming of this ordered sequence of activities, much of which will depend on cellular interactions later in development, is initiated by that first interaction of the male and female pronuclear genes, about which we know very little so far. It certainly merits further work.

In conclusion, we have seen in this chapter that the process of fertilization in animals involves a long series of cellular interactions. First, there is the interaction between cells of the pituitary gland and those of the gonad, leading to the maturation and release of the gametes; next, interactions between specific molecules carried on the male gametes and in the genital tract of the female. The egg also carries antigenic and antibody-like substances which activate the spermatozoa and facilitate the approach of genetically compatible sperm to the egg. During these interactions an acrosomal reaction occurs in the sperm which releases its lytic enzymes, enabling it to penetrate to the egg surface. At this surface, membrane interactions help to draw the sperm into the egg. This is followed by reactions in the egg which may prevent the entry of further sperm, and which initiate orientation and metabolism in the future embryo. Meanwhile, the male and female pronuclei interact both genetically and structurally, though we know very little about these last events.

4

Interactions of cells
during embryonic
cleavage

The first phase of embryonic development in animals is *cleavage*, when the fertilized egg divides by mitosis into a number of cells called 'blastomeres', which will later pursue various paths of differentiation to from the different tissues and organs. The pattern of cleavage is usually constant and characteristic for each species, but there are some forms, particularly among the vertebrates, which show individual minor variations in the cleavage pattern and yet produce normal embryos. Generally speaking, the pattern of cleavage is much more rigidly constant in embryos with relatively few cells, the fates of which are determined at a very early stage. This is true of many invertebrates, particularly the molluscs and the annelids which show 'spiral' instead of radial cleavage (cf. Fig. 4.1). In embryos with spiral cleavage the pattern is so invariable that detailed cell-lineages have been worked out on several species, and we know exactly which tissues will arise from each blastomere and its progeny. In conditions as rigid as this, it follows that any accident causing disruption of the normal cleavage pattern, or loss of a cell, will result in an abnormal embryo, lacking whatever organ the lost cell would have formed. The inability to adapt to abnormal situations implies also that the differentiation of the individual blastomeres is *not* governed by adaptable interactions between them. This is borne out by experiments (see later, Section 4.3) in which isolated blastomeres of spirally-cleaving embryos are found to carry on differentiating quite successfully on their own, forming exactly the part of the embryo that they would have if left *in situ*.

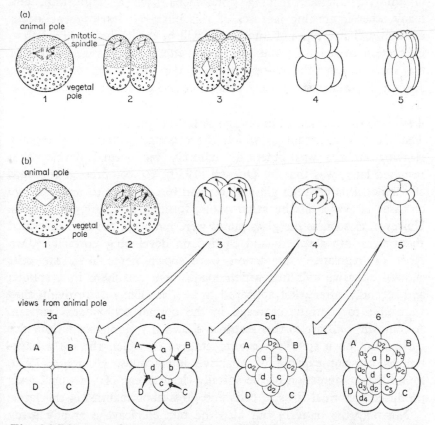

Fig. 4.1 Diagrams of external views of a radially cleaving and a spirally cleaving embryo, for comparison. Stages up to 16 cells shown. (a) Radial cleavage. (b) Spiral cleavage.

In radially cleaving embryos with large numbers of cells – a category that includes all the vertebrates – there is a much less rigid determination of the fates of individual blastomeres. Experiments in which the cleavage pattern is altered, or cells removed, added or transplanted, show that there is a remarkable ability for the embryo to form normally in all these unusual circumstances. This great adaptability, or power of 'regulation' as it has been called, has fascinated observers for a long time. It clearly results from the ability of the cells to interact in new ways according to the changed circumstances, and to reorganize both the numbers of divisions that they undergo and the course of differentiation that each cell pursues.

We must now consider some of the ways in which cleavage cells of embryos interact, making phenomena such as regulation, and many other interesting features of this phase of development, occur efficiently. Three types of interaction will be distinguished here: first, interaction to maintain the rate and pattern of cleavage typical of the species; secondly, co-operation in the mechanism of cleavage; and thirdly, influence on each other's course of differentiation.

4.1 Maintenance of cleavage rate and pattern

The classic experiment in which the cleavage pattern of a radially cleaving embryo was drastically altered, yet normal development resumed later, was that by Driesch (1910). He compressed the eggs of echinoderms between glass plates and forced them to undergo two or three cleavages in the same plane, forming a monolayer of cells. However, as soon as the glass plates were removed, the cells reformed themselves into a sphere and carried on developing normally. Two types of 'regulative' interaction were shown here: first, the cells showed adhesive affinities which made them re-adhere in a sphere, and secondly, after what appeared to be a random reassortment, they were able to continue cleavage in the normal echinoderm pattern. A complete contrast to this result is shown if a similar experiment is carried out on a spirally cleaving embryo in which the early blastomeres are not capable of adaptive interactions. Guerrier (1970) found that compression of the spirally cleaving egg of the slug *Limax* produced abnormal sizes of cells and gross abnormalities in the larva.

Not only the pattern, but also the rate of cleavage at any given temperature is constant for each species. It appears to be determined by factors orginally present in the egg cytoplasm, for when eggs of one species are fertilized by sperm of another species, it has been found in both echinoderms and amphibians that the cleavage rate is typical of the species to which the egg belongs (see Raven, 1954 for review). These observations refer to the period of cleavage when all cells are dividing synchronously. Normally, however, there is a definite stage when divisions start to be asynchronous, some cells (usually the less yolky ones) dividing faster than others – and this stage is characteristic of the species also. In teleost fish, for instance, there are regularly twelve synchronous divisions before asynchrony sets in. In the sturgeon and in amphibians, asynchronous divisions usually start just before gastrulation, and Chulitskaia (1970) has shown that at higher temperatures, they begin earlier, and gastrulation may take place with smaller numbers of cells than usual. Nevertheless, normal larvae of normal size result. This shows that the cells must be able to adjust the number of divisions that they undergo at some later stage, so that the normal number of cells is restored. The most interesting and puzzling facet of this phenomenon is how the cells are informed about their total number, so that they can adjust this if necessary. We shall see examples later of the ability of embryonic cells to adjust their number and also the numbers of certain organs like somites, under experimental conditions. Clearly they must be able to transmit information to each other about their size and relative positions; this applies also to the re-forming of a sphere by echinoderm cells after Driesch's compression experiment.

There are a number of more subtle problems if we try to explain how synchronous divisions are maintained in an early embryo. It is fairly easy to envisage that the length of each mitotic cycle might be the same in all cells at first, but not so easy to explain why there should also be identical intervals between successive mitoses in all cells. Van den Biggelaar (1971) followed in detail the timing of each phase of the cell cycle in the spirally-cleaving snail *Lymnaea,* but so far no one has studied a radially cleaving egg in such detail, perhaps because the much larger number of cells in these embryos makes such a study far more difficult. In *Lymnaea,* Van den Biggelaar showed that there was less strict synchrony of events than had hitherto been supposed: the phases of the mitotic cycle differed in length in cells of different regions of the embryo, even during the first four cleavages. At 16- and 24-cell stages there was a distinct lengthening of the whole mitotic cycle in all cells, and at this stage too, nucleoli appeared, having been absent at earlier cleavage stages. This is an interesting finding, since in amphibian embryos too, nucleoli appear just after the onset of asynchronous divisions. There is no evidence to suggest that nucleoli directly affect the rate of cell divisions, though; their presence is an indication that new ribosomal RNA can be synthesized, and hence new ribosomes for protein synthesis. An interesting conclusion of Van den Biggelaar was that in *Lymnaea* the pattern of cleavage at the asynchronous phase was controlled by the '3D' blastomere. This is a cell which receives a special, basophilic cytoplasm from the egg, which forms a projection known as the 'polar lobe' (Fig. 4.2). The 3D cell itself becomes mesoderm, but if it is removed or if the polar lobe is excised from it, neither the cleavage pattern nor, as we shall see later, the differentiation of other cells can proceed normally. So here is at least one example of cellular interaction in a spirally cleaving embryo. It is probable that when the technical difficulties of experimenting on these very small and fragile invertebrate embryos are overcome, evidence of further cellular interactions during their cleavage will accumulate, and it will become apparent that they do have some limited powers of regulation (see also a recent study on isolated blastomeres of molluscs, by Verdonk and Cather, 1973).

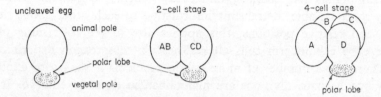

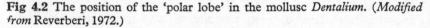

Fig 4.2 The position of the 'polar lobe' in the mollusc *Dentalium*. (*Modified from* Reverberi, 1972.)

4.2   Cellular interactions in the mechanism of cleavage

As each cleavage takes place, in the early embryo, a very obvious furrow appears and deepens gradually until two separate cells are formed. The mechanism of this furrow-formation has interested embryologists for a long time. It clearly requires a steady contraction in a localized zone at the cell surface, and this contraction must be maintained equally in the two cells as they form.

Recent electron microscope work has revealed certain structures which could be responsible for the contractions in the furrow region of cleaving cells. For instance, in amphibians it has been shown that there are microfilaments orientated transversely to each cleavage furrow (Fig. 4.3), in the cortex of the blastomeres on either side of the furrow (Selman and Perry, 1970). It is thought that these micro-fibrils are contractile and may be made of material similar to actomyosin. But experiments with cytochalasin B, an agent which blocks the action of actomyosin, have given equivocal results (Blue-mink, 1971). Cleavage in embryos treated with this agent is abnormal, but the microfilaments appear undamaged. There is still considerable controversy too about the action of cytochalasin on living cells, (cf. Forer *et al.*, 1972), so we must await much further work, before the nature of the microfilaments becomes clear. A general point that should be emphasized is that if the microfilaments are contractile and are responsible for furrow-formation, they must contract to the same extent and with the same timing in both daughter cells, to make a symmetrical furrow. This requires an interaction between the two cells.

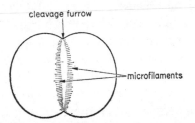

Fig. 4.3 Diagram to show the position of microfilaments in relation to the cleavage furrow in Amphibia. The first cleavage is shown, viewed from the animal pole.

There is no problem about postulating that cleavage cells in vertebrate embryos must be able to communicate and to co-ordinate the activity of their contractile structures during each cell division. The membrane between them forms only very gradually, and for some time there are wide open areas where adjacent cells' cytoplasms are continuous (cf. Fig. 1.9b). Even after the cell membrane is complete, which may not be until each cell has undergone further divisions, there are certain types of junctions between them (see p. 73) which would allow the passage of small molecules and ions. In some species, the earliest intercellular membranes at first consist entirely of structures which could play some part in intercellular communication: for instance, in the zebra fish *Brachydanio rerio*, Thomas (1968) has observed a gradual coalition of a chain of vesicles across the gap between two daughter cells. These vesicles may themselves contain some of the intermediaries of cell communication.

The mechanisms of cleavage in some invertebrates have also been studied ultrastructurally in recent years. Microfilaments have also been seen in these, but instead of lying transversely to the cleavage furrow they may be found in a circumferential band parallel to the furrow. This band of microfilaments has been beautifully illustrated by Szolossi (1970) in eggs of a polychaete annelid, *Armandia brevis*, and also of a jellyfish, *Aequoria aequoria* in which the cleavage furrows pass through only one side of the egg. The filamentous band is circular in *Armandia*, but arcuate in *Aequoria*, since it coincides exactly with the plane of the furrow in each case (Plate 4.1). It looks as if in *Armandia* the band forms a contractile ring which gradually constricts the cells apart – a mechanism for cleavage which was postulated some years ago by Marsland and Landau (1954). A similar

63

circumferential band of filaments has been seen in cleavage stages of the squid, *Loligo* (Arnold, 1969).

It may be partly because of these differences in orientation of the filaments in vertebrate and invertebrate eggs that whereas the blastomeres of most vertebrates remain in close contact during cleavage, the cells of many invertebrates are more loosely adherent and at the end of contraction of each furrow the daughter cells may have a noticeable space between them (cf. Plate 4.2). In cases like these there is clearly no extensive cytoplasmic continuity between the cells as there is in many vertebrates. This may be one reason why, with the exception of the 3D cell in molluscs, there seems to be less cellular interaction during cleavage in invertebrates than in vertebrates. In amphibian embryos, the transverse filaments tend to draw together the margins of the cleavage furrows as they deepen, so that the daughter cells are held in fairly close contact as they form. At the very base of each furrow, however, a small gap persists, forming a hollow crypt, between cells in the animal half of the embryo (Plate 4.3). Kalt (1971 a, b) has shown that these crypts later amalgamate and form the beginnings of the blastocoel cavity.

4.3 Interactions controlling the differentiation of cleavage cells

There is both direct and indirect evidence that even at very early cleavage stages, the blastomeres of many kinds of animal embryos interact to influence each other's differentiation. The indirect evidence comes from isolation of blastomeres. If, after isolation, a blastomere develops to form exactly the same tissue as it would have if left *in situ* – no more and no less – one can deduce that *no* essential interaction occurs between it and other blastomeres in the intact embryo; its development is evidently entirely independent of them. This situation is true of a few groups of invertebrates: for instance ctenophores, some annelids (e.g. the marine worms *Sabellaria* and *Nereis*) and to some extent in ascidians. Perhaps the most spectacular demonstration of this cellular independence was Costello's experiment (1945) in which he showed that each of the first 32 blastomeres of *Nereis* would develop on its own into exactly what it would have formed if left *in situ*. But in many other invertebrates and in all vertebrates in which blastomeres have been separated, there is evidence that they *do* interact with other cells, for they *do not* develop into the same tissues when isolated, as they would have if left *in situ*. Often, a single blastomere from as late as the 8-cell stage is capable of regulating to form a whole embryo when isolated. This is true of Nemertine worms, of echinoderms and of mammals. The classic experiments in which these regulative powers were demonstrated were carried out at the two-cell stage by Driesch (1891) on echinoderms, by Spemann (1901) on the newt and by Seidel (1960) on the rabbit. Separation of the first two blastomeres of an echinoderm embryo by shaking them in calcium-free sea water produced two separate, complete larvae; ligaturing the 2-cell stage newt embrko to separate the cells produced two complete tadpoles; killing of one cell by cautery in the rabbit and returning the surviving cell to the uterus of another doe, produced a normal rabbit, born with the rest of her litter (Fig. 4.4). From all these results, it is evident that full potencies to produce a whole embryo arc present in each blastomere, but they are partially suppressed when another blastomere is present. This is an example of the suppression of certain gene activities which

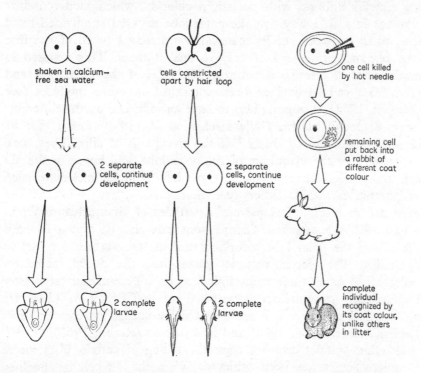

Fig. 4.4 Examples of the adaptability of embryonic cells at the 2-cell stage. Results of separating the first two blastomeras in (a) Echinoderms (Driesch, 1891); (b) amphibians (Spemann, 1901); (c) the rabbit (Seidel, 1960).

is thought to be a general feature of cell differentiation, as was discussed in Chapter 1. It shows, however, that the suppression is not permanent and can be reversed.

Direct evidence of the interaction of blastomeres on each others' differentiation is not very easy to obtain. What is needed are experiments in which the relative positions and numbers of blastomeres are altered. This is technically very difficult to do in spirally cleaving embryos, which are the ones about which we need information most. Experiments that have been done on these forms relate mainly to the functions of the polar lobe (e.g. see Cather and Verdonk, 1974) mentioned on p. 61. As long ago as 1904, E. B. Wilson showed in embryos of two marine molluscs, *Dentalium* and *Patella*, that only those groups of blastomeres to which the polar lobe was attached were capable of forming the internal tissues of the embryo. If the polar lobe was removed, a purely ectodermal larva with cilia and epidermis only, was formed. Novikov (1940) showed the converse type of effect by treating annelid embryos with potassium chloride, which modified their cleavage in such a way that the polar lobe material subdivided more than usual and some of its material was attached to all of the first four blastomeres instead of only to one of them. This resulted in duplications of internal structures. The roles of the polar lobe and of the 3D cell have since been confirmed in other molluscs (see Clement, 1952). It appears also to have an influence on the differentiation of ectoderm, for Geilenkirchen *et al.* (1970) found that its removal could lead to absence of the apical tuft of cilia. They have suggested that the cytoplasm of the polar lobe, and hence of the 3D cell and its derivatives, contains some informational molecules such as extranuclear DNA. Others (e.g. Spirin, 1966) have suggested the existence in it of 'informosomes' – particles of ribonucleoprotein to which mRNA is attached. Comparisons may also be drawn between the role of the polar lobe cytoplasm and the grey crescent cortex in Amphibia. This cortex material passes into the dorsal mesoderm cells, which exert their main influence on differentiation at a later stage, during gastrulation, as we shall see in Chapter 5. There are also cortical plasms which are essential for normal differentiation in Cephalopods (Arnold, 1968) and in *Lymnaea* (Raven, 1967).

In other spirally cleaving eggs, the killing of certain blastomeres by microcautery has been achieved. When the 4d cell in annelids, which later forms mesoderm, is cauterized, cleavage is abnormal and the larva is incomplete, lacking mesodermal tissues. This has been shown in the oligochaete *Tubifex* by Penners (1938) (see Fig. 4.5) and in the leech *Glossiphonia* by Mori (1932). The 4d cell is known to give rise to mesoderm, and this mesoderm may influence the differentiation of other larval tissues later (cf. Chapter 5).

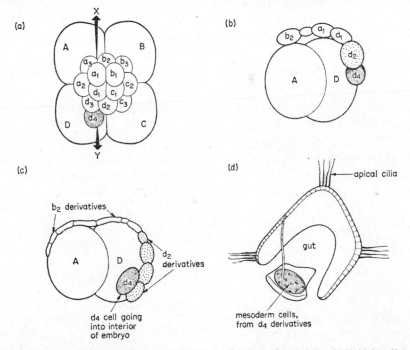

Fig. 4.5 Diagrams to show the position and future fate of the '4d' (d_4) cell in annelid embryos. (a) Upper surface view at 16-cell stage. (b) vertical section in the plane marked XY in *a*. (c) Section in similar plane, at gastrula stage. (d) Section of larva.

There is some direct experimental information about the interactions of cleavage cells in the radially cleaving embryos of echinoderms. We owe this information mainly to the many years of careful work by Hörstadius and his associates in Sweden (for a recent review see Hörstadius, 1973). He was able to dissociate the cells of 32-cell stage echinoderm embryos, by placing them in calcium-free sea water, and then to recombine them so that there were abnormal proportions of cells from animal and vegetal hemispheres. He obtained a graded series of different kinds of larvae, by a series of proportionately different combinations of cells. The larvae ranged from extremely 'animalized' individuals with cilia and ectodermal structures only, to non-ciliated, yolky 'vegetalized' larvae in which only endoderm was represented. Because of the apparently quantitative relationship between the numbers of animal or vegetal cells and the degree of animalization or vegetalization in the larvae, Runnström and Hörstadius produced their 'double gradient' theory, which stated that an 'animalizing' and a 'vegetalizing' influence emanated from the opposite poles of the embryo, in concentrations which were highest at the pole of origin and declined in a gradient towards the opposite pole. All the varying forms of larvae were then explained by interaction of the two gradients, at different concentrations accord-

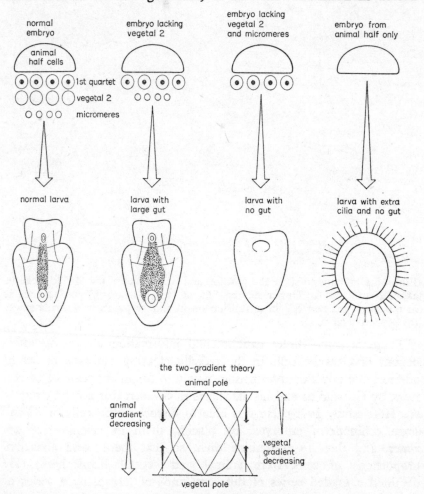

Fig. 4.6 Diagrams to illustrate blastomere recombination experiments and their results in echinoderms, and the double-gradient theory put forward to explain these. (*Modified from* Hörstadius, 1973.)

ing to the regions from which the blastomeres came. Whatever materials were in the postulated gradients would of course be contained in the blastomeres – so the cellular interactions were explained on the same basis. The theory, and some of Hörstadius' classic results, are illustrated in Fig. 4.6. Recently Hörstadius and Josefsson (1969, 1972) have been able to extract several factors with animalizing and vegetalizing properties from unfertilized eggs and from 8-16 cell stages of the sea-urchin *Paracentrotus lividus*, using chromatography and Dowex-50 resin to separate the substances in solution. Czihak and Hörstadius (1970) have also been able to show by labelling experiments that RNA passes from the micromeres (small cells at the vegetal pole which have a particularly strong vegetalizing action) into adjacent blastomeres. They suggest that RNA is the vehicle for the induction of vegetal structures – i.e. endoderm and ventral mesoderm. So far, however, no gradients in concentration of these substances have been found which could indicate any connection with the postulated animalizing or vegetalizing gradients.

It is not so well known that in ascidians, too, recombination of blastomeres have been carried out (see Reverberi, 1972) and that the cells here also influence each others' differentiation, according to the relative proportions of cells from different regions that are combined. This is surprising in view of the earlier findings (cf. p. 64) that isolated blastomeres of ascidians at early cleavage stages form partial embryos and seem to develop quite independently. It suggests that the original, indirect evidence which led people to believe that many types of invertebrate embryos were totally incapable of regulation may be refuted when further experimentation becomes possible. The results of recombinations of ascidian blastomeres are shown in Fig. 4.7. Similar recent experiments in Nemertine worms (Reverberi, 1972) indicate that in these, however, the cells do not influence each others' differentiation in specific ways until the gastrula stage.

Because of the difficulty of separating the closely-knit blastomeres of most vertebrates at early cleavage stages, we have no evidence about their interactions except in mammals, whose cells do not have continuous cytoplasm between them like those of amphibians, birds and reptiles. Graham (1971) and Hillman, Sherman and Graham (1972) have been able to manipulate the cells of 4-, 8- and 16-cell stage mouse embryos and to see how their positions in relation to other cells affect their differentiation. They have found that placing a cell from even as early as the 4-cell stage onto the outside of

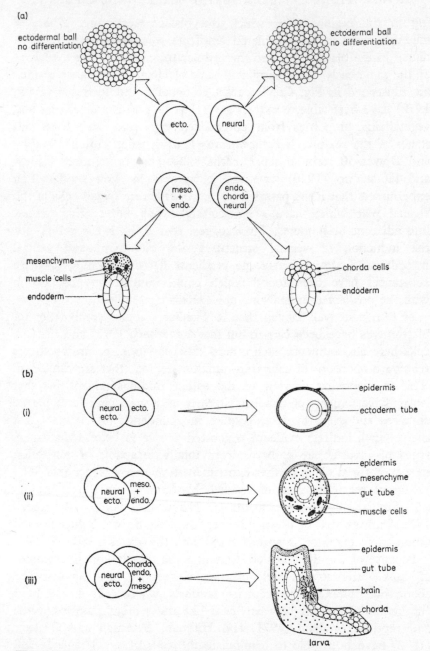

Fig. 4.7 Results of isolating and recombining blastomeres in ascidians at the 8-cell stage. (a) Pairs of cells dissociated from the normal 8-cell stage embryo; (b) neural ectoderm combined with other pairs of cells. (*Modified from* Reverberi, 1972.)

another 4-cell embryo makes it form trophoblast (extraembryonic membrane tissue) in the majority of cases. By contrast, if the cell is placed in the centre of another 4-cell stage, it usually forms part of the embryonic body (Fig. 4.8). A whole 4-cell stage embryo, too, can be made to contribute all its cells to the embryonic body, by surrounding it with other 4-cell stages which then fuse with it. So evidently the decision between differentiation into either trophoblast or embryonic body is made by some factors resulting from either an outside or an inside position. It could be that the cells on the outer surface have more rapid access to oxygen and nutrients than those inside, and begin already to function in a nutritive capacity towards the inner cells. But we have no evidence yet as to what precise 'signal' informs the cell of its new position when it has been transplanted experimentally, and causes it to switch to the appropriate course of differentiation.

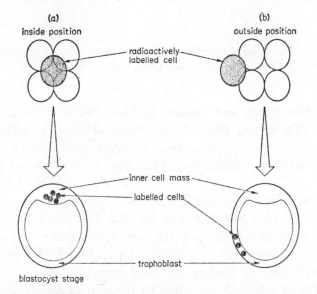

Fig. 4.8 Effects of inside or outside position on the fate of 4-cell stage blastomeres of the mouse embryo (see Graham, 1971). (a) Inside position leads to destination in inner cell mass, to form embryonic tissue. (b) Outside position leads to destination in trophoblast, contributing to placenta.

Several recent experiments on cleavage stages of mouse embryos have provided indirect evidence that the cleavage cells interact and control each others' differentiation, though the method of interaction is not yet known. It is possible to do the converse of

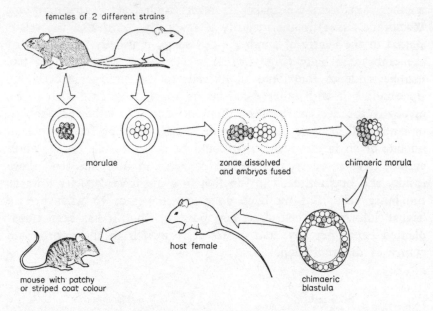

Fig. 4.9 The production of chimaeric mice by fusion of two morulae from genetically different parents. (See Tarkowski, 1964; Mintz, 1964.)

isolating blastomers, and instead to fuse several cleaving embryos together, (Tarkowski, 1964, 1965; Mintz, 1964). It is found that when such composite (chimaeric) embryos are inserted into the uterus of another female, they will develop into normal mice, of normal size (Fig. 4.9). Embryos from different strains of mice have been fused in this way, and if they carry genetic 'markers' such as differences in coat colour, it can be seen that more or less equal numbers of cells from each genotype are represented in the chimaeric mouse. So evidently cells of different genotypes have been able to 'co-operate', first in moderating their rate of cleavage so that a normal size is achieved, and then in forming chimaeric tissues and organs. There is some evidence that the cells of different genotypes may segregate themselves into 'clones' (groups derived by mitosis from one progenitor cell) in each organ, however, for coat-colours sometimes appear in stripes, suggesting that each segment of the body contained neural crest cells of only one genotype. We shall return to this point later when considering axial differentiation, in Chapter 6.

4.4 Possible mechanisms of cell-interaction during cleavage

We have seen that cellular interactions during cleavage may have at least three kinds of effect: to regulate the pattern and timing of mitoses, to assist in the mechanism of cytoplasmic division and cell-membrane formation and to control the course of differentiation in the blastomeres. We have also noted that cells of spirally cleaving invertebrates, as far as present evidence goes, seem to show less interaction than do those of radially cleaving invertebrates and vertebrates. A further point already remarked on is that cells of vertebrates, which show the most interaction of all, are usually very closely adherent and have continuity between their cytoplasms for prolonged periods during the process of cleavage. An exception to this are very early mammal embryos, but even these already show desmosome-like junctions between the trophoblast cells at the blastocyst stage. The more information we can glean, either by ultrastructural study or by experimental methods about the possible materials and signals that can be exchanged between cleavage cells, the nearer we can come to understanding the mechanisms by which they interact. So far, unfortunately, very little clear information is available. Electron microscope work has revealed a number of close junctions between adjacent cell membranes, some of these were previously designated 'septate junctions' and were believed to allow the passage of molecules of molecular weight up to 69 000. So far it has not been practicable to follow the passage of any normal metabolites through these junctions. As autoradiographic and labelling methods for electron microscopy improve, however, it should be possible to establish with more certainty how much chemical exchange can take place between cleavage cells, in a variety of animal species. Presumably during early phases of cleavage in amphibians, reptiles and birds, when the cell cytoplasms are still confluent, there can be a complete intermingling of cytoplasmic constituents between each pair of daughter cells.

Besides, molecules, ions or electrical charges can act as communicating signals between cells. Furshpan and Potter (1968) drew attention to the passage of ions between the cells of a number of animals, both in adult and in embryonic stages and recently Sheridan (1973) has discussed the conditions under which either electrical signals, ions or molecules could pass through the various types of intercellular junctions. In late cleavage cells of the frog *Xenopus*, Palmer and Slack (1969) have demonstrated that there are electrical

73

discharges – i.e. drops in voltage which can be detected with micro-electrodes placed in two adjacent cells – passing from cell to cell. These discharges evidently play some essential role in maintaining cleavage, for if the embryos are treated with halothane, an agent which causes electrical uncoupling between the cells, cleavage is blocked (Slack and Palmer, 1969).

It is probable that to maintain a control over the speed and synchronous timing of cleavage, some periodic signal is required to 'trigger off' each set of mitotic cycles in the embryo. The electrical discharges we have just mentioned above could provide a suitable mechanism for such a signal, since they travel very rapidly and are easily transferred from cell to cell, unlike macromolecules. Cinematograph films of cleavage in amphibian embryos (e.g. Hara, 1971) give a very striking impression that periodic waves of movement pass across the whole embryo just before each cleavage occurs. These waves of movement look remarkably like the responses of smooth muscle cells to an electrical stimulus, though on a very much smaller and slightly slower scale.

In the absence of any other positive evidence about the mechanisms by which cleavage cells interact, except when special cytoplasms such as those of the 3D cell and the cortex in molluscs are involved, the idea that electrical impulses play a part has some attractions. It is amenable to further experimentation, on invertebrates as well as vertebrates, and it involves no major theoretical difficulties as to how the ions could pass from cell to cell. The mechanism could work in the loosely-knit cells of invertebrates and mammals, just as effectively as in cells which are closely bound together. It is unlikely to offer any wide variety of types of signal, however, so the number of different processes it could control must be limited. When we go on to consider more complex events at later stages of embryonic development, in the following chapters, other types of signal or stimulus will have to be considered, to explain the variety of cellular interactions that are later involved.

5

Cellular
interactions during
gastrulation

Gastrulation is the phase of embryonic development when extensive cell-movements start, resulting in the formation of distinct layers of tissue from which the embryonic organs develop. Generally there is less cell movement in the embryos of invertebrates than in those of vertebrates at this stage. This is particularly true of those invertebrate embryos that have relatively few cells with which to achieve a layout of separate tissues. There are only some thirty cells in the gastrula of an annelid worm, for example, compared with over thirty thousand cells in the early gastrula of a frog. It would be difficult for the integrity of the annelid embryo to be maintained, if several of its rather large and few cells became mobile.

No one can fully appreciate the co-ordinated and dramatic movements of cells that occur during gastrulation, unless he has been lucky enough to see a cinematograph film of the process. It has been filmed very successfully in the transparent embryos of sea-urchins, by Gustafson and Kinnander (1956). Several other experimenters have made films of the movements in amphibian gastrulae and in the primitive streak of the chick. Some of the range and varieties of gastrulation movements in both vertebrates and invertebrates are illustrated in Figs. 5.1 and 5.2. The 'fate maps' of the future destiny of different embryonic regions, which are also given in these figures, have been devised from marking experiments carried out by a number of workers in the 1930s and 40s, in which they coloured groups of cells with vital dyes and then observed where they moved to during gastrulation. Recently Anderson (1973) has suggested new

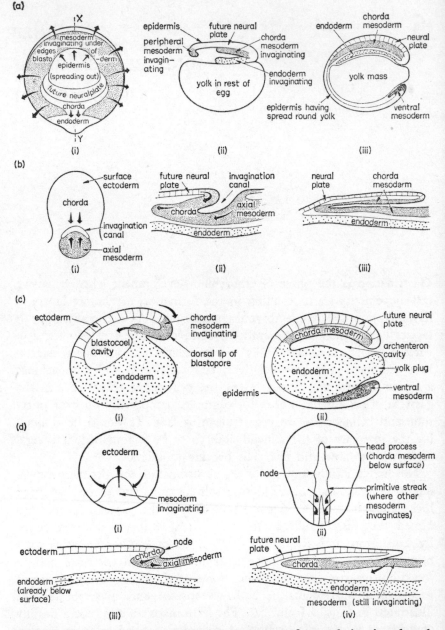

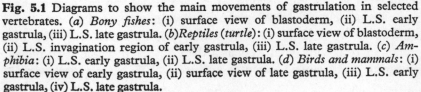

Fig. 5.1 Diagrams to show the main movements of gastrulation in selected vertebrates. (*a*) *Bony fishes*: (i) surface view of blastoderm, (ii) L.S. early gastrula, (iii) L.S. late gastrula. (*b*)*Reptiles* (*turtle*): (i) surface view of blastoderm, (ii) L.S. invagination region of early gastrula, (iii) L.S. late gastrula. (*c*) *Amphibia*: (i) L.S. early gastrula, (ii) L.S. late gastrula. (*d*) *Birds and mammals*: (i) surface view of early gastrula, (ii) surface view of late gastrula, (iii) L.S. early gastrula, (iv) L.S. late gastrula.

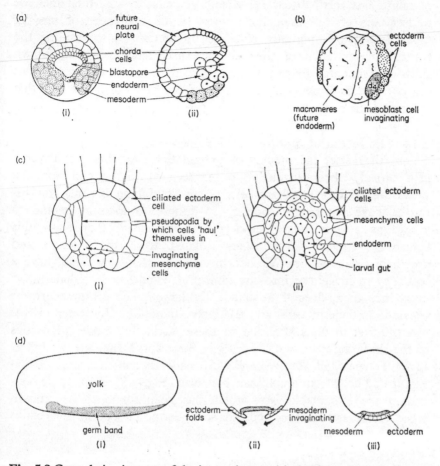

Fig. 5.2 Gastrulation in some of the invertebrates. (*a*) *Ascidians*: (i) ventral view of gastrula, (ii) L.S. gastrula (based on Reverberi, 1972). (*b*) *Annelids*: L.S. early gastrula (cf. Fig. 4.5). (*c*) *Echinoderms*: (i) early gastrula, (ii) late gastrula. (*d*) *Insects*: (i) view of whole egg with ventral germ band, (ii) T.S. at gastrula stage, (iii) T.S. at end of gastrulation. (*Based on* Waddington, 1956.)

fate maps for some of the invertebrates; his ideas are not included in Fig. 5.2.

Whatever the numbers of cells involved, their range of possible contacts with one another is enormously increased during gastrulation. Consider one cell only: if it passes along a row of ten or a hundred cells instead of remaining stationary, it will increase its possibilities of cell-interaction tenfold or a hundredfold. In vertebrate embryos, whole sheets of several hundred cells are involved

77

in migrations relative to other sheets of cells: so the total increase in numbers of cell-interactions must be of the order of tens of thousands. The importance of these interactions in controlling the future differentiation of the cells in vertebrates has been known since experimental grafting work began, in the first decades of this century. We shall now consider some of the kinds of interaction that occur and their effects.

5.1   Differential mobility and adhesions

Before they start the process of gastrulation, vertebrate embryonic cells already show distinctive differences in motile and adhesive behaviour. Future ectoderm cells of an amphibian blastula, for instance, have an innate tendency to adhere to each other and to spread as a sheet over mesoderm or endoderm; they show this spreading tendency even when isolated from the embryo and grown in tissue culture. Mesoderm cells, on the other hand, have a tendency to invaginate into any clump of ectoderm or endoderm, or even into themselves if a sufficiently large piece of mesoderm is isolated. Endoderm cells are relatively immobile. Holtfreter (1944) was the first to draw attention to these distinctive motile properties of the three different types of cells in newt and salamander embryos. Later Townes and Holtfreter (1955) made random mixtures of disaggregated cells from amphibian blastulae (Fig. 5.3) and they demonstrated the remarkable fact, which has been found to be true in other vertebrate embryos too, that ectoderm, mesoderm and endo-

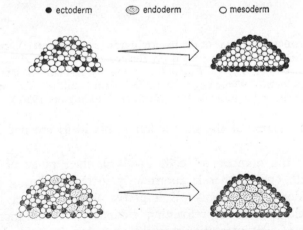

Fig. 5.3 'Sorting out' of ectoderm, mesoderm and endoderm cells from random mixtures, derived from amphibian gastrulae (cf. Townes and Holtfreter, 1955).

derm are able to sort out from the mixture and to form separate layers, appropriately placed with the ectoderm outside, the meso-derm in the middle and the endoderm on the inside. This reassort-ment is explicable if one assumes that cells of like type tend to adhere to one another in preference to cells of other types. It is also of course essential that the cells should be motile, to be able to initiate the sorting out process (Curtis, 1961).

This difference in adhesive properties represents an early stage of differentiation in the gastrula cells. It implies that there are some surface differences between ectoderm, mesoderm and endoderm. The motility is, in addition, essential for their survival and further development. Elsdale and Jones (1963) pointed out that amphibian embryonic cells in tissue culture would only continue to develop if they had first gone through a 'germination' phase, characterized by changes in shape and amoeboid movements. Cells which did not 'germinate' failed to consume their yolk, showed no histological changes and eventually died.

It is not known what causes the onset of movement in embryonic cells at this particular phase in development known as gastrulation. Their ability to change shape, however, which is an essential part of their motility, is thought to be a function of the surface charge on their membrane. This surface charge is also thought to play some part in controlling preferential adhesions to cells of their own type (see Curtis, 1960, for discussion of the surface forces governing cell movement and cell adhesions). McMurdo and Zalik (1970) were able to measure the electrophoretic mobility, which is related to

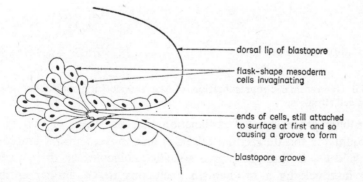

Fig. 5.4 Diagrammatic longitudinal section through blastopore lip region of early amphibian gastrula, to show shape of invaginating cells. (*Modified from* Holtfreter, 1944.)

surface charge, in gastrula cells of amphibians. They found that mesoderm had a lower surface charge than either ectoderm or endoderm. This lower charge should, they thought, tend to make the cells more easily deformable, and hence might explain the ability of dorsal mesoderm cells to squeeze their way into the blastopore at the beginning of gastrulation (cf. Fig. 5.4). What has still not been explained by any of the several theories on this subject, however, is why the invagination should begin at a particular place on the surface of the embryo and at a particular time, characteristic of each species. For instance, it always starts at the dorsal lip, in amphibians, and at the lateral edges of the early primitive streak in birds. Cohen and Morrill (1969) obtained evidence that there is a unidirectional flow of sodium ions into the amphibian blastula, making the blastocoel 30-40mV positive with respect to the cells and to the external medium. They suggested that the unidirectional electric field resulting from this could initiate invagination of cells. Even if this theory is true, it still does not explain why invagination should start only at one particular point, the dorsal lip.

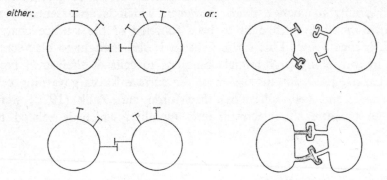

either: _or:_

identity of surface molecules   complementarity of surface molecules

Fig. 5.5 Diagrammatic representation of the 'interlocking molecules' idea to explain cell affinities.

The only other type of explanation that has been offered so far, to account for the differential affinities and movements of amphibian gastrula cells, is that they have specific molecules at their surfaces which interlock by a mechanism analogous to an antigen-antibody reaction (Fig. 5.5). This idea was put forward by Weiss (1947) and attracted many supporters. It has since been shown by Spiegel (1954) that treating cells of newt embryos with antiserum specific to that species abolishes their preferential affinity to cells of their own species and makes them equally liable to stick to cells of another newt species. Spiegel did not extend his observations to the different cell types in gastrulae of one species, however. What other side effects the antiserum may have had in his experiments was not clear either. In fact we have no certain evidence yet of the existence of *surface* antigens in amphibian gastrula cells, though there is quite a lot of evidence showing that the cells differ in their total complements of antigens. Clayton (1953) demonstrated antigenic differences between ectoderm, mesoderm and endoderm of the newt, *Triturus alpestris* by serological methods (Fig. 5.11). Others have confirmed this on other species of amphibians (e.g. Vainio, 1956; Takayanagi, 1958; Inoue, 1961, Stanisstreet, 1972). But so far no distinction has been made between possible surface antigens and those that are internal to the cell. A suitable approach might be to look for tissue-transplantation antigens, by investigating whether the cells are able to evoke cell-bound antibodies when injected into rabbits. If so, they would then be rejected rapidly if subsequently used as grafts into the rabbit, or if grown in media containing the rabbit's lymphocytes.

Before leaving the subject of differential cell affinities, a few general points should be added. The first is that one would expect similar properties to be shown by the cells of invertebrate embryos, but experimental data on these is scarce. Another point is that the affinities in vertebrates cannot be detected *before* the late blastula or early gastrula stage. Curtis (1957) was unable to get cells of pre-blastula stages of the frog, *Xenopus* or the newt, *Triturus* to show affinity or to reaggregate after they had been separated in a calcium- and magnesium-free medium. By the late blastula stage, however, reaggregation and sorting out into ectoderm, mesoderm and endoderm do occur, even if the whole embryo is disaggregated *en masse* (Landesman and Gross, 1968). A third point of general interest is that the differential affinities, which are stable and characteristic for each species, are much impaired in hybrid embryos of amphibians (obtained by fertilizing eggs of one species with sperm of another). Gastrulation is also abnormal in these hybrids and may fail altogether. Johnson (1969) showed that there is a correlation between the extent to which the cells are able to reassort after disaggregation, and the viability of the hybrids. The more successful hybrids were able to gastrulate, and in these the disaggregated gas-

81

trula cells were also able to reaggregate and to sort out into ectoderm, mesoderm and endoderm successfully. Other hybrids which were unable to gastrulate showed an inability of their cells to reaggregate or to sort out. There were also unusually large gaps between the mesoderm cells (Johnson, 1972). So it is clear that these evidences of differential cell affinity, although often investigated in what seem to be artificial conditions quite unlike those in the intact embryo, are nevertheless important indicators for normal development. Most embryologists believe that this is a general phenomenon and not peculiar to amphibian embryos.

5.2   Interactions that exert major control over embryonic differentiation

Most of the work on tissue interactions has been carried out on embryos of vertebrates, because in these, grafting experiments are easier to do than in invertebrates. So we shall consider first the findings in vertebrates and then go on to mention such evidence as there is from invertebrates, in Section 5.3.

An interaction between cells of vertebrate gastrulae that has occupied the attention of embryologists for longer than almost any other event in development is *neural induction*. In this process, the dorsal mesoderm acts on the ectoderm as it invaginates underneath it, and causes it to form a neutral plate, which then differentiates into the brain and spinal cord. The discovery of this induction was first made by Spemann and Mangold during the years 1916-22, in embryos of the newt, *Triturus*. The grafting experiments by which they proved that dorsal mesoderm would induce ectoderm, even of another species of animal, to form neural tissue, were epoch-making. They have been described frequently in textbooks af embryology, so de-

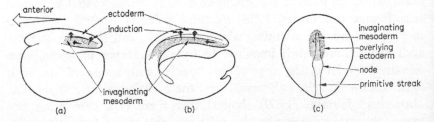

Fig. 5.6 Neural induction: the inducing and reacting tissues in (a) fish, (b) amphibians and (c) birds and mammals. *a* and *b* show longitudinal sections, and *c* a surface view, of gastrula stages.

tails will not be given here; Fig. 5.6 shows examples of the tissues that have been shown to take part in neural induction in various vertebrates.

Spemann's findings started a veritable gold-rush of experimentation in the 1930s and 40s, to try to discover the nature of the inducing stimulus that passed from mesoderm to ectoderm cells. Dead mesoderm, and a wide variety of organic chemicals and tissue extracts were found to give neural inductions when implanted into amphibian embryos. At that time it seemed that *the* inductor would soon be discovered. But the enormous variety of experiments that has ensued since seems only to have broken down the original broad theories and generalizations, leaving us without any unifying concept by which to explain all the results. There are several good reviews of the main trends in theory and experimentation from the 1930s onwards (e.g. Saxèn and Toivönen, 1962; Tiedemann, 1967; Balinsky, 1970), so we shall confine ourselves here to quoting some of the more recent work which exemplifies the three lines of approach that have characterized all investigations. These are: (a) descriptions of structural detail under normal and experimental conditions, in the hopes of finding evidence of interactions between mesoderm and ectoderm; (b) biochemical work on the identification of materials both from embryonic and foreign tissues which act as inductors; and (c) studies of biochemical changes that take place in gastrula ectoderm which has responded to neural induction.

5.2.1 *Descriptions of structure.*
Electron microscopy has re-awakened interest in the histological and cytological changes that occur both in the inducing mesoderm and in the induced ectoderm of the embryo during gastrulation. Most of the new ultrastructural studies have been made on amphibian and chick embryos. For instance, Tarin (1971) has re-examined the mesoderm/ectoderm interface, both histochemically and by electron microscopy, in gastrulae of the frog *Xenopus*. Besides re-describing the changes in shape of the ectoderm cells as they form neural plate (cf. Fig. 6.2) he confirmed the existence in *Xenopus* of two layers of dorsal ectoderm, the outer one of which according to Nieuwkoop and Faber (1956) forms epidermis and the inner one neural plate. Ave, Kawakami and Shameshima (1968) had also drawn attention to these two cell types, and they had found that only the inner cells responded to inductors *in vitro*. Schroeder (1970) however believes

that *both* ectoderm layers form the neural tube. It is not clear yet which authors' interpretations are correct. As we shall see in Chapter 6, however, the induction stimulus can in any case be passed on from one group of ectoderm cells to others which have not had direct contact with mesoderm.

Electron microscope work has also helped to elucidate the results of experiments in which filters of known pore sizes were placed between ectoderm and mesoderm, in attempts to find what size of molecule may be involved in neural induction. In the chick, Gallera *et al.* (1968) found that induction can occur even when a millipore filter is placed between the two tissues. With the electron microscope it could be seen that microvilli had penetrated from the mesoderm cells a short distance into the filter, but that there was definitely no contact between mesoderm and ectoderm cells. So the inducing stimulus must have been able to pass through the filter in soluble form. England (1969) observed materials exuded from the mesoderm cells within the pores of the millipore filter. These materials were not identified. In a similar study on *Xenopus,* Nyholm *et al.* (1962) found that here too, insertion of a millipore filter did not block neural induction. In this case, no cellular processes could be seen penetrating the filter, but a granular material which appeared to emanate from the mesoderm was seen to a depth of 2-4 μm in the filter. This is interesting in view of the observations by Kelley (1969) and by Tarin (1971) that granules accumulate in the mesoderm/ectoderm interspace at the late gastrula stage in *Xenopus* (cf. Plate 5.1). Neither of these authors has been able to find clear evidence that these granules either originate from the mesoderm or enter the ectoderm, so it cannot be said whether or not they play any role in induction. Kelley did report that lobes containing the granules were found *only* on inducing mesoderm cells and not on ventral mesoderm, however. Van Gansen and Schram (1969) saw chains of similar granules, apparently being discharged from extracellular vesicles lying between mesoderm and ectoderm, but the origin of the vesicles was unfortunately not clear. Histochemical tests by several investigators have indicated that at least some of these granules contain RNA, while others contain glycogen. The glycogen granules could perhaps be partly a remainder of the contents of the blastocoel cavity, for Kalt (1971) observed glycogen granules passing into the blastocoel from surrounding cells at earlier stages (cf. Chapter 4). Any RNA-containing granules would seem likely candidates for the role of induction, since so much biochemical work (Section 5.2.2) suggests that ribonucleoproteins are effective neural inductors. But without dynamic evidence of the fates of these granules, the question of their role must remain open. So far one of the main limitations of electron microscope work is that it gives only a static picture of events.

The work with millipore filters placed between inductor and induced tissue has led to a further conclusion, namely that whatever molecule is concerned in induction is able to pass effectively through pores of 0.8 μm diameter, and to cross distances of 20 μm, the thickness of the filters (Saxèn, 1961). Earlier, Brahma (1958) had found that 0.4 μm was the smallest pore size through which an induction stimulus could pass, when Gradocol membranes of known pore sizes were tried in axolotl and *Xenopus* embryos. In both cases, in order not to block induction the filter would presumably have to allow a sufficient concentration of inductor to pass across, in a time period equivalent to that normally taken for gastrulation, since this is the minimum period for which the ectoderm normally needs to be exposed to inducing mesoderm if a normal nervous system is to form. Arrest of gastrulation, or premature removal of the dorsal mesoderm, results in little or no neural differentiation.

There have been a number of studies aimed at determining the minimum period needed for induction of neural tissue. Ectoderm has been explanted for varying times in contact with an inductor, then removed and allowed to continue differentiation on its own. Goettert (1966) found, as did many others, that the longer the contact allowed, the more complete and normal was the neural tissue that formed. Johnen (1964) found that the minimum time of contact between chorda-mesoderm and ectoderm needed for induction to occur in the newt *Triturus vulgaris* was 4h. During this time the chorda-mesoderm invaginated into the explanted ectoderm, as in an intact embryo. Shorter times than four hours gave no neural inductions, and longer times produced progressively more posterior neural structures. This corresponds to the normal sequence of events in gastrulation, for the mesoderm passes under the future posterior end of the neural plate first and for the longest time, whereas it only reaches future cranial regions at the end of gastrulation and has a relatively short time passing under these regions (cf. Fig. 5.7). Other data on the time required for induction have been obtained by Gallera (1965) in the chick. He found that 7-8h contact between inductor and ectoderm was necessary in this case. This time corresponds

t.i. = tail inductor region
tr.i. = trunk inductor
h.i. = head inductor region

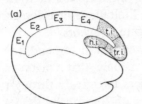
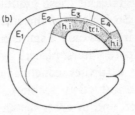
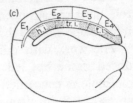

E₄ beginning to receive induction
from head inductor region

E₂ beginning to receive head
induction: E₄ now receiving
trunk inductor

E₁ now receiving induction for
the first time. E₂, E₃ and E₄
receiving successively more
caudal induction

Fig. 5.7 Diagram to compare the relative times of exposure of different regions of ectoderm to neural induction in amphibians. (a) Early gastrula; (b) mid-gastrula; (c) late gastrula; seen in L.S.

with that normally taken for the notochord mesoderm to invaginate and form the head process, at 37°C. (See Hamburger and Hamilton's Normal Table, 1951, of the development of the chick).

From the rather imprecise but nevertheless indicative findings outlined in this section, it can be concluded that since the neural inducing agent is able to pass through filters, without there being direct contact between mesoderm and ectoderm, it is probably a diffusible material. It is not likely to be an entire structural component such as the granules seen in electron micrographs, but it could be some soluble molecule which is at first contained in granules or other structures and is then released into the interspace between mesoderm and ectoderm. To have some effect on the differentiation of the ectoderm, it possibly need not enter the cells directly, but may influence the nature of the extracellular material, which then in its turn may pass on a further chemical change to the ectoderm cells. It seems likely that the inductor substance emanating from the mesoderm is a slow-moving and therefore perhaps a large molecule, since it takes some hours to exert its effect, in amphibians and birds, and it cannot pass through pores of diameter less than 0.4 μm. As we shall see in the following section, the most favoured candidate for the role of neural inductor is some form of ribonucleoprotein.

5.2.2 Materials that may pass between mesoderm and ectoderm and may act as inducing agents.

Attempts to find out what kind of agent may be responsible for the inductive interaction between mesoderm and ectoderm in gastrulae of vertebrates have consisted mainly of extracting known, purified components from quite foreign tissues and testing the effects of these on gastrula ectoderm of amphibians. This kind of work was started as soon as it became clear, in the 1930s, that dead mesoderm was an effective inductor, i.e. that a chemical rather than a biological agency was involved. Bautzmann *et al.* (1932) obtained neural inductions in amphibians with implants of mesoderm that had been killed by boiling, and Waddington and Schmidt (1933) showed similar effects with heat-killed primitive streak in the chick. So it seemed likely that in normal development the inducing agent was a material exuded from mesoderm. This suggestion was supported by experiments of Niu and Twitty who placed pieces of ectoderm in drops of medium in which dorsal lip mesoderm had previously been cultured. This 'conditioned' medium caused neuralization of the ectoderm. When analysed, the medium was found to contain RNA (Niu and Twitty, 1953). However, subsequent work has shown that pure RNA is not an effective neural inductor; on the other hand proteins and ribonucleoproteins are. Furthermore, in a recent follow-up of Niu and Twitty's work, Ajiro (1971) found that several foreign tissues are far more effective at 'conditioning' the medium for neuralization than is dorsal lip tissue itself. Probably the foreign tissues could be used in greater bulk than the dorsal lip and therefore exuded higher concentrations of products into the medium. It is clear that the conditioning substances are by no means peculiar to mesoderm tissue.

Biochemical extraction of purified components is much more readily possible from adult tissues than from embryonic material, simply because the adult tissues are available in greater bulk. So until very recently, most of the extraction work has been done using tissues other than the inducing mesoderm of the gastrula. Tiedemann and his collaborators, who have persisted longest and with most success in work of this kind, have compromised by using whole 9-day chick embryos for extractions of inducing materials. Their earlier work is reviewed by Tiedemann (1967). With some of the latest analytical methods such as column chromatography, electrophoresis and electro-focusing devices, they have been able to purify a number of protein and nucleoprotein components from homogenates of the chick embryos. Each of their isolated fractions induces a characteristic range and proportion of different types of neural tissue (Fig. 5.8). Some fractions also induce mesoderm, or even endoderm and germ cells

87

D

(Kocher-Becker and Tiedemann, 1971). This was unexpected at first, but only because in the overriding enthusiasm for discovering neural inductors, embryologists have tended to forget the likelihood that all cell types in the gastrula – ectoderm, mesoderm and endoderm – may be capable of producing materials that influence the differentiation of tissues adjacent to them.

The fact that successful neural inductions have been obtained with certain nucleoprotein materials extracted from foreign tissues is not proof that inducing tissues of gastrulae produce the same materials. The experiments by Tiedemann and others represent only a first step towards trying to identify inducing substances extracted from the

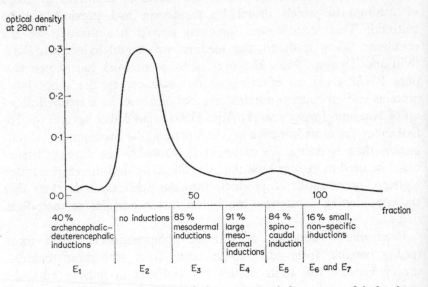

Fig. 5.8 Graphic representation of the range and frequency of inductions produced by different fractions of a 'mesoderm inductor'. (*Redrawn from* Tiedemann, 1967.)

gastrula itself. I tried a crude, mitochondria-rich extract from homogenates of different parts of *Xenopus* gastrulae (Deuchar, 1967) and found that the extract from dorsal lip mesoderm gave much higher percentages of neural inductions than did extracts from other regions of the gastrula (see Table 5.1). Recently Faulhaber and Geithe (1972) have gone much further in chemical analysis, using whole embryos, and have succeeded in isolating two active fractions from whole gastrulae. One is a protein and the other a high-molecular-weight ribonucleoprotein. Both induce brain and spinal cord, but in different proportions. So we are a little nearer the goal of being able to say what components of the living cells may be responsible for this particular cellular interaction in the vertebrate gastrula.

Table 5.1 Numbers of inductions obtained with small-particle extracts of *Xenopus* embryos (see Deuchar, 1967). (Figures in brackets are percentages)

	Total explants	Total inductions	Neural tube	Neuroid	Epithelial thickenings
Whole blastula					
expt. (a)	87	18 (20.7)	17 (19.5)	1 (1.1)	—
expt. (b)	51	5 (9.8)	4 (7.8)	—	1 (1.96)
Whole gastrula					
expt. (a)	58	12 (20.7)	6 (10.3)	—	6 (10.3)
expt. (b) '	30	7 (23.3)	—	—	7 (23.3)
Whole tailbud stage					
expt. (a)	105	32 (30.5)	15 (14.3)	1 (0.9)	16 (15.2)
expt. (b)	41	21 (51.2)	9 (21.9)	1 (2.4)	11 (26.8)
Dorsal lip of gastrula					
expt. (a)	38	22 (57.9)	6 (15.8)	3 (7.9)	13 (34.2)
expt. (b)	32	18 (56.3)	3 (9.4)	3 (9.4)	12 (37.5)

Another promising, but so far inconclusive approach which has been used to investigate the chemical nature of the neural inductor is to trace the movement of labelled molecules during induction. For this purpose, radioactive labelling and immunofluorescence techniques are ideal. It has been found, for instance, that when amphibian gastrula mesoderm is labelled with radioactive amino acids such as glycine or methionine, the ectoderm induced by it also acquires label (Ficq, 1954; Friedberg and Eakin, 1949; Waddington and Sirlin, 1964). When the proteins of bone marrow, a successful foreign inductor tissue, are labelled by immunofluorescent methods (Clayton and Romanovsky, 1959; Vainio, Saxèn and Toivönen, 1960), fluorescence is picked up by the ectoderm on which it acts. But it has not been certain in any of these experiments so far, whether the passage of labelled materials has any immediate or essential connection with neural induction. There is some passage of materials even in circumstances when there is no neural induction. So we are still very ignorant as to what kinds of chemical exchange normally take place between tissues of the gastrula, and even more ignorant of the extent to which their survival and development depends on such exchanges.

It remains to consider the third line of investigation in studies on neural induction, namely, studies of biochemical changes in the induced ectoderm.

5.2.3 *Biochemical events in induced ectoderm*

It is often easier to detect the results of cellular interactions than to identify the events that have caused them. Thus, there are fewer difficulties in detecting biochemical changes that occur in gastrula ectoderm after it has been induced, than in recognizing what exchanges if any, occur between mesoderm and ectoderm during the induction process. One hopes to be able to deduce somthing about the earlier processes, from what one sees after the cellular interaction is over.

Any tissue which is starting a new phase of differentiation will be expected to show evidence of mRNA synthesis and then of new proteins. These two features have been looked for in amphibian ectoderm at the late gastrula stage, just after induction has occurred. Some preliminary observations on RNA and protein synthesis in the induced, dorsal ectoderm show that it has different characteristics from epidermal or from uninduced ectoderm. Earlier, Bachvarova *et al.* (1966) had observed an increased synthesis of high-molecular-weight, heterogeneous RNA (probably mRNA) in dorsal halves of *Xenopus* embryos at the gastrula stage, but it was not clear what proportion of this synthesis was occurring in the ectoderm. Waddington and Perkowska (1965) obtained RNA profiles which suggested that more high-molecular-weight and ribosomal RNA was being

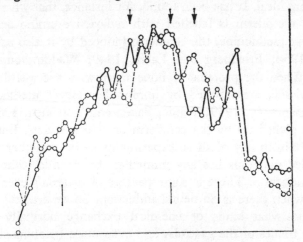

Fig. 5.9 Differences in the size distribution of polysomes extracted from neural and non-neural ectoderm of late gastrulae of *Xenopus laevis*. Profiles from an agarose-gel electrophoresis experiment of Evans (1969). (————) neural ectoderm; (- - - -) epidermis.

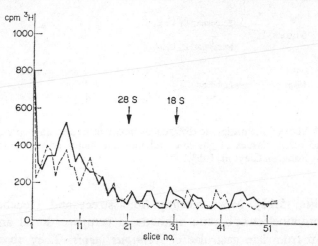

Fig. 5.10 Comparison of RNA extracted from neural and non-neural ectoderm of *Xenopus laevis* at the end of gastrulation. Profiles from an acrylamide gel electrophoresis run, with uridine-3H-labelled RNA. It can be seen that more high-molecular weight RNA is synthesized in neural (————)than in non-neural (– – – –) ectoderm. (*From* Thomas and Deuchar, 1971.)

synthesized in neural ectoderm than in mesoderm or endoderm in newt embryos at the neurula stage, but of course this is some time after induction has taken place. Evans (1969) was able to use earlier stages and to show that new polysomes, differing from those found in epidermis, appeared in dorsal ectoderm of late gastrulae of *Xenopus* immediately after induction (Fig. 5.9). Thomas and Deuchar (1971) found that the dorsal, induced ectoderm of the late gastrula in *Xenopus* had an apparently higher rate of RNA synthesis, and that it synthesized a greater proportion of high-molecular-weight material (?mRNA) than the uninduced, ventral ectoderm (Fig. 5.10). All of these observations were based on rather few data owing to the difficulty of obtaining enough ectoderm material for RNA analysis, but at least they are indicative of changes that can be investigated further in the future when new methods are available for dealing with the small quantities of embryonic material.

There is also evidence that synthesis of new proteins occurs in dorsal ectoderm of amphibian gastrulae, immediately after neural induction has taken place. It has already been mentioned (p. 81) that Clayton (1953) demonstrated antigenic differences between the tissues of gastrulae of the newt, *Triturus alpestris*. She showed some differences between neural and epidermal ectoderm too at the neurula

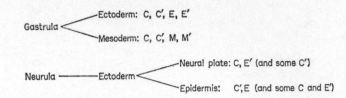

Gastrula — Ectoderm: C, C', E, E'
Mesoderm: C, C', M, M'

Neurula — Ectoderm — Neural plate: C, E' (and some C')
Epidermis: C', E (and some C and E')

Fig. 5.11 Plan of the antigenic differences between neural and epidermal ectoderm and other tissues of gastrula and neurula stages of the newt, *Triturus alpestris*. (*Based on* Clayton, 1953.)

stage (Fig. 5.11). More recently Stanisstreet and Deuchar (1972) used immunoelectrophoretic methods to compare dorsal and ventral ectoderm from late gastrulae of *Xenopus laevis*. They observed that three antigenic components were present in higher concentrations in the neural than in the ventral ectoderm. It is also possible to extract the protein from ectoderm samples and to separate a number of protein components by electrophoresis on acrylamide gel. It can be shown by radioactive labelling experiments which of these components are being synthesized actively in the different samples. Thomas and Cordall (1972) applied this procedure to extracts of late gastrula ectoderm from *Xenopus,* and found that different proteins were being synthesized in neural and in ventral ectoderm, and that the overall protein synthesis rate in dorsal ectoderm was 3-4 times that in ventral ectoderm.

It can be seen from the examples given, that there is plenty of evidence that new mRNA and proteins are formed in ectoderm of amphibian embryos as soon as it has responded to neural induction, and that these products differ from those formed in ventral or uninduced ectoderm at this stage of development. We do not yet know, however, the precise relevance of any individual nucleic acid or protein changes to the future differentiation into central nervous tissue, nor do we know whether they are first responses to the neural induction stimulus, or are preceded by some other, as yet undetected changes. Studies of the biochemical events in ectoderm after it has been treated with artificial inductors, such as some of Tiedemann's purified extracts, might throw more light on these points.

Because of the very long-standing interest of vertebrate embryologists in the process of neural induction, studies of cellular interactions at the gastrula stage have been confined almost exclusively to those between mesoderm and ectoderm. The possible influences of ectoderm and endoderm on each other or on mesoderm, have been somwhat neglected. However, Nieuwkoop has recently re-directed attention to these other possible interactions by showing that in amphibian blastulae, mesoderm will not form at all unless proper proportions of ectoderm and endoderm are present. He recombined dorsal and ventral tissues in various proportions, and concluded that the mesoderm was itself 'induced' by the endoderm (Nieuwkoop, 1969). Sudarawati and Nieuwkoop (1971) found that a minimal quantity of ectoderm had to be present also, for mesoderm to be induced successfully, and that in some recombinations of dorsal and ventral parts, the inner layers of the ectoderm were converted into mesoderm. Recently Nieuwkoop and Ubbels (1972) have suggested that both mesoderm and endoderm originate from tissue that is potentially ectoderm at the blastula stage. So we seem to be faced with a circle of so-called 'inductions' in which no one tissue necessarily acts first. How, then does the differentiation into germ layers start? Is it all really decided long before, at fertilization, when in amphibians the grey crescent marking future dorsal mesoderm is first seen? Grafting experiments by Curtis (1962a) in which he showed that the grey crescent could 'induce' a second embryonic axis to form, suggest that this may be so. But all our knowledge of the great adaptability of the embryonic cells of vertebrates in the courses of differentiation they take in response to interactions with other cells, argues against there being any *permanent* 'determination' of their future fates at pre-gastrulation stages.

Unfortunately the techniques so far used in demonstrating cellular interactions in vertebrate grastulae are far too crude to show up events that may occur at a molecular level. If the interactions are chemical processes, they may take place in milliseconds or less. Some very rapid and sensitive recording process is needed to show up events as small and speedy as this. But before being able to design any such instrument, we first have to know what kind of event is the crucial one to record; it is a vicious circle of needs and ignorance.

5.3 Cellular interactions in gastrulae of invertebrates

It may seem strange that embryologists have not spent more time looking first at cellular interactions in simpler embryos such as those of some invertebrates which have fewer cells and less complex paths of differentiation than vertebrates. Invertebrates have not been used as often as vertebrates for experimental work, however, because of their small size and limited availability. Another reason is that earlier evidence suggested that the cells of many invertebrate embryos have a rigidly determined course of development from the start (cf. Chapter 4), so that no interactions with other cells could be expected at so late a stage as the gastrula. More recent work has refuted this idea, however. At least a few groups of invertebrates show evidence of one gastrula tissue 'inducing' another to differentiate. We see this in some insects, molluscs and lower chordates.

5.3.1 *Insects*
Krause has led much experimental work on cleavage and gastrula stages in insects (see Counce and Waddington, 1972; Krause, 1953; Seidel *et al.*, 1940). He showed that invagination of the mesoderm (see Fig. 5.2) is essential for the formation of the ectoderm. In addition he and Haget (1953) found independently in different species of insects that once the ectoderm has started to form, the further development of the mesoderm and endoderm are dependent on it. Removal of pieces of ectoderm leads to defects in the mesoderm underlying it, and also to defects in the gut endoderm. So the main induction in insect gastrulae seems to pass from ectoderm to mesoderm and then to endoderm. This looks like the reverse of its direction in vertebrates. Whether the reversal of trend bears any relation to the apparent 'inversion' of the germ layers in insects before the embryo rotates in blastokinesis (Fig. 5.12), is an interesting point for speculation.

5.3.2 *Molluscs*
In this group, the innermost tissues appear to have an inductive action on the ectoderm, as in vertebrates. Raven (1958) showed in the freshwater snail *Lymnaea,* and Hess (1956) showed in the marine snail *Bithynia* that the tip of the foregut endoderm induces the ectoderm with which it comes in contact at the end of gastrulation to form the shell gland. If gastrulation is incomplete so that the endoderm does not make contact, no shell gland and therefore no shell will form. It was also shown that when combinates of foregut endoderm and gastrula ectoderm were explanted, a shell gland formed from the ectoderm. It is not certain how far *Lymnaea* and *Bithynia* are representative of the situation in other molluscs, however. Cather

94

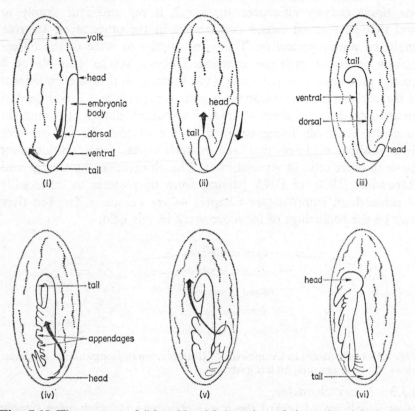

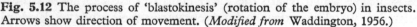

Fig. 5.12 The process of 'blastokinesis' (rotation of the embryo) in insects. Arrows show direction of movement. (*Modified from* Waddington, 1956.)

(1971) pointed out that in some species the gut endoderm invaginates too late for it to be involved in inducing the shell gland, which has already begun to form. He suggests that in these, the 4D macromere cell may be the inducer. Polar lobe material, and the 4d micromere (which forms the mesoderm) are also apparently effective inducers of shell gland. With mesoderm as an inductor, these molluscs fall more into line with the situation in vertebrates.

Cephalopods have a very different mode of development from other molluscs. Gastrulation takes place at one end only of the heavily-yolked egg, forming three distinct layers of cells, the innermost of which is the yolk endoderm. Arnold (1968) has shown that this endoderm layer is essential for the formation of the mesoderm and ectoderm: if the yolk-epithelial cells are damaged or deleted,

95

the rest of the embryo is abnormal. However in order to show that one tissue actively influences another it is not sufficient simply to find that its removal causes deficiencies in the other layer. Its role might be purely protective. The same applies to some of the observations on insect gastrulae mentioned above. As an example of a purely protective role, Zwilling (1963) showed that the endoderm of the coelenterate *Cordylophora* is able to regenerate if it is protected by an outer layer of dead, irradiated ectoderm, though it cannot do so if the ectoderm, living or dead, is not present. So far we have little positive evidence that any materials pass from an inducing tissue to other cells, in invertebrates. The observations by Czihak and Hörstadius (1970) of RNA passing from micromeres to other cells of echinoderm embryos (see Chapter 4) are unique so far, but they may be the beginnings of far more work in this field.

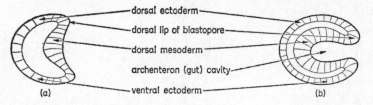

dorsal ectoderm
dorsal lip of blastopore
dorsal mesoderm
archenteron (gut) cavity
ventral ectoderm
(a)   (b)

Fig. 5.13 Gastrulation in 'Amphioxus' (*Branchiostoma*). Longitudinal sectional views: (a) early gastrula, (b) late gastrula.

5.3.3 *Lower chordates*

Most people would regard the lower chordates as being more closely related to vertebrates than to invertebrates in their mode of development. For instance, the protochordate commonly known as *Amphioxus* (or *Branchiostoma*) gastrulates with its mesoderm turning in at the dorsal lip of the blastopore and coming to lie under the future neural plate ectoderm, just as in the Amphibia (cf. Figs. 5.2 and 5.13). Moreover, Tung *et al.* (1962) showed that dorsal lip tissue from *Amphioxus* induces a secondary nervous system when grafted into another embryo, just as does the dorsal lip of amphibians. In the urochordates (ascidians) too, a similar situation seems to obtain at the gastrula stage. The cells equivalent to the dorsal lip are the chorda cells and the anterior endoderm (prechordal plate) which migrate inwards and come to lie at the anterior end, beneath the ectoderm. It has been shown (see Reverberi, 1972) that in the absence of these underlying cells, no brain will form. However, transplantation of the cells beneath other areas of ectoderm does *not* induce an extra brain. So evidently in ascidians either the induction process is more limited, or there is a limited capacity of ectoderm to respond to it, implying that other regions of ectoderm are not as 'competent' to respond to neural induction as they are in vertebrate embryos. In view of the previous work on ascidians (cf. Chapter 4), which suggested that each cell had a strictly limited fate from very early cleavage stages, it is not surprising that limitations of competence should show later. What *is* surprising is that cellular interactions *do* occur in these embryos that used to be thought to have set fates for each cell. It seems likely that as more experiments are done on invertebrates, more evidence of the versatility of their cells will come to light, and some of the old impressions of rigid determinations in this group will be withdrawn.

5.4   Conclusions

We might end with a few words about the general significance of cellular interactions in the gastrula. Medawar (1954) pointed out that they provide for flexibility in developmental processes. They also help towards selection of the most viable genotypes in a species. If neural induction has to occur before a nervous system can form, then only those embryos whose genes are normal enough to produce normal gastrulation movements will have a nervous system and be able to survive. We have seen the converse situation, in the amphibian hybrids studied by Johnson (f. p. 81) whose gastrulation movements were sometimes blocked because of genetic incompatibilities. It can be said in general, that when development occurs by a series of steps in which cellular interactions occur, it can only be successful if all those cells show at least a minimum of normal behaviour, movement and metabolism. Geneticists would probably agree with embryologists that the first rigorous environment to which the genome is exposed during the life of the organism, with a large number of selective pressures, is the environment of its own cells during embryonic development. One of the early 'crises' which the genome has to surmount is that of gastrulation. If it can get through this stage successfully, it has a good chance of producing a mainly normal organism, since most of the cellular interactions that occur at later stages take place within individual organs or tissues, and the rest of the organism could be relatively unaffected by them. It is these later interactions that we shall go on to consider in the following chapters.

97

6

Cellular interactions
during
axis-formation

The previous chapter dealt with the phase of embryonic development
when many cells become mobile and change their relative positions,
thus extending greatly the range of their interactions. After this
phase they settle down to form more compact groupings which will
become distinct tissues. It is this new phase, the beginnings of axial
organization and tissue development, which we now have to consider.
It is one about which very little is known, in terms of cellular interactions, in invertebrates. This is partly because the life-histories of
invertebrates often involve transitory larval stages which do not
achieve permanent organization into tissues until metamorphosis, so
they are difficult to study. Another feature of invertebrate larvae
which makes them difficult material is that their symmetry is sometimes radial, with no single head-to-tail axis at first.

In the present chapter we shall concern ourselves with the axial
organization of vertebrate embryos only. In these, a head to tail
axis with bilateral symmetry appears during or immediately after
gastrulation. The first external signs of this craniocaudal axis become
obvious during the phase of development which is known as
neurulation; this we must now consider.

6.1   Cellular interactions during neurulation
One of the earliest and clearest evidences of axis-formation in vertebrates is the appearance on the dorsal surface of the embryo of a
thickened plate of ectoderm, the neural plate, which as we have seen
is induced by the underlying mesoderm. The lateral edges of this

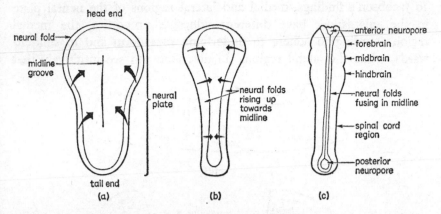

Fig. 6.1 Surface views of the neurulation process in a vertebrate embryo. (a) Early neurula; (b) mid-neurula; (c) late neurula. Diagrammatic.

plate soon rise up and meet in the midline, fusing to form a hollow neural tube which will become the brain and spinal cord. This conversion of the neural plate into a tube is known as neurulation (cf. Fig. 6.1). The cell-interactions involved have been a focus for interested study for many years. The rolling up of the sides of the neural plate seems to be due at least partly to changes in shape of its component cells. During neurulation, the cells change from a cuboidal to

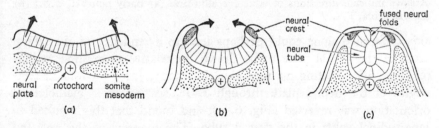

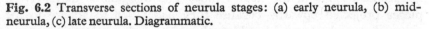

Fig. 6.2 Transverse sections of neurula stages: (a) early neurula, (b) mid-neurula, (c) late neurula. Diagrammatic.

a columnar shape, as seen in transverse section (Fig. 6.2), and their upper surfaces become narrowed as the neural tube closes. Earlier investigators thought that these changes in shape could be due to differential absorption of water, but no convincing evidence has been found in support of this view. The problem has been taken up again more recently by Jacobson (1968), who tried to explain neurulation in salamander embryos on the basis of differential cell affinities, similar to those observed in gastrula cells (cf. Chapter 5). According to Jacobson's findings, medial and lateral regions of the neural plate in the salamander have different adhesive properties, the medial regions tending to adhere to underlying mesoderm and to sink inwards, while the lateral regions do not. One can see intuitively that

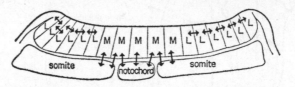

(a)

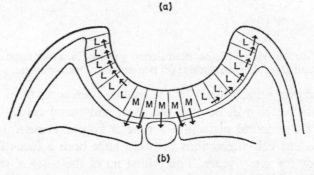

(b)

Fig. 6.3 Diagrams attempting to explain the movements of neurulation on the basis of affinities between contiguous cells of the neural plate and between midline neural plate cells and the underlying notochord and somite mesoderm. Arrows indicate directions of adhesive affinities. (a) Early neurula, T.S.; (b) Late neurula, T.S.

this difference alone could perhaps lead to a tendency of the lateral areas of the neural plate to be drawn up relative to midline regions (cf. Fig. 6.3). Testing his idea experimentally, Jacobson rotated rectangles of the neural plate through 180°, so that their mediolateral orientation was reversed (Fig. 6.4a), and found that this resulted in longitudinal splits in the neural tube. The margins of the reversed pieces projected upwards because they had failed to adhere to the areas of neural plate with which they found themselves in contact after the operation. Another of Jacobson's operations was to exchange a medial strip of neural plate with a lateral strip from the neural fold (Fig. 6.4b). In its new position laterally, the medial strip still tended to sink inwards, as it would have if it had been left in contact with the notochord in the midline. The neural fold strip, placed in the midline, detached itself later and joined up with the neural folds above it at the end of neurulation. Thus it is evident that there are innate differences in cellular adhesive affinities, leading to dif-

100

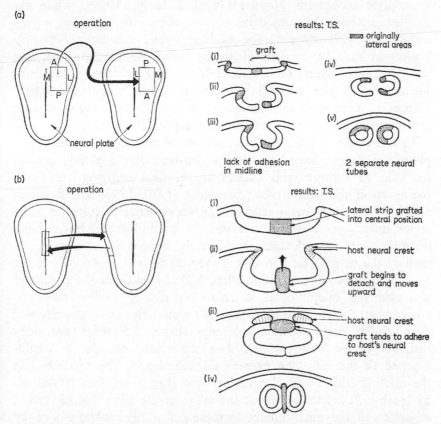

Fig. 6.4 Jacobson's experiments to demonstrate differential adhesive affinities in medial and lateral regions of the neural plate in *Ambystoma*, and their effects on neurulation movements. (a) anteroposterior and mediolateral inversion of pieces of neural plate lead to failure of adhesion in the midline of the neural groove, and eventually to formation of two separate neural tubes. (b) exchanging a piece of neural crest and a piece of midline neural plate, shows a tendency of the grafted crest to break away and fuse to the crest of the host: doubling of the tube may also result here. (*Drawings based on* Jacobson, 1968.)

ferences in the movements of different regions of the neural plate. Whether or not these alone could account for the process of neurulation, remains to be established by further work. The mesoderm may also play an important role, however, in controlling neurulation. Dorsal mesoderm has at this stage a tendency to segregate from the neural ectoderm except in the midline where the notochord is forming. The dorsolateral mesoderm, which forms the somites and which tends to increase in height at this time, undergoes a mainly centripetal movement (Jacobson and Löfberg, 1969), while the midline notochordal mesoderm shows some co-ordination of its movements with those of the midline neural ectoderm. This co-ordination may be brought about by means of junctions between the two tissue layers, for Flood (1970) has seen neuro-notochordal junctions with the electron microscope. They have some resemblances to neuromuscular junctions, but it is not known whether any contractions of cells occur in this region (Plate 6.1).

In a number of ultrastructural studies of neural plate cells, cytoplasmic inclusions have been seen which may play a part in causing the cells to contract and change shape, thus bringing about the rolling up of the neural tube. Schroeder (1970, 1973) and Burnside (1971, 1973) have drawn attention to the presence of microfilaments at the dorsal, outer ends of the cells in Amphibia and birds; these microfilaments are orientated horizontally as seen in transverse sections of the neural plate, so that if they contracted, the apices of the cells would become narrower (Plate 6.2). In addition these authors also observed microtubules, orientated vertically, and it could be argued that if these expanded, the cells would elongate vertically and thus achieve their eventual columnar shape. Schroeder concluded that the gradual folding of the plate into a tube was due to a combination of the following events: (i) adhesion to the notochord in the midline (this could also involve the special affinities postulated by Jacobson); (ii) thickening at the edges of the plate due to vertical expansion of the microtubules in these cells; (iii) contraction of the filaments so that the cell apices narrowed; and finally (iv) an increase in the height of the somite cells alongside the rising neural folds, preventing any drag or lateral resistance to the upward movement of the folds. The centripetal movement of the mesoderm, already referred to, would also aid in reducing resistance to the changes in shape of the neural plate.

Several other workers have suggested that microfilaments and microtubules play a part in morphogenetic movements of embryonic tissues (e.g. Spooner, 1973; Hilfer, 1973). Burnside (1971) stressed that the vertical orientation of the microtubules in neural plate tissue contrasted with their random array in epidermal cells. Karfunkel (1971) thought it possible that the tubules later broke down and gave rise to microfilaments. There is no conclusive evidence about the functions of either of these ultrastructural components so far however, as it is difficult to investigate ultrastructures experimentally.

102

Jacobson (1970) found, for instance, that β-mercaptoethanol, an inhibitor of the activity of sulphydryl (-SH) groups, will block neurulation, but Messier (1972) saw no effects of this agent on either microtubules or microfilaments in the neural plate cells. Inhibitors of -SH activity would be expected to alter the tertiary structures of proteins, and perhaps the activities of enzymes, since the -SH groups help to bind the molecules in a particular configuration. It is conceivable that changes in shape of large molecules may, if they occur in sufficiently large numbers, alter the shapes of embryonic cells and hence of tissues. This is a subject on which we have no clear evidence yet, though it has been reported that the cells of the neural plate in chick embryos show birefringence in polarized light, in a direction parallel to the tube (Hobson, 1941). This may mean that some of the molecules at the surfaces of the uppermost cells become reorientated, but whether this is the cause or the effect of the neurulation process is as yet uncertain.

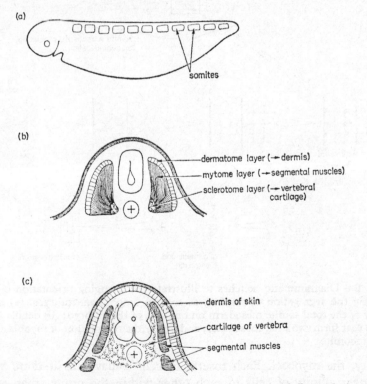

Fig. 6.5 The somites and their derivatives in vertebrates. (a) Lateral view; (b) transverse section of axis; (c) later stage.

6.2   The formation of somites

Another major event in the formation of the embryonic axis in bilaterally symmetrical animals is metameric segmentation. In vertebrate embryos, one of the earliest segmented structures to show are the somites. These paired blocks of dorsal mesoderm cells will give rise to segmental musculature, to the dermis of the skin and to the cartilage of the vertebral arches (cf. Fig. 6.5). They are remarkably constant in number, for any given species. The mechanism by which they form is not understood except in purely descriptive terms, but is a fascinating subject for study. From longitudinal sections through amphibian embryos (Fig. 6.6) it can be seen that the somites arise from a continuous strip of cuboidal cells on each side of the notochord. Successive groups of cells in this strip elongate and form rosette-like clusters, radially orientated around a central

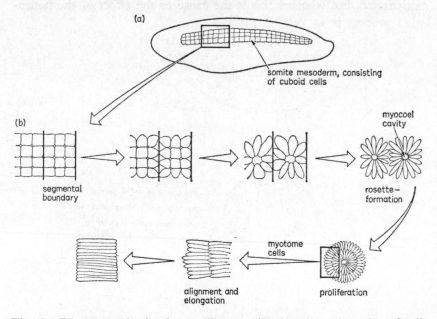

Fig. 6.6 Diagrammatic sketches to illustrate the changing orientation of cells during the segregation of somites, as seen in amphibian embryos. (a) lateral view of the total somite mesoderm on one side of the embryo; (b) details of the cells that form two adjacent somites, and finally the orientation of myoblasts in a single somite.

cavity, the myocoel. Each rosette of cells behaves as if there was a stronger affinity of cells to each other within the group, than to adjacent cells in the strip of mesoderm. So this appears to be an example of differential affinities appearing in cells of the *same* tissue type, just as with the neural plate cells. We do not know how these differential affinities arise, but from earlier work (Holtfreter, 1938) they seem in amphibians to be dependent in the first place upon the presence of the notochord. Holtfreter showed that there was an inductive influence emanating laterally from the notochord to the adjacent somite mesoderm. In explants, segmented somites would not form in this mesoderm unless notochord was also present. Further, it was possible for more lateral parts of the mesoderm such as the prospective kidney and mesenchyme to be converted into somite tissue and to undergo segmentation, in the presence of the notochord. The notochord itself elongates, and its cells become vacuolated, as the somites segment. If there are any adhesions between the elongating notochord cells and adjacent somite tissue, these might result in some of the somite cells being stretched and perhaps breaking their adhesions with each other; this could perhaps lead to segmentation if the stretched regions came at regular intervals along the embryonic axis. But we have no evidence so far as to whether there is any regular 'zoning' of the vacuolating and stretching regions in the axial tissues of amphibian embryos; it is a subject well worth investigating.

In embryos of birds, there has been much interest in the possible control of somite segmentation by a 'somite center' on each side, just lateral to the node where cells invaginate to form the notochord. It is difficult in birds to follow the details of the rearrangements and changes in shape of the somite cells, because these are so small compared with those of amphibians: they are also difficult to culture in isolation *in vitro*. Whole chick embryos can, however, be cultured *in vitro* successfully at early stages, and in such cultures Spratt (1957) removed the node and its environs from contact with posterior regions, by cutting across the blastoderm behind the node on one side. He found that this prevented somites from forming posterior to the operated region. Spratt's work was discredited by Bellairs (1963) however, since she showed that it was possible to obtain somite differentiation in blastoderms from which Spratt's 'somite centers' had been removed, provided the culture medium used allowed regression of the node to occur normally, as in normal gastrulation. Bellairs concluded that it was the mechanics of node regression, rather than any 'inductive' influence from the somite centre, which was essential for normal somite-formation. Nevertheless

(a)

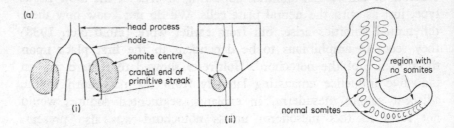

head process
node
somite centre
cranial end of
primitive streak

(i)

(ii)

region with
no somites

normal somites

(iii)

(b)

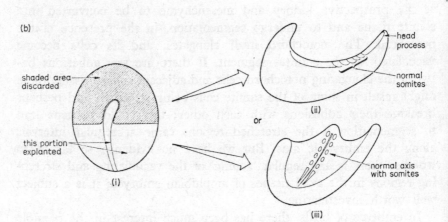

shaded area
discarded

this portion
explanted

(i)

head
process

normal
somites

(ii)

or

normal axis
with somites

(iii)

(c)

this area
excised

no somites form in
posterior regions

(i)

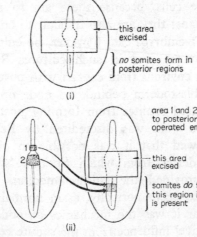

area 1 and 2 grafted
to posterior axis of
operated embryo

this area
excised

somites *do* form in
this region if graft
is present

(ii)

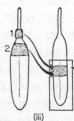

this axial
portion
explanted onto
area opaca

(iii)

Nicolet (1970) concluded that some kind of 'induction' was involved, since he could restore the ability of transected blastoderms to form somites, by transplanting node tissue into the posterior regions later. These transplanted nodes may perhaps have undergone regression, however, which would support Bellairs' view. Fig. 6.7 compares the experiments of these three authors and their interpretations. Whichever interpretation one accepts, it is clear that we have in the chick, as in amphibians, evidence of cellular interactions between axial tissues in both mediolateral and craniocaudal directions. It is still not clear how essential the notochord is in the chick for initial somite segregation, though (see Lipton and Jacobson, 1974).

Three further suggestions have been made recently as to other factors that influence the onset of segmentation and the shape of the somites in chick embryos. Lawrence and Burden (1973) have shown by fluorescent staining methods that the neural tube, and more especially the notochord, become rich in catecholamines during the process of neurulation. They suggest that these substances are involved in initiating movements of the neural folds, and this idea is supported by the finding that inhibitors of catecholamine synthesis inhibit neurulation and result in brain malformations and spina bifida. Lipton (1973) believes that the notochord by its expansion causes the somite mesoderm to become detached from the neural plate, and that this enables it to break up into segments. In experiments in which chick embryos had been transected behind the node, he was able to restore powers of segmentation to posterior regions, by dissecting apart the somite mesoderm and the neural plate. Packard (1973) has shown that the lateral plate mesoderm has an influence on the shape of chick somites: if it is removed, they become expanded laterally and do not acquire their normal rectangular outline.

Fig. 6.7 (*opposite*) Experiments on the control of somite-formation in the chick embryo. (a) Removal of the 'somite center' prevents somite formation caudal to this point (Spratt, 1957). (i) Position of the 'somite centers' lateral to the node; (ii) explant with somite centre removed from one side; (iii) result, 22 h later. (b) Removal of areas corresponding to the somite centre, but leaving node attached to rest of primitive streak so that it can regress normally, does not prevent somite-formation (Bellairs, 1963). (i) Operation; (ii) and (iii) types of result obtained. (c) Explants lacking node and adjacent parts of axis, do not form somites, unless tissue from the node or head process is transplanted into them (Nicolet, 1970). (i) Control explant; (ii) and (iii) explants with transplanted node and head process pieces.

6.3 Theories about the mechanism of control of metameric segmentation

A number of theories have been put forward to account for the formation of segmented series of structures in embryonic development. One of the earlier theoreticians in this field was Turing (1952) who devised a formula applicable to a ring of cells in which substances were being produced at given rates and diffusing from cell to cell. He was able to show mathematically that what had originally been a random distribution of a metabolite could eventually acquire a wave-like distribution of concentrations through the ring of cells, with a series of maxima and minima, and could become stabilized in this state. Several embryologists have been inspired by Turing's arguments to suggest that it would be possible for wave-like maxima and minima of concentrations of morphogenetic substances to exist along the axis of an embryo, causing certain structures to form at regular intervals, which coincided with the maxima or minima of the concentration curves. However, Turing's formulae would first have to be modified to make them applicable to linear arrays of cells rather than to rings. Maynard-Smith (1960) has developed Turing's ideas further and has shown that they could help to explain the arrangements of serially repeated parts, in animals ranging from insects to mammals. Some of the variations in bristle pattern in the fruit-fly *Drosophila,* as well as the control of the numbers of segments formed in both vertebrates and invertebrates could, he argues, be based on the number of 'waves', or maxima in concentration of some substance or signal which caused these structures to form.

It has already been remarked (6.2) that the number of somites is constant for each vertebrate species. This is true even if tissue is removed from, or added to, the somite mesoderm. Waddington and Deuchar (1953) added strips of somite mesoderm to late gastrulae of the newt, *Triturus alpestris,* thus doubling the thickness of the original tissue, and found that the resultant size of the somites was increased proportionately in all three dimensions, while the total number of somites remained unchanged. This three-dimensional adjustment in size presents a problem if one is trying to explain it in terms of a control mechanism in one dimension only. But it may be that the practical observers and histologists, rather than the theoreticians, will find an explanation of how changes in size in one dimension can effect other dimensions as well. For instance, Hamilton (1969) has observed in *Xenopus* that as the somites form there

108

is a rotation of cells from a vertical to a longitudinal orientation. In this case, then, any cells added to the thickness of the tissue would, without any other adjustment mechanism being involved, eventually add to the length of the somites when the cells rotate. What still needs to be explained, however, is how in embryos in which tissue has been added or removed, the divisions between the somites then come to be spaced at either longer or shorter intervals than normally, so that the total number of somites remains unchanged (see Waddington and Deuchar, loc. cit.). This phenomenon does seem to demand some fairly elaborate control mechanism. At present we have no explanations based on observations of cellular interactions, partly because it is very difficult to see the normal behaviour of somite cells during segmentation. Present methods for looking at living cells are successful only with monolayers; cells in thicker layers such as the somite mesoderm cannot be seen clearly and are not suitable for time-lapse cinematography with the equipment available at present. So the tendency has been for people not to try to describe what happens in any more detail, but to go on thinking in theoretical terms of some overall control system which could span the whole length of the embryo. Since the somites form in craniocaudal sequence, it would seem likely that some stimulus or 'signal' passing in a craniocaudal direction within the mesoderm could control their sequential formation.

A few years ago Ann Burgess and I carried out a series of operations on amphibian embryos, designed to test whether or not some influence controlling somite segmentation passed from cranial to caudal regions of the somite mesoderm. We obtained only negative results, however (Deuchar and Burgess, 1967). By various operations (see Fig. 6.8) we isolated posterior, unsegmented regions of the mesoderm from the anterior regions which had already begun segmentation, at late neurula and tailbud stages in *Xenopus* and in *Ambystoma*. After all of these operations, segmentation still proceeded normally in posterior as well as anterior parts of the embryo. So it had to be argued either that the signal had passed from the cranial to caudal regions already, at a stage before our operations were performed, or that segmentation is *not* subject to a control or signal from the cranial end of the embryo, but can start from any point along the axis independently of regions cranial to it. If the latter is true, it does not rule out the possibility that in any isolated section of the embryonic axis, a *new* craniocaudal control mechanism

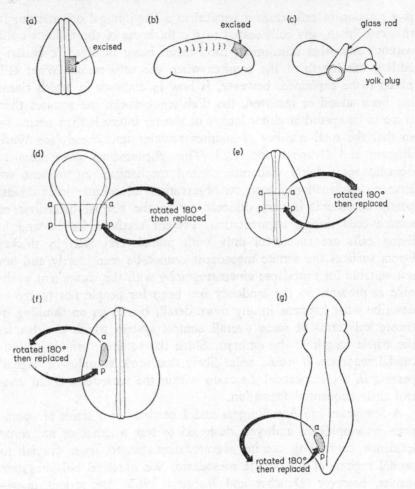

Fig. 6.8 Experiments on somite-formation in amphibian embryos (see Deuchar and Burgess, 1967). (a) Excision of mesoderm and overlying ectoderm, at neurula stage. (b) Similar excision, from unsegmented region, at tailbud stage. (c) Separation of posterior from anterior parts at late gastrula stage, by means of a glass rod. (d) Rotation of mesoderm and overlying neural tissue, at early neurula stage. (e) Similar operation at late neurula stage. (f) Rotation of somite mesoderm only, at late neurula stage. (g) Rotation of unsegmented somite mesoderm only, at tailbud stage. All of these operations resulted in normal segmentation of somites in and posterior to the operated region.

is initiated as soon as the cranial end of the posterior isolate recovers from the operation and (for want of a better concept) 'realizes' that it is now the cranial end of the piece.

6.4 Current theories about the control of axial differentiation

In the last few years Wolpert (1971) and Goodwin (1971) have developed theories attempting to account for the control of axial differentiation. Neither of these theories is as yet supported by enough experimental evidence for it to be acceptable, but they have aroused much interest and are worth serious consideration here. Wolpert expresses his theory at the cell level, and discusses mechanisms by which cells might obtain information about their position in relation to the rest of the cells in an organism: 'positional information' as he calls it. He envisages this potential information as a gradient with, for instance, highest values at the cranial end and lowest values caudally. Each cell has an appropriate course of differentiation according to where it is in this gradient. Obviously there would have to be a complex receptive machinery in the cell – perhaps in its genes – which is able to interpret the effects it receives at any one point in the gradient. Wolpert has not so far offered any suggestions as to how this receptive machinery might work. But he has gone on to suggest, using the results of operations on the coelenterate *Hydra* to support his ideas, that when the original gradient is interrupted by a transection operation, there is an adjustment of levels of the gradient at the boundaries of the cut, which leads to a *raised* value at the anterior end of the posterior piece. The principles of this theory are set out very briefly in Fig. 6.9. It has been used to explain why when a *Hydra* is transected, a new head will form at the anterior boundary of the posterior portion of it, although in an intact animal it would not be possible for a head to form at this level. Similarly one might argue that in our transections of somite tissue in amphibians, a new high value for the controlling gradient

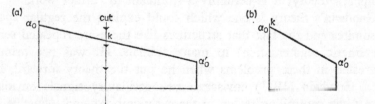

Fig. 6.9 'Positional information' and the effects of removal of part of an axis in a developing organism, according to Wolpert's gradient theory. (a) Simplified representation of the original gradient in the intact axis, and the values assigned to cells at different points along the gradient which provide them with information about their position. (b) If the anterior portion is removed, it is postulated that there is a steepening of the gradient, k now having the same value as α_0.

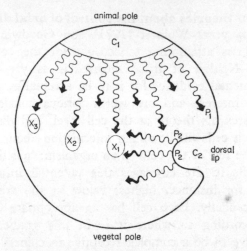

Fig. 6.10 Some of the points of Goodwin's theory (1971) illustrated in a diagram of an amphibian embryo, and the determination of its axes of symmetry. Two successive signals, S and P_1 are considered to emanate as oscillations from the animal pole region C_1. A further signal P_2 emanates from the region of the future dorsal lip of the blastopore, which is the centre of the grey crescent (C_2). At points labelled X_1, X_2 and X_3, respectively, there are increasing time intervals between signals P_1 and P_2. It is postulated that these differing time intervals provide information which leads to differentiation along the craniocaudal axis of the embryo.

was established at the cranial end of the caudal portion of mesoderm, so that this now became able to initiate segmentation, although in an intact animal it would not have been able to do without first receiving a stimulus or signal from the cranial end. The reader will see that there are still many flaws and unproven points in these arguments, and many more experiments need to be devised to test Wolpert's theory; it is certainly a stimulant to further work.

Goodwin's theory is one which could explain the regular spacing of somites and the fact that structures like these are repeated serially (metameric segmentation) in many animals. He was not primarily interested in these problems when he put the theory forward, however. Goodwin (1971) envisaged two oscillatory signals emanating from some controlling region in the early embryo, and following each other with a definite time interval between them. If these signals are imagined to be passing along an axis of symmetry in the embryo, and if one oscillation is at a higher frequency than the other, then the time interval between the two kinds of signal will differ according to what point along the embryonic axis is reached (cf. Fig. 6.10).

112

This difference in time-relations could cause different genes to be activated at different points along the axis, and result in the formation of different organs: it could perhaps be the basis for craniocaudal differentiation. There is also another effect that could arise if the two signals were of sufficiently high frequency to have several points along the axis where they showed the *same* interval of time between them, or even *coincided* (Fig. 6.11). This could result in the same genes being activated, at equidistant points along the axis (assuming that the signals travelled equally rapidly along the whole length of

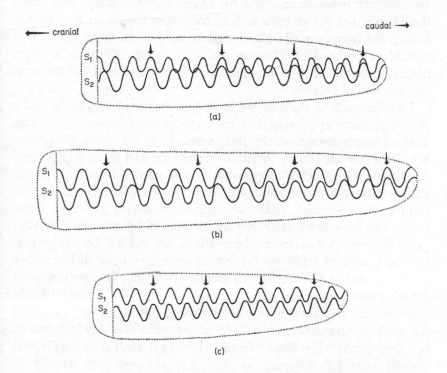

(a)

(b)

(c)

Fig. 6.11 An attempt to relate Goodwin's theory to the formation of a fixed number of somites in a vertebrate embryo and the constancy of this number when tissue is either added or removed. (a) Two oscillatory signals, S_1 and S_2, are shown passing from cranial to caudal along the somite mesoderm. At points where maxima of the two signals, which are of different frequencies, coincide, their combined effect could be the stimulus for formation of a somite. To produce the same number of somites when mesoderm is either added (b) or removed (c), the frequencies of the two signals would have to be either reduced, or increased, respectively. Alternatively, the difference in frequency between S_1 and S_2 could be increased in b and reduced in c, so that their maxima coincided less, or more often within the same distance of tissue.

113

the embryo). Thus a series of *similar* structures might form at regular intervals along the embryonic axis. The simplest analogy is that of two sound waves of different frequencies which produce 'beats' when their maxima coincide, at regular time intervals. Goodwin did not go so far as to suggest this kind of effect in his original argument, but it seems to me that such a repetitive reinforcement in the two signals, occurring at regular intervals along the embryonic axis, could provide a mechanism by which the formation of repeated segmental structures such as somites could be explained. In the cases where tissue is removed or added experimentally, one could then argue that the embryo is able to restore the *status quo* by adjusting the frequency of its two signals, so that the same number of coincident peaks occurs within the new length of tissue as in a normal embryo. (*How* this adjustment is made, would be the next big question of course.)

Theories such as the two we have just discussed may seem to be over-speculative at present, but they are valuable in suggesting new lines of experimentation. In fact research in all fields progresses best by a combination of intuitive speculation and experimental test. To support Goodwin's theory, what is now needed is some evidence that oscillatory events do occur in embryos, then we need a demonstration that artificial oscillatory signals made to simulate those events can have predictable effects on axial differentiation. To support Wolpert's ideas, more observations are needed on the extent to which cells in embryos receive information about their relative positions, during axis-formation. Cooke (1972) has interpreted some experimental results on differentiation in amphibian gastrulae, in terms of this 'positional information' theory, but he has not extended the work to neurulation or tailbud stages when the axis is forming. We have plenty of evidence from behavioural studies on amphibian gastrula cells (cf. Chapter 5) that they respond very readily to changes of position at this stage, when they are still mobile. Much of their information is apparently obtained actively, by throwing out pseudopodia and exploring the surfaces of adjacent cells (cf. Plate 1.4). In the less mobile cells of later stages, it is more difficult to observe whether, and how, they respond to changes of position. But they may still carry out small-scale explorations of adjacent surfaces by means of microvilli, which can be seen at several points on their surfaces when they are viewed with the electron microscope. Information about position may also be exchanged between cells wherever junctions form between their membranes. There is a great need for further observations and experiments in this field.

6.5 Cell-interactions during the further differentiation of somite cells

The cells of vertebrate somites are very versatile in their developmental potentialities. They have at least three main paths of differentiation open to them: to form tissues of the dermis; to form segmental muscles, or to form cartilage (later ossified) in the vertebral column. Normally, if left *in situ,* the outer, dermatome layer of the somites forms dermis, the middle, myotome layer forms muscle and the inner, sclerotome layer forms vertebral cartilage. When isolated in culture, however, the paths of differentiation taken by somite cells is seen to be dependent on what other tissues are present with which they can interact. One tissue which can influence their differentiation is the spinal cord, and this provides a good example of the reciprocity of interactions between embryonic cells. For the spinal cord, which depended for its formation on an induction from somite as well as notochord cells, is in its turn responsible for causing the somite cells to form cartilage. Sclerotome cells isolated, will only form cartilage if some spinal cord, or alternatively notochord tissue is put with them (Lash, Holtzer and Holtzer, 1957). Myotome cells too, fail to continue differentiation in culture unless some spinal cord is put with them (Muchmore, 1968).

Lash, Holtzer and their collaborators have worked extensively on the differentiation of somite cells of chick embryos *in vitro* (see review by Lash, 1968). They showed by labelling experiments that although precursors of cartilage, such as glucosamine, may be taken up and metabolized by isolated sclerotome cells, the last steps in synthesis of chondroitin sulphate (Fig. 6.12) do not take place in these cells unless either spinal cord or notochord tissue is also present in the culture. Another component of cartilage is collagen, and the source of this was previously assumed to be the mesoderm cells, since fibroblasts in tissue culture readily form collagen. But recent ultrastructural and biochemical studies by Cohen and Hay (1971) show that at the time when the spinal cord and the somite cells interact in the chick embryo, the neuroepithelial cells of the cord have deposits of collagen on their basement laminae. Whether this is the source of the collagen later found in sclerotome cells, or whether it somehow induces these cells to start their own synthesis of collagen, is not yet clear.

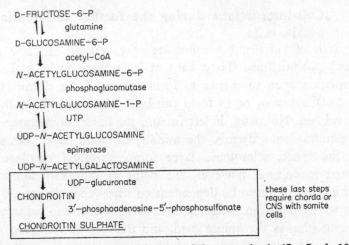

Fig. 6.12 Chart of the steps in chondroitin sulphate synthesis. (See Lash, 1968.)

The precise stage in development at which interaction occurs between spinal cord and somites, to determine that the sclerotome cells shall form cartilage, has been investigated in detail in chick embryos by O'Hare (1972 a-c). His procedure was to take out groups of somites from different levels of the embryonic axis, at a series of developmental stages, and to test by either explanting them on the chorioallantoic membrane, or implanting them into another embryo, whether they were able to form cartilage. He found that stage 11 of Hamburger and Hamilton (1951), i.e. at 40-45 h incubation, when 13 pairs of somites are present, was the earliest stage at which any somites were able to form cartilage when removed from the vicinity of the spinal cord. At this stage, only the anterior somites were able to do so. The more posterior somites could not form cartilage independently until stage 16 (51-56 h incubation: 27 pairs of somites present). O'Hare also tested whether tissues other than the spinal cord could 'induce' cartilage-formation. He found that dorsal ectoderm from the 3-4 day embryo was able to do so. He also found that other cells besides sclerotome cells could respond to induction by the spinal cord: lateral mesoderm cells were induced to form cartilage, if explanted together with spinal cord tissue. Unlike the situation in neural induction (see Chapter 5), dead spinal cord was not effective as a cartilage inductor. This rather suggests that the collagen production by the living neuroepithelium may indeed play some active role in stimulating cartilage-formation by the mesoderm cells.

Even in the absence of inducing tissues, the versatile somite cells may follow one or other path of differentiation in culture, according to conditions in the medium. Ellison, Ambrose and Easty (1969) have shown that even mature sclerotome cells which are competent to form cartilage *in vitro,* will not do so unless there are optimal culture conditions. An adequately large colony of cells must be established before they will differentiate, but overcrowding must be avoided. The same is true of the differentiation of cartilage-forming cells isolated from the sternum region of the chick embryo: Lavietes (1970) showed that optimal cell density was essential for successful chondrogenesis by these cells in tissue culture. So not only cell and tissue interactions, but also interactions between cells and the culture medium have an influence on differentiation. This means that we must accept the earlier work on tissue inductors of cartilage with caution; perhaps the spinal cord and notochord are simply good 'conditioners' of the culture medium, and the substances they liberate into it may not be specific to these tissues only.

In further work on cells of the sternum, Eguchi and Okada (1971) showed that colonies of cartilage-forming cells can be built up from *one* original chondrocyte in isolation. Similarly Watanabe (1970, 1971) showed that colonies can be obtained from single somite cells of chick embryos. The colonies tend to regress, if the cell is derived from an embryo of earlier than stage 35 ($8\frac{1}{2}$-9 days incubation), but after this stage, viable colonies can be established from single cells. Quite elaborate arrays of collagen-forming cells can be obtained when single fibroblasts derived from the somites or from the mesenchyme are isolated at these late stages when organogenesis has begun. Elsdale and Bard (1972) have also obtained such arrays from single cells of human fetal lung. This 'cloning' work, forming colonies derived from single cells of mammal or bird embryos, is a useful method for detecting genetic differences between individual cells in early tissues of the embryonic axis. It is another promising field for future research.

The examples given in this chapter represent just a few of the types of cellular interaction that may take place during the formation of the embryonic axis in vertebrates. Their mechanisms, though still not fully understood, include differential cell affinities within different areas of each single tissue, and may involve ultrastructural organelles such as microfilaments, microtubules and microvilli. Experimental work indicates that there must also be some overall control mechanism acting along the craniocaudal axis. Perhaps it is a series of oscillatory signals, such as Goodwin envisaged. Perhaps there are other clues too by which cells recognize their positions along the embryonic axis and can thus set up new gradients of metabolite concentrations, or initiate new signals, when their normal arrangement is upset by the experimenter. Besides craniocaudal controls, there are also evidently medio-lateral influences, both from the notochord and from the spinal cord, which influence the segmentation and the further differentiation of somite cells. Whether these influences should be termed 'inductions', or whether they are less specific, like the variations in culture conditions which also influence cell differentiation, still has to be made clear by further experimentation.

7

Cellular interactions in organogenesis: I. Organs formed from endoderm and mesoderm

In the growth and development of organs, cell movements, adhesions and de-adhesions still play some part, although they are not so extensive as at earlier embryonic stages such as gastrulation. Changes in cell shape also occur, though these may not be as rapid or obvious as in neurulation. Some cells continue to divide, too, but others cease division and begin to lay down special proteins and other structural materials, either within their cytoplasm or as an extracellular 'matrix' between them and other cells. All of these cellular activities must of course be co-ordinated within each organ, to bring about its normal development as a whole. To achieve this co-ordination, a chain of interdependent interactions occurs between the layers of cells that compose each organ.

Many of the organs in animal embryos are composed of two main layers of cells, which originate from two of the three basic germ layers of the embryo; i.e. either from endoderm and mesoderm, or from mesoderm and ectoderm. In the development of the organs, these pairs of embryonic cell layers interact, often in a series of reciprocal steps which can be analysed experimentally. We shall consider in this chapter some of the organs formed from endoderm and mesoderm, then in Chapter 8 shall deal with organs composed of mesoderm and ectoderm.

Much of the work on which our knowledge of cellular interactions in organogenesis is based has been carried out on cells of vertebrate embryos grown *in vitro*. Recent work on cultures of invertebrate organs has also begun to yield interesting results, which will be referred to briefly here: for a fuller review, see Lutz (1970). It has

119

often been found convenient for studies of organogenesis *in vitro* and the behaviour of the cells involved, to dissociate these first in a medium lacking calcium and magnesium, and then to follow the details of their re-adhesion and reorganization, while at the same time adding cells of another tissue layer to see how these interact with them. It must be emphasized at the outset that the events one sees under these experimental conditions cannot be assumed to be identical to what occurs *in vivo*, without a great deal of further study. One important point is that the disaggregation of the cells may destroy their intercellular matrix, which may have important morphogenetic properties (see 7.3.2). At the same time, it can be admitted that some interesting and important facts have come to light from studies *in vitro*, which could never have been discovered *in vivo*. For instance, those who work with dissociated cells have found that it is possible to produce chimaeric organs that are a mixture of chick and mouse cells, because evidently the adhesive affinity between cells of like tissue type is greater than that between cells of the same species. If, for example, dissociated heart and retina cells of mouse and chick embryos are mixed, all the heart cells tend to group together and form chimaeric mouse/chick heart muscle, while all the retina cells group to form a chimaeric mouse/chick retina. Several other instances of this type of phenomenon have been observed (see Moscona, 1962 for review). Roth (1968) made a quantitative estimate of the affinities between cells of like tissue type (isotypic) in mixtures of cells from mouse and chick embryos, by labelling some of the samples with radioactive tracers and then estimating the radioactivity of cell clumps resulting from mixtures of labelled and unlabelled cells. He found that the affinity between isotypic cells of mouse and chick was just as high as that between isotypic cells of the same species.

Of course, in normal circumstances such mixtures of tissue cells from different species do not occur in embryonic development; this is a highly artificial situation created by the experimenters. But the general finding that there are strong adhesive affinities between cells of like tissue and organs is an important one, which we shall bear in mind as we go on to consider examples of cellular interactions in endodermal and mesodermal organs. Adhesion is the first, important step towards many types of short-range interaction, by passage of molecules or signals as we saw in Chapter 1, enabling adjacent cells or cell layers to influence each other's differentiation.

7.1 Organs formed primarily of endoderm

7.1.1 *The gut tube in the early embryo*

The primitive gut tube, or 'archenteron' is one of the earliest organs to form, either during or immediately after gastrulation in vertebrates. It is evident in lower vertebrates before the neural tube and axial mesoderm have differentiated, since the gut forms as the endoderm invaginates during gastrulation. But in reptiles, birds and mammals where the endoderm is only a single layer under the embryonic disc at the end of gastrulation (cf. Fig. 7.1a), further foldings of tissue have to occur to form the primitive gut tube. These provide an example of the role of cell-movements and adhesions in organogenesis.

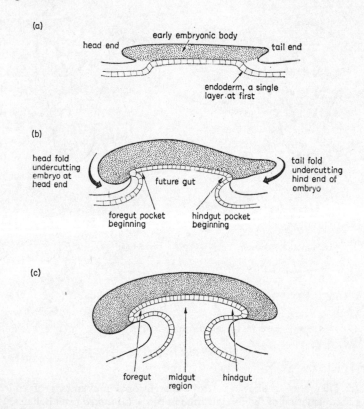

Fig. 7.1 Diagrams to show the formation of the foregut and hindgut tubes in a reptile, bird or mammal embryo. Longitudinal sections. (a) Early embryonic disc stage (neurula). (b) Formation of head and tail folds. (c) Deeper undercutting of folds.

121

The foregut and hindgut in birds and mammals form as blind-ended tubes, when the head and tail folds undercut the embryonic disc (cf. Fig. 7.1b and c). The movements of the cells in the foregut region of the chick embryo have been studied by Bellairs (1953) and Stalsberg and DeHaan (1968) using either carbon or ferric oxide particles as markers. The latter authors suggested that regression of the axial mesoderm helps to pull the lower folds of the endoderm caudalwards, while Bellairs laid more emphasis on the co-ordinated movements of the endoderm cells during the folding process. These movements take place in three dimensions: craniocaudally, latero-medially and vertically (Fig. 7.2). The co-ordination so that the whole sheet undergoes symmetrical folding would not be possible unless the cells, besides making appropriate individual movements

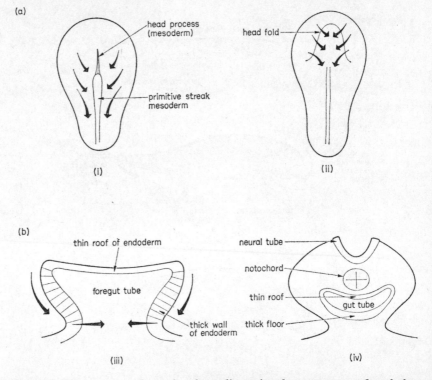

Fig. 7.2 Diagrams to show the three-dimensional movements of endoderm during the formation of the foregut tube in birds. (*Adapted from* Bellairs, 1953). (a) Ventral surface views of the blastoderm, (i) at the head-process stage, (ii) at a later stage when the head fold has begun to form. (b) Transverse sectional views: (iii) at same stage as (ii), and (iv) when the foregut tube is complete (cf. Fig. 7.1c). Arrows show directions of movements in the endoderm.

and changes of shape, also adhered firmly to one another by some tissue-specific affinity.

Later in its development, the primitive gut tube of vertebrates becomes asymmetrically placed, with the region of the future stomach deflected to the left of the midline and the first coil of the future intestine lying to the right. Work carried out on amphibians by Von Woellewarth (1969) shows that this asymmetry is dependent on the mesoderm. If the dorsal mesoderm overlying the gut roof is damaged, or if pieces of it are excised, a reversal of the asymmetry occurs in many cases. It is not yet understood what kind of interactions between mesoderm and endoderm are responsible for this control of asymmetry. Purely mechanical factors could be involved, since mesoderm conrtibutes to the muscle and connective tissue throughout the gut tube and also to the mesentery which supports it. In mammals too, it is known from organ culture work that the gut endoderm depends for its normal development on interactions with mesoderm at an early stage. David (1972) showed that the whole endoderm of the rabbit embryo cannot differentiate in isolation, but that if mesoderm is placed with it, development proceeds normally in culture. The mesoderm, on the other hand, has a considerable degree of independence in its development, for if it is cultured alone, it is able to form gut muscle in the absence of endoderm.

In the development of derivatives of the gut tube, there are many other examples of the endoderm depending for its developmnt on an interaction with mesoderm cells, whereas the mesoderm can develop quite well alone. In several organs that are made up of endoderm and mesoderm, the mesoderm appears to play a dominant role in controlling what type of differentiation shall take place. It has been shown in organ culture work on the chick embryo by Wolff and his associates (see Wolff, 1968 for review) that the mesoderm of a given organ is often capable of imposing its own type of differentiation on endoderm from another region. For instance, mesoderm from the gizzard of the chick embryo combined with endoderm from the stomach, will cause that endoderm to form the type of mucosal layer characteristic of gizzard, rich in glycogen and mucus. Conversely, stomach mesoderm combined with gizzard endoderm causes this endoderm to form a folded epithelium with gastric glands (Table 7.1).

Table 7.1 Effects of different types of mesoderm on stomach endoderm of the chick embryo (see Wolff, 1968)

Proventricular mesoderm + proventricular endoderm→normal gastric glands·
Proventricular mesoderm + gizzard endoderm→normal gastric glands.
Gizzard mesoderm + gizzard endoderm→glycogen-rich, gizzard-like mucosa.
Gizzard mesoderm + proventricular endoderm→glycogen-rich, gizzard-like
mucosa.

A more complicated series of mesoderm/endoderm interactions takes place in the development of the liver, a specialized diverticulum of the gut tube, which we will now consider separately in the following section.

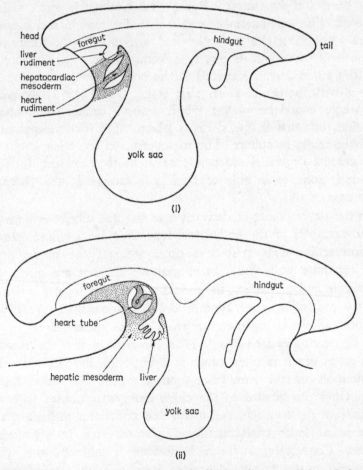

Fig. 7.3 Diagrams to show the origin of the liver rudiment and its relations to the cardiac mesoderm. Longitudinal sections at successive stages, (i) and (ii).

7.1.2 *The liver*

The sequence of steps by which the differentiation of the endodermal layer of the liver is controlled by interactions with mesoderm has been worked out in birds by Wolff, Le Douarin and their associates (see Croisille and Le Douarin, 1965; Wolff, 1968). The liver rudiment originates at the anterior edge of the yolk sac at the embryonic disc stage (Fig. 7.3i) and at first lies cranial to the cardiac mesoderm. Later, however, it passes back through the cardiac mesoderm as the head fold forms (Fig. 7.3ii). Wolff *et al.* showed that this contact with the cardiac mesoderm is essential for the first step in differentiation: the formation of hepatic cords. Liver endoderm cells isolated *in vitro* will not form hepatic cords, unless heart cells are also present. Also if an obstacle is placed in the path of the liver rudiment in the embryo so that it cannot pass caudalwards under the cardiac cells, it will not form hepatic cords. To complete their differentiation, these endodermal hepatic cords next require contact with the mesoderm layer of the liver; mesoderm from other sources such as the head region, the somites, or the general mesenchyme do not promote differentiation of the hepatic cords. Surprisingly, though, metanephric kidney mesenchyme does promote some further differentiation of the hepatic cords (but not the deposition of glycogen in these liver cells: that functional stage requires interaction with liver mesoderm). We shall meet further examples later, of the ability of metanephric mesenchyme to promote differentiation of cells belonging to other organs. This particular kidney mesenchyme is an exception to the usual situation that mesoderm specifies differentiation according to its own origin, for it does *not* always convert other cells into kidney tissue. One may note that there is no endoderm layer in the kidney, so kidney mesenchyme does not control the differentiation of any endoderm during its own organogenesis. Perhaps this is the reason why it has not acquired the same specific control over endodermal differentiation as some other mesenchymes have.

The steps in differentiation of the liver endoderm and the interactions that control these, according to Wolff's findings, are set out in Table 7.2. We do not know yet what precise mechanisms of cellular interaction are involved, but at least we can say that cell-contacts, and movements of one cell layer over the other, are essential, so that probably some chemical interchanges take place between the endoderm and mesoderm layers.

125

Table 7.2 Steps in the control of differentiation of liver endoderm (See Wolff, 1968).

Hepatocardiac mesoderm

Pre-hepatic endoderm

1st induction

Hepatic mesoderm   Cardiac mesoderm

Hepatic endoderm

2nd induction

Other mesenchyme (metanephric)   Hepatic cords

3rd induction

Endothelium of liver sinusoids

Differentiation but no glycogen.   Further differentiation. Glycogen in cord cells.

7.1.3 *The lung*

Like the endoderm of the stomach and gizzard, the lung endoderm of the chick embryo can be influenced in its mode of differentiation by mesoderm from another region. Spooner and Wessels (1970) have shown that lung endoderm is not capable of differentiating in isolation. In the presence of either chick or mouse embryonic mesenchyme, however, it will differentiate according to the region from which the mesenchyme came. With stomach mesoderm (from either mouse or chick) the lung endoderm will form gastric glands; with intestinal mesoderm, it forms epithelium with villi; with liver mesoderm, it forms hepatic cords. Metanephric mesenchyme allows lung buds to form, but these do not branch further. Only *lung* mesenchyme will cause the lung endoderm to differentiate normally, into lung epithelium. Alescio and Dani (1971, 1972) have shown that the further growth of the lung bud and the deposition of glycogen in its cells also depend upon the presence of lung mesoderm. Both of these last processes are inhibited by other mesenchymes, including metanephric mesenchyme.

Alescio and Dani (loc.cit.) and Wessels (1970) went on to show that at later stages of development of the lung, mesenchyme from bronchial and tracheal regions have different effects on the endoderm layer. Bronchial mesenchyme induces bronchial buds to form, even from tracheal endoderm, and it does not lay down collagen. Tracheal mesenchyme, on the other hand, does lay down collagen and it inhibits morphogenesis of bronchi by the endoderm.

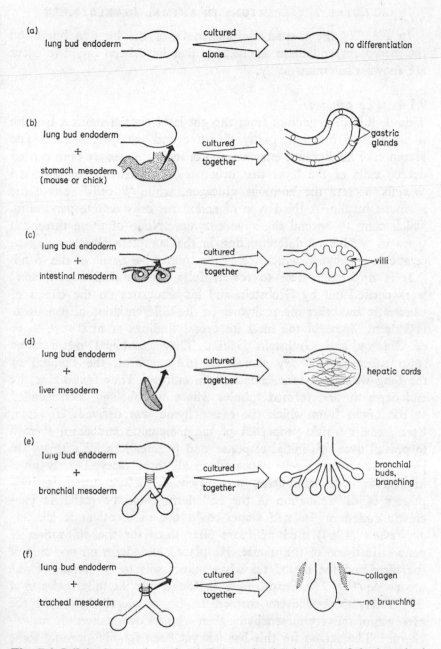

Fig. 7.4 Cellular interactions that influence the development of the lung bud in birds. (a) Lung endoderm is unable to differentiate in isolation, and its development may be modified if it is combined with other types of mesoderm, as in (b), (c) and (d). In (e) and (f), the different effects of bronchial and tracheal mesoderm are shown.

The labels within the figure read:

(a) lung bud endoderm — cultured alone — no differentiation

(b) lung bud endoderm + stomach mesoderm (mouse or chick) — cultured together — gastric glands

(c) lung bud endoderm + intestinal mesoderm — cultured together — villi

(d) lung bud endoderm + liver mesoderm — cultured together — hepatic cords

(e) lung bud endoderm + bronchial mesoderm — cultured together — bronchial buds, branching

(f) lung bud endoderm + tracheal mesoderm — cultured together — collagen, no branching

In Fig. 7.4 the normal morphogenesis of the lung in birds and the mesoderm/endoderm interactions that have been found to occur are shown diagrammatically.

7.1.4 *The pancreas*

This double diverticulum from the gut is of special interest because it has a dual function, partly digestive and partly endocrine. The secretion of its digestive enzymes and of its hormones are both carried out by cells of the layer that originates from endoderm. So-called 'α cells' secrete the hormone glucagon, while 'β cells' secrete the hormone insulin. A third type of endoderm cells secrete pancreatin, which contains several digestive enzymes. None of these three cell types is capable of differentiating in the complete absence of pancreatic mesenchyme, which forms the connective tissue of the gland.

It is impossible here to review fully the detailed experimental work carried out by Grobstein and his associates on the effects of pancreatic and other mesenchymes on the differentiation of pancreatic endoderm. Some of the most important findings at first were those of Golosov and Grobstein (1962). They combined mesenchyme from either the salivary glands, the metanephros, the stomach or the lung with pancreatic endoderm in culture. They found that the endoderm at first formed tubules whose morphology corresponded to the organ from which the mesenchyme was derived. However, later some intrinsic properties of the pancreatic endoderm seemed to prevail over this initial response, and pancreatic acini with typical pancreatic secretory cells appeared, in all these cases. Yet an initial interaction with mesoderm was essential, before these intrinsic powers of differentiation in the endoderm could be realized: pancreatic endoderm isolated alone, could not differentiate at all.

Wessels (1964) used millipore filter discs for the cultivation of pancreatic tissue of the mouse. He placed endoderm on one side of the filter and the mesoderm whose effect was to be tested, on the other side (Fig. 7.5). Surprisingly, he found that the differentiation of the pancreatic endoderm seemed to be promoted better in the presence of salivary mesenchyme than with its own pancreatic mesenchyme! The reason for this has not yet been found: possibly some quite non-specific conditions in the cultures in the two cases caused this unexpected result. What Wessels' work has shown, however, is that whatever mesodermal influence promotes the differentiation of the pancreatic endoderm is capable of passing through a millipore filter. This implies that it is some relatively small, diffusible molecule. No subcellular particles have been seen crossing the filters, when these are examined by light and electron microscopy.

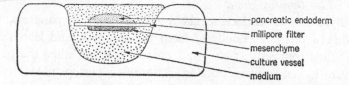

pancreatic endoderm
millipore filter
mesenchyme
culture vessel
medium

Fig. 7.5 Method of explantation to test the effects of diffusible substances from one tissue on the differentiation of another (cf. Wessels, 1964). Pancreatic endoderm, for example, may be placed on one side of a millipore filter disc, and the mesoderm whose effects are to be tested is placed on the other side. The combinate is grown in a suitable chamber of culture medium, under sterile conditions.

Some recent work by Dieterlen-Lièvre (1970) showed that metanephrogenic mesenchyme placed in combination with pancreatic endoderm can also promote the differentiation of secretory cells. Both α and β cells form normally, but the size of the pancreatic acini and ducts is reduced.

The changes in overall shape that take place in the pancreatic rudiment of the chick as it develops, either *in vivo* or *in vitro*, are normally brought about by two forms of cellular interaction within the endoderm. These are, changes in shape of the endoderm cells, and movements of the cells round the base of the original pancreatic diverticulum. These events have been described in some detail by Wessels and Evans (1968) who also studied the ultrastructural changes in the endoderm cells. Some of these morphogenetic events take place in the absence of mesoderm, which confirms the impression already gained from Golosov and Grobstein's findings described above, that the pancreatic endoderm is to some extent independent of interactions with mesoderm during its differentiation. In his earlier studies of the interactions between salivary mesenchyme and pancreatic endoderm on either side of a millipore filter, Wessels (1964) observed that the cells forming the acini migrated inwards from the periphery of the endoderm, having evidently first received some induction stimulus from the mesoderm, and were then able to differentiate in the centre of the endoderm tissue mass. So it is possible that a quite brief exposure of peripheral cells to the influence from mesoderm is sufficient to allow further differentiation of the pancreatic acini to continue independently.

Other experiments by Wessels and his colleagues on the factors influencing development of the pancreas in the mouse have been reviewed by Kratchowil (1972).

7.1.5 *The thyroid gland*

The thyroid is an endocrine gland which, like the pancreas, begins as a diverticulum of endoderm from the gut tube and later acquires mesodermal connective tissue and blood vessels. Unlike the pancreas however, its endocrine function does not develop unless it is in continuous association with mesoderm. Unlike the pancreas, again, no mesoderm other than that of the thyroid region can promote full and functional differentiation of the endodermal cells. Hilfer, Hilfer and Iszard (1967) tried the effects of combining other mesenchymes with chick thyroid endoderm *in vitro*. If they used mesentery, the thyroid endoderm formed only epithelial islands. Heart mesoderm caused highly vascularized follicles to develop, and perichondrial mesoderm caused follicles to form in a matrix of cartilage. In these last two cases, the follicles formed by the thyroid endoderm appeared normal, but did not become functional in secretion; the rest of the gland had a morphology which reflected the origin of the mesoderm.

7.1.6 *General features of the cellular interactions in endodermal organs.*

Two striking features emerge from the work on cultured rudiments of organs derived primarily from endoderm of embryos of vertebrates. One is that the cells secreting either enzymes or hormones are always derived from the endodermal layer, not the mesoderm associated with it. The other is that two rather remote mesodermal tissues, salivary mesenchyme and metanephric mesenchyme, are sometimes able to promote differentiation in unrelated endoderm, either when placed in direct contact with it or when it is on the opposite side of a millipore filter.

Pursuing the first of these features further, Adelson (1971) has suggested a broad generalization about the secretory roles of endoderm cells. He has pointed out that in both vertebrates and invertebrates, endoderm has evolved the ability to secrete proteins which are functional in some cases as digestive enzymes and in other cases as hormones. Some of Adelson's examples of these biochemical similarities between enzyme and hormones are given in Table 7.3. One can envisage that an evolutionary step from an enzyme to a hormone could have wide-ranging effects on embryonic development and

Table 7.3

Amino-acid sequences compared, in some enzynes and hormones.

Caerulein: Glx-Glu-Asp-Tyr-Thr-Gly-Trp-Met-Asp-Phe

Cholecystokinin: Asp-Tyr-Met-Gly-Trp-Met-Asp-Phe

Gastrin: Glx-Gly-Pro-Trp-Leu-Glu-Glu-Glu-Glu-Ala-Tyr-Gly-Trp-Met-Asp-Phe

Growth Hormone: Ala-Phe-Asp-Thr-Tyr-Glu-Phe-Glu-Glu-Ala-Tyr-Ile-Pro-Lys-Glu-Glu

Insulin B: Ser-His-Leu-Val-Asp-Ala-Leu-Tyr-Leu-Val-Cys-Gly-Asp-Arg-Gly-Phe-Phe-Tyr-Asn-Pro-Lys

Growth Hormone Ser-His-Asn-Asp-Asp-Ala-Leu-Leu-Lys-Asp-Tyr-Gly-Leu-Leu-Tyr-Cys-Phe-Arg-Lys-Asp-Met growth, for it would provide a new long-range interaction between certain endoderm cells and one or more target organs in the embryo (see Chapter 10). However, it should not be assumed that the endoderm has the monopoly of this type of protein secretion: Pearse (1969, 1971) has pointed out that many important polypeptide hormones in mammals and birds are secreted by cells which are derivatives of the neural crest, an 'ectomesenchymal' tissue (see Chapter 8).

Referring now to the second of the general features mentioned above: we shall see further examples later of the ability of salivary and metanephric mesenchyme to promote differentiation, not only of endoderm but also of other mesoderm and of ectodermal organ rudiments. These two mesenchymes are so different both in derivation and morphology that it seems likely that some quite unspecific factor (such as their ability to survive well *in vitro* and to make extensive contact with other tissues) is the basis of their success in promoting differentiation. Neither of these tissues contains any constituent which is unique to them or which could be expected to have any special power to promote differentiation, as far as we know. Nor is their action species-specific, for Le Douarin and Houssaint (1968) showed that metanephrogenic mesenchyme from the mouse as well as the chick would stimulate the proliferation of chick liver cells. It needs stressing again that organ and tissue culture work may sometimes produce results which cannot be related easily to known events of normal development. Some of the results may turn out to be 'red herrings', diverting attention from more important matters.

When we go on now to consider certain organs which are derived entirely from mesoderm, we shall find that essentially the same types of cellular interaction are involved as those we have seen above: namely, changes in cell shape and in their degree of adhesion to one another, bringing about changes in shape of the whole organ; cell movements, sometimes resulting in migration of the whole organ to a different position, and biochemical exchanges between cells of different parts of the organ.

7.2 Organs and tissues derived from mesoderm

The emphasis so far has been on the role of mesoderm in controlling the differentiation of other layers of tissue. But in some tissues and organs that are derived from mesoderm alone, such as the vascular system, it is often the case that their early differentiation depends on some control from the endoderm. We do not know what kind of influence passes from the endoderm to the mesoderm rudiments, but since endoderm cells are often nutritive in function, their main role could be to supply essential metabolites, perhaps.

7.2.1 Blood vessels

Work on early chick embryos by Miura and Wilt (1969) shows that the blood-forming layers of the mesoderm will not develop successfully if the endoderm is removed from beneath them. Heat-treated endoderm is not effective in restoring the ability of the mesoderm to develop, so evidently some active function of living endoderm is essential. This could, as we have already suggested, be a nutritive function. Deuchar and Herrman (1962) showed that the yolk sac endoderm of the chick embryo is the main route of uptake of amino-acids into the embryo from a nutrient medium. Miura and Wilt (loc.cit.) went on to show that the influence from the endoderm was not entirely blocked by inserting a millipore filter between it and the mesoderm, though rather fewer blood islands formed than when the filter was absent. The filter would not have prevented passage of small metabolites such as amino-acids from endoderm to mesoderm.

Many complex interactions are involved during the formation of blood cells in various embryonic organs in vertebrates. This process of haemopoiesis occurs successively in yolk sac mesoderm, liver, spleen and bone marrow. In organ cultures it has been shown that haemopoiesis in liver, spleen and bone marrow may be enhanced by adding further embryonic liver. Salvatorelli (1969) showed an enhancement of haemopoiesis in organs of the duck and the guinea pig, when chick embryonic liver cells were added to the cultures. It has been suggested that the liver produces some agent with similar effects to erythropoietin, a hormone from the metanephric kidney which stimulates the formation of red blood cells both prenatally and post-natally in mammals (Cole and Paul, 1966). So far, there is no evidence that the development of the blood vessels, as distinct from the blood cells, is controlled by any hormonal influence, however. It seems simply to require metabolites from the endoderm at early stages, as we saw above.

After the blood cells have formed in haemopoietic organs, they then have to enter the blood vessels. Those that develop initially within blood islands in the yolk sac, or near the open ends of capillary vessels in the liver, have easy access to the vascular system. But at later stages, haemopoiesis occurs extravascularly also. Jones (1970) has described blood cells entering vessels in mouse embryos by a process of diapedesis similar to that shown by germ cells in birds (cf. Chapter 2). These early blood stem cells have pseudopodia by which they penetrate in between the endothelial cells of the blood vessel walls.

7.2.2 *The heart*

The heart, a highly specialized part of the vascular system in vertebrates, is also highly independent in its development and does not appear to require interaction with other tissue layers in the embryo. However, in conditions of culture the development of isolated heart rudiments may be enhanced if endoderm is also present. Both Jacobson and Duncan (1968) and Fullilove (1970) have shown that the differentiation of heart rudiments from newt embyros *in vitro* is more rapid and more complete if anterior dorsal endoderm is present too. It is possible that here, as in the previous cases we have discussed, the endoderm plays a mainly nutritive role. Endoderm cells of amphibian embryos still contain yolk during early larval stages, and this could be a source of nutrition for other cells in contact with the endoderm.

The morphogenesis of the heart in vertebrates includes a number of changes in shape of the originally tubular structure. It has been shown in chick embryos that both intra-tissue and intra-organ interactions of cells are involved, in three dimensions. Before the heart tube forms, cells from a pair of rudiments on either side of the primitive streak have to move to the midline and fuse with each other. This process has been followed by DeHaan (1965) using cinematographic methods. He has been able to describe the course of the movements and the future fates of different groups of marked cells, but so far the mechanism by which the originally paired rudiments fuse in the midline, and then give rise to an asymmetrically curved heart tube, has not been explained. It seems probable that the fusion is due to tissue-specific adhesive affinities between the heart cells. The asymmetrical single tube, which in birds projects to the right of the midline, results in a heart that eventually lies slightly to the right of the midline and has a larger right ventricle with the aorta leading from it. (This is in contrast to mammals, which have a larger left ventricle and a left aorta.) Stalsberg (1969), and Stalsberg and De Haan (1969) found that a larger number of cells from the right-hand rudiment contributed to the heart than from the rudiment on the left side, in the chick embryo, so this could be the basis for the original asymmetry of the heart tube. It is still not known, however, why there should be different numbers of cells in the heart rudiments of each side. Some have suggested that differences in mitotic rate on the two sides of the heart tube may account for its later asymmetrical curvature, but attempts to find evidence of such differences have so far failed. Another factor controlling the shape of the heart is that the initial asymmetry of the tube leads to a spiral twisting of the

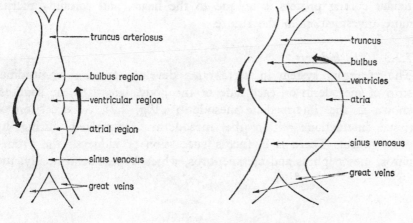

(i) before coiling

(ii) beginning of twist

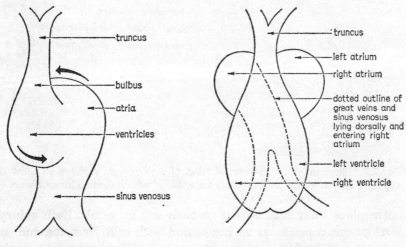

(iii) further twisting

(iv) final positions of heart chambers

Fig. 7.6 Diagrams, ventral view, of the mammalian heart tube at successive stages of its coiling process. (i) Before coiling begins, (ii) beginning and (iii) and (iv) later stages of coiling.

muscle fibres that develop from the myoblast cells in its outer layer. These twisted muscles begin to contract at a very early stage, and help to promote further twisting of the heart as it grows.

As is evident from the rather speculative arguments above, we cannot yet explain the gross morphogenesis of the heart tube of vertebrates in terms of any precisely defined cellular interactions. The changes in shape that occur are shown in Fig. 7.6. This particular coiling process is unique to the heart, and certainly merits more investigation in the future.

7.2.3   *The kidneys*

The excretory system in vertebrates develops from a longitudinal strip of mesoderm on each side of the body, lateral to the somites, known as the 'intermediate mesoderm' (Fig. 7.7). A succession of tissue interactions within the mesoderm is involved during its development. There is in fact a succession of kidneys: the pronephros, mesonephros and metanephros, which appear sequentially, the

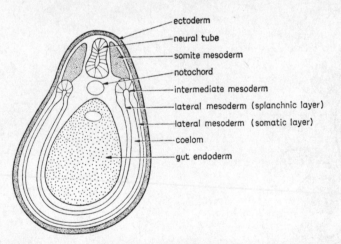

ectoderm
neural tube
somite mesoderm
notochord
intermediate mesoderm
lateral mesoderm (splanchnic layer)
lateral mesoderm (somatic layer)
coelom
gut endoderm

Fig. 7.7 Diagrammatic transverse section of a vertebrate embryo to show the original position of the 'intermediate mesoderm' which forms kidney tissue.

metanephros being present only in birds and mammals. Each kidney's development depends on an interaction with cells from the one that preceded it (Fig. 7.8). It was first shown by Waddington (1938) that in the chick embryo the pronephric duct which grows caudalwards from the pronephric kidney interacts with the intermediate mesoderm caudal to the pronephros and causes this mesoderm to develop into the mesonephros. If the pronephric duct is removed, no mesonephros develops. In its turn, that part of the pronephric duct which has now become the mesonephric duct initiates the development of the metanephros. A bud, the ureteric bud, grows out from near the caudal end of the mesonephric duct to make contact with intermediate mesoderm caudal to the mesonephric kidney. This mesoderm, known as the metanephrogenic mesenchyme, is then induced to start forming the metanephros. This kidney in turn influences the

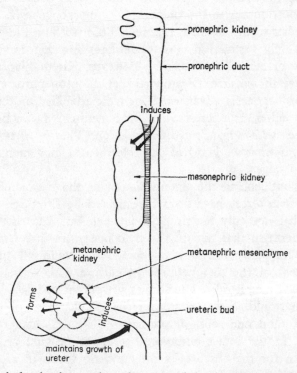

Fig. 7.8 The inductive interactions that control the development of successive kidneys in vertebrate embryos. The pronephric duct induces mesonephric tissue to form, and the ureteric bud from the mesonephric duct induces metanephric mesenchyme to form the metanephros. In its turn, the metanephros maintains the development of the ureter.

further development of its duct, the ureter. The interdependence of these metanephric components has been demonstrated *in vitro* as well as by operations on chick embryos *in vivo* (see Romanoff, 1960 for review). There is also evidence from mutants in mice: the 'Danforth short-tail' mutants lack a metanephros, and it has been shown that in their embryos, the ureteric bud does not grow far enough to make contact with the metanephrogenic mesenchyme (Gluecksohn-Schoenheimer, 1945). Hence the deficiency is apparently due to a failure of induction by the ureteric bud. Although the term 'induction' has been applied to these interactions between the vertebrate kidneys and their components, we are still ignorant of whether any material that one could regard as an 'inductor' passes between the cells: this is a matter for future investigation.

Some of the cellular interactions which occur *within* each type of kidney, contributing to the details of its morphogenesis, have come in for further study. There is relatively little work on the pronephros, since it is only vestigial in most vertebrates and degenerates early in embryonic or larval life. Recently, however, Kiremidjian and Kopac (1972) have shown that the properties of the pronephros cells change as its morphogenesis advances. When the mobility of these cells is tested by cell-electrophoresis in a fluid medium, it can be seen that it decreases with advancing differentiation. This is what one would expect from what we know of the roles of cell movements in earlier embryonic stages.

Factors influencing the differentiation of the mesonephros in the newt, *Taricha torosa* have been investigated by Etteridge (1969). He showed that not only the pronephric duct, but also tissues such as the endoderm of that region, the notochord, the early somite mesoderm or the early lateral plate mesoderm could all promote the development of the mesonephros in culture. Other tissues such as neural crest, neural tube, *late* somite mesoderm and *late* lateral plate mesoderm would, however, inhibit the development of the mesonephros. If (and one must emphasize the *if*) the same relationships hold true in the intact embryo, this might provide an interesting explanation for the absence of mesonephric tissue in more anterior regions of the body, for here the somite and lateral plate mesoderm would be in a more advanced state of differentiation than in posterior regions (as axial differentiation proceeds craniocaudally) and they may already be exerting their inhibitory action. Similarly, limits could be set to the time for which the mesonephric tubules continued to proliferate in posterior regions, for they would eventually be inhibited as the development of somites and lateral plate mesoderm in these regions became more advanced. It seems an interesting general possibility, which has so far received little attention by investigators, that inhibitory cellular interactions at later stages in development may

138

set limits to the period of morphogenesis in some embryonic organs.

Metanephrogenesis, too, may be influenced by a variety of other tissues *in vitro*. Ectodermal as well as mesodermal and endodermal tissues have been found to affect it. Grobstein (1953a, 1957) and colleagues working with him have shown that in mouse and chick embryos, spinal cord and brain were particularly effective in promoting the differentiation of the metanephros in culture. The 'induction' process from the spinal cord was found to be able to act across a millipore filter. In the mouse, Unsworth and Grobstein (1970) have tried the effects of various mesenchymes, finding a difference in the rates at which they induce the formation of metanephric tubules. Jaw mesenchyme must be left in the culture for at least two days to be effective, bone needs four days, and salivary or somite mesenchyme need five to six days. It is puzzling that the types of mesoderm most unrelated to the metanephros seem to act fastest, and suggests that these findings are unlikely to throw much light on interactions that normally occur during development. The kind of information emerging from such studies as these which could be important, would be any indications of what substances pass from the 'inducer' tissues to the metanephros and any observations of the behaviour of these cells as they interact.

7.2.4 *The genital system in vertebrates*
We shall be dealing in Chapter 10 with a number of hormonal influences that control the differentiation of the genital system. In this section here, only a few points about the interaction of the gonads with their ducts and about cellular interactions in the development of the genital system will be discussed.

The gonads and gonoducts in vertebrates develop from mesoderm very near to that which forms the kidneys. This mesoderm moves nearer to the midline than the kidney mesoderm, and primitive gonoducts then develop medially to the excretory ducts. Interactions between the gonads and their ducts have been studied in both vertebrates and invertebrates by means of grafting *in vivo* and culture *in vitro*. In extensive investigations using organ cultures from chick embryos, Wolff and his associates (see Wolff, 1969; also Hamilton and Teng, 1965) have shown that both the ovary and the testis interact with the gonoduct rudiments and control their differentiation, so that either an oviduct or a vas deferens is formed, as appropriate. The testis, in addition to stimulating the growth of the vas deferens, also inhibits the development of the Mullerian duct, which in the female forms the oviduct (Akram and Weniger, 1969). So the testis cells play a dual role, as stimulators of male gonoducts and inhibitors of female gonoducts. The same is true of the mammalian testis: this inhibits the growth of the Mullerian duct and stimulates development of the vas deferens from part of the Wolffian (mesonephric) duct. Jost (1948) showed that this action is local to only one side of the body and perhaps should not therefore be regarded as hormonal. A testis grafted into one side of a female rat fetus causes masculinization of the gonoducts on that side, but leaves the ducts of the other side unaffected. In mammals, unlike birds, the ovary has no positive interaction with gonoduct rudiments: female ducts will develop just as well if the ovary is removed as if it is present.

In the chick, male and female gonoducts have been shown to interact, *in vivo*, when they are both at a rudimentary stage. Didier (1971, 1973) showed that if the backward growth of the early Wolffian ducts was blocked, this could prevent development of the Mullerian ducts in the female. This was true if the blockage lay between the levels of somites 9 and 13, but not if the blockage was further caudal, between somites 13 and 19. This difference may be because by the time the Wolffian duct has grown as far caudal as somite 13, it has already exerted an inductive action on the Mullerian duct, so this can already develop independently. It is perhaps surprising that the Wolffian duct, which is the potential vas deferens of the male, should at first stimulate the growth of a potentially female duct together with it. But this is no more strange than the fact that in vertebrates each embryo always has two sets of ducts at first and goes through a bipotential stage before sexual differentiation occurs.

The development of the gonads and gonoducts and their interactions, in birds and mammals, are summarized in Fig. 7.9.

7.2.5 *The genital system of invertebrates*

Although relatively little experimental work has been done on the genital system in invertebrates, we do know of some cellular interactions that occur in the developing gonads and gonoducts in insects. In the 'holometabolous' insects which metamorphose from larva to adult via a pupal stage, the gonads do not develop until the last larval stage becomes a pupa. So there is only a very limited period of the life history in which cell-interactions in the genital system can be studied. Bodenstein (1946) showed, however, that in the fruit-fly

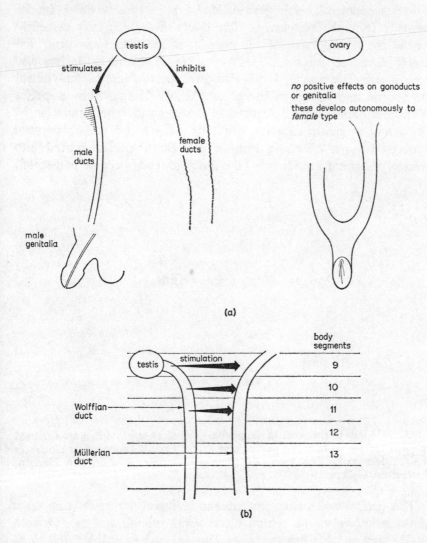

Fig. 7.9 Some early interactions between gonads and gonoducts in mammals and birds. (a) Mammals. (b) Birds.

Drosophila, female genital rudiments differentiate into normal ovaries and oviducts even when transplanted into males, so they are evidently unaffected by the testes of the host. Moreover, if the oviducts from the graft come close enough to make contact with the host's testes, they cause these to degenerate. So in this system (as in the inter-

141

actions we noted during gastrulation) certain insects differ funda-
mentally from the vertebrates. The ducts of some insects evidently
control the development of the gonads, rather than vice versa, and
female elements can dominate over those of the male. It has also
been shown (Bodenstein, 1955) that gonoducts of insects can control
the general morphology of the gonad. Vasa efferentia from a species
of *Drosophila* which has spirally-shaped testes will, when transplanted
into a host of another species, cause its testes to become spiral also.
Conversely, vasa efferentia from a non-spiral species grafted into
hosts of the spiral species cause the hosts' testes to become non-spiral.

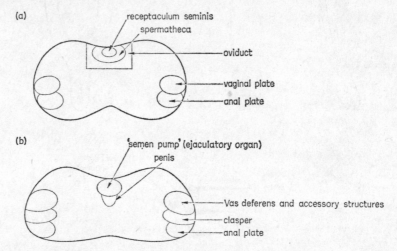

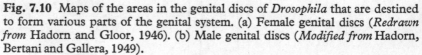

Fig. 7.10 Maps of the areas in the genital discs of *Drosophila* that are destined
to form various parts of the genital system. (a) Female genital discs (*Redrawn
from* Hadorn and Gloor, 1946). (b) Male genital discs (*Modified from* Hadorn,
Bertani and Gallera, 1949).

The genital rudiments are present in insect larvae as tiny discs
of tissue known as the genital discs. Some minute and careful work
by Hadorn and his associates on *Drosophila* has enabled the prospective components of the genital system represented in these discs
to be 'mapped' very accurately (cf. Fig. 7.10). In a series of experiments in which portions of these discs were transplanted into pre-
metamorphic larvae to test their potentialities (Hadorn & Gloor,
1946; Hadorn, Bertani and Gallera, 1949) it was found that these
areas were predetermined and would develop independently, without
requiring any interaction with surrounding cells. There was some
overlapping of the predetermined areas, however, so that each

isolated portion might form more than it would if it had been left *in situ*. This implies that *in situ* there are interactions which limit the course of differentiation of the parts of the disc. We do not yet know what kind of interaction is involved here. The behaviour of the individual cells in culture has not yet been followed in detail, nor has anyone yet isolated any materials that might be passing between the cells: it would be very difficult indeed to do this with such tiny areas of tissue.

One of the exciting recent discoveries which has emerged from the work by Hadorn and his colleagues on genital and other imaginal discs is the process that they have named 'transdetermination'. This will be referred to again in Chapter 8. Studying leg, wing, eye and antennal discs as well as genital discs, they showed that a whole disc of one kind might, after a series of transplantations into immature larval hosts, change its course of development and form one of the other organs when it was finally transplanted into a metamorphosing larva. As far as is known, no alterations of the potentialities of individual regions within the genital disc occurs during experiments of this kind, however.

The patterns of development in endodermal and mesodermal organs that we have reviewed so far have involved at the cellular level much the same kinds of interaction that we saw governed earlier embryonic development. The main types of event which can always be recognized are: changes in cell shape; adhesive affinities of cells within their own organ; movements of cells over each other's surfaces which both alter the shape of the organ and allow a greater range of cells to interact; and active growth and proliferation of cells so that extremities of the organ, such as the excretory and genital ducts, meet and make contact with other parts of the organ system, influencing its development. What substances or signals pass between the cells during all these processes, remains unknown so far.

We must now go on to consider a rather different type of effect in morphogenesis, in which cellular interactions mould the structure of hard, supporting tissues in various regions of the body. These events occur particularly among mesodermal cells.

7.3 Cellular interactions that mould structure during organogenesis

So far we have been dealing mainly with interactions that initiate differentiation in a particular layer of cells in an organ rudiment.

There are a number of later interactions between cells which complete the structure and maintain the integrity of each organ: these usually occur towards the end of its developmental period. In skeletal and connective tissues these late structural moulding processes are most important. It is not appropriate here to describe the details of chondrification and ossification in vertebrates, since these have mostly been studied at post-embryonic stages and are covered adequately in text books of histology. The types of events that we shall mention here, which do occur at embryonic stages, are those in which some cells are eroded, while others proliferate and fuse, and yet others lay down matrices to complete the form of the organ or tissue, before chondrification or ossification take place.

7.3.1 Cell death and epithelial fusion

Glücksmann (1951) and Saunders (1966) have stressed that cell death is an important component of morphogenesis. We shall see in the next chapter that in some organs such as the spinal cord, the retina, the inner ear and the limb ectoderm, certain cells are predetermined to die, and their death is essential to the moulding of the final shape of the organ. Mesodermal and endodermal organs show this phenomenon too. A good example of the kind of events involved is seen in the development of the palate, in mammals. The fusion of the two sides of the palate and its adhesion to the lower edge of the nasal septum (see Fig. 7.11) involve the breakdown of epithelial cells along all these borders, so that the underlying mesenchyme cells can interdigitate and fuse. Recent histochemical and ultrastructural work by a number of investigators (Angelici and Pourtois, 1968; Farbman, 1968; Smiley and Dixon, 1968; Mato *et al.*, 1969) has shown that the two halves of the palate fuse by, first, adhesion between epithelial cells in the midline, then the projection of mesenchyme cells through gaps in the basement membrane of the ectoderm, to join with mesenchyme tissue of the opposite side. Granules have been seen in the midline cells, and microvilli projecting from them also, but there is no evidence that the cells secrete any sticky material that could account for their initial adhesion. It must be presumed that they have intrinsic adhesive affinities such as we have seen in cells of other organ rudiments. Phagosomes have also been seen in these epithelial cells, and it is thought that these particles are responsible for their breakdown by an autolytic process.

144

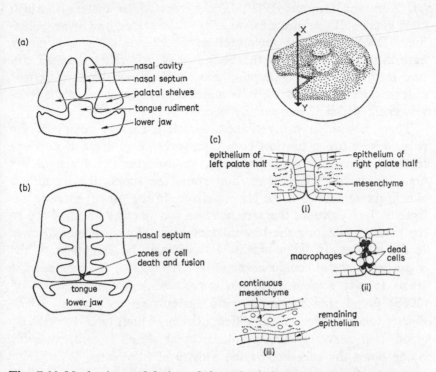

Fig. 7.11 Mechanisms of fusion of the palatal shelves in mammal embryos. (a), (b), coronal sections (in the plane XY, in inset figure) through nasal and buccal cavities before and after fusion of palatal shelves. (c) Detail of the processes of epithelial breakdown and mesenchymal fusion, shown diagrammatically in three stages, (i)–(iii).

In the nasal septum too, signs of lysis appear in the epithelium at its lower edge, and interestingly this breakdown starts *before* the septum fuses to the palatal shelves; it is evidently 'programmed' into the genetic events in these cells and does not depend on whether or not they make contact with the palate. Recently Smiley and Koch (1972) have shown that there is an intrinsic programming for cell breakdown in the palate epithelium too: one half of the palate, grown alone *in vitro*, will show epithelial degeneration along its midline surface at the appropriate stage when it would have fused with the opposite palatal half if it had been left *in situ*.

7.3.2 The laying down of matrix material

At many stages in embryonic development, cells exude materials which help them to remain attached to other cells in the same tissue (cf. Kean and Overton, 1969). These extracellular materials, which may include RNA, proteins, mucopolysaccharides and even sometimes DNA, may have important morphogenetic roles and, as we have mentioned already at the beginning of this chapter, organ culture methods and disaggregation procedures which tend to destroy at least some of the extracellular material may therefore produce development which is not entirely normal.

There is also in many tissues, including the mesoderm of the palate whose fusion has just been discussed, a final stage of development in which supporting material, to maintain the shape of the organ or tissue, is laid down. Sometimes the supporting tissue also aids in the moulding of the organ's shape. In the palatal shelves, collagen is laid down in the mesenchyme and imparts an elasticity to the tissue, facilitating the upward movement of the shelves to meet in the midline. In other organs, collagen may be laid down at the stages when tissue components are interacting, and in these cases it seems to play some part in the interactions. Wessels and Cohen (1968) found that treatment with the enzyme collagenase, which digests collagen, blocked the differentiation of both lung and salivary gland. It is almost always the mesenchyme which is responsible for laying down the collagen, by the activity of fibroblasts. Whether or not materials are passed through the collagen from mesoderm to endoderm, or vice versa, in the organs we have been discussing in this chapter, is not yet known. It should be possible to investigate this point by means of radioactive tracers and ultrastructural observations in the future.

The matrix materials laid down by cells of invertebrates differ from those of vertebrates, and are often concentrated at the external surface of the organism, rather than between layers of tissue in its internal organs. There are some embryos and larvae of invertebrates which develop an internal skeleton of calcified material, however: sponges and echinoderm larvae have spicules, for example. Gustafson and Wolpert (1961) followed the growth of spicules in echinoderm larvae by cinematographic methods. They showed that the spicules form at the gastrula stage, as a result of interaction between clusters of mesoderm cells. These cells send out pseudopodia as they invaginate, and the skeletal foci between them also elongate, apparently being laid down by the extended pseudopodia (Fig. 7.12). At a later stage, the skeletal rods themselves play a morphogenetic role, for their position determines the direction and placing of the arms of the larva, which form by local proliferation of the ectoderm layer.

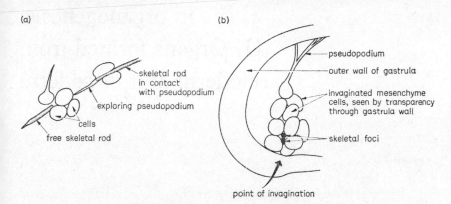

Fig. 7.12 Semi-diagrammatic sketches to show the process of spicule-formation in echinoderm embryos. (*Based on* Gustafson and Wolpert, 1961.) (a) Clusters of mesenchyme cells with pseudopodia and skeletal rods forming between them. (b) Invaginating mesoderm cells at the gastrula stage (cf. Fig. 5.2), showing some foci for spicule-formation already present at this stage.

These few examples serve to illustrate the general principle that extracellular material, as well as the cells themselves, can play an important role in the morphogenesis of organs and tissues. There are some types of cell such as the mesodermal fibroblast, whose chief mode of interaction with other cells appears to be by laying down collagen. The important role of this material in the development of embryonic organs has been discussed recently by Hay (1973). Chondrocytes also produce collagen in response to conditioning factors produced by the cells themselves (Solursh *et al.* 1973). But collagen is only one of many kinds of extracellular, or perhaps one should say '*inter*cellular' material which are now being observed in much more detail with the electron microscope. In addition, the basement membranes of epithelial cells in many organ rudiments contain essential materials (including glycosaminoglycans) and if these are removed, further differentiation is prevented (Bernfield *et al.*, 1973). We shall see further examples of special materials that lie between tissue layers when, in the following chapter, we consider organs formed from ectoderm and mesoderm.

147

8 Cellular interactions in organogenesis: II. Organs formed from ectoderm and mesoderm

It was emphasized in the last chapter that in vertebrates the mesoderm often plays a dominant role in controlling the type of differentiation that occurs in other layers in contact with it during organogenesis, just as it does during neural induction at the gastrula stage. When we go on now to consider some of the organs that form by interactions between mesoderm and ectoderm, we find that the mesoderm again appears to dominate, initially at least. Later, however, reciprocal interactions occur in which ectoderm and mesoderm are mutually dependent.

The major part of the embryonic ectoderm in both vertebrates and invertebrates becomes the epidermis, with its several appendages, glands and other surface structures. A layer of mesoderm lies close beneath this epidermis, in all three-layered animals, and contributes to some of the appendages as well as to the skin. The skin in vertebrates is composed of the epidermis, derived from ectoderm and the dermis, derived from the outer, dermatome layer of the somite mesoderm (Fig. 8.1a).

There is a special feature of the skin in humans which is known to be controlled by the general layout of the dermis: the finger-ridge pattern. This gives rise to our characteristic fingerprints. The pattern of ridges develops before birth, and is initiated by patterns of folds in the dermis, which cause a corresponding set of ridges to form in the epidermis above it. Other features which characterize different regions of the skin, in animals as well as man, are the numbers and kinds of appendages such as hairs, feathers, bristles, scales, teeth and glands. These also are determined by mesoderm of the dermis layer. Grafting of either epidermis or dermis between different regions of the body in mammals (Billingham and Silvers, 1967) has shown that even at postnatal stages, the epidermis takes on some minor characteristics that are appropriate to the region of dermis that lies beneath it. In embryos of birds, the same is true with regard to major characteristics, provided that the epidermis has not already become determined in its characteristics by the dermis over which it originally lay (see later examples).

Most of the skin appendages in vertebrates have an outer layer derived from ectoderm and an inner core of mesoderm. The limb bud too is, in its early stages, an appendage of the same type, with a core of mesoderm covered by ectoderm. In all the vertebrate appendages whose development has been studied experimentally, it has been found that normal differentiation depends on reciprocal interactions between these two layers (cf. Fig. 8.1b).

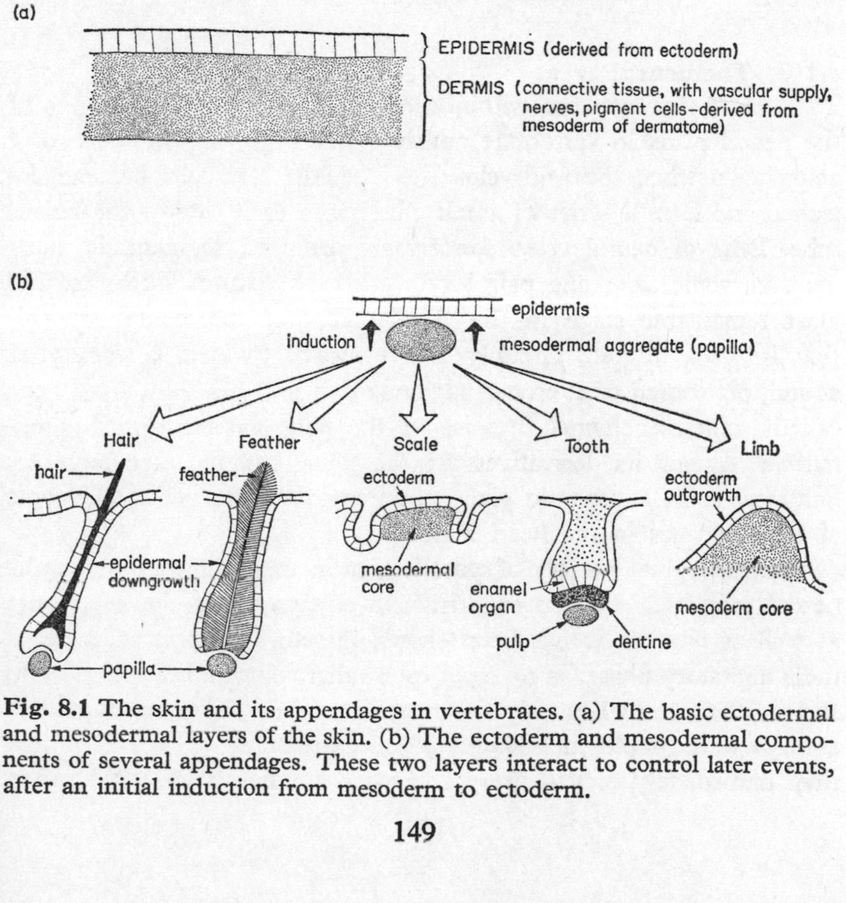

(a)

} EPIDERMIS (derived from ectoderm)

} DERMIS (connective tissue, with vascular supply, nerves, pigment cells—derived from mesoderm of dermatome)

(b)

Fig. 8.1 The skin and its appendages in vertebrates. (a) The basic ectodermal and mesodermal layers of the skin. (b) The ectoderm and mesodermal components of several appendages. These two layers interact to control later events, after an initial induction from mesoderm to ectoderm.

In invertebrates, many of the appendages are primarily purely ecto-dermal, and their development cannot therefore be discussed in terms of interactions between ectoderm and mesoderm. Brief con-siderations of experimental work on the integument and on the limb discs of insects are given later in this chapter, however, since these show interesting examples of interactions between different group-ings of ectoderm cells. Intra-ectodermal interactions may occur in vertebrate appendages also, but less attention has been paid to this, since the main interest of most of the work on vertebrates has been in ectoderm-mesoderm interactions.

There is one tissue in vertebrates which is famous for the remark-able migrations and interactions that its cells undergo, and which is difficult to assign definitely either to ectoderm or mesoderm, since it forms such a variety of structures. This is the neural crest. It seems appropriate to consider this tissue first, before going on to the organs that are made up of distinct ectoderm and mesoderm com-ponents.

8.1   The neural crest

This tissue first appears within the neural folds at the margins of the neural plate, in vertebrate embryos (see Fig. 8.2). At the end of neurulation when the folds close, the neural crest cells become de-tached and form a series of small clusters on each side of the neural tube. Pairs of neural crest clusters are arranged segmentally along the embryonic axis; one pair level with each somite. The next and most remarkable stage in the development of the neural crest is that its cells migrate laterally and ventrally to form a variety of neural, pigmented or even cartilaginous tissue. It has sometimes been called 'ectomesenchyme' because of the wide variety of its poten-tialities. Among its derivatives are the spinal ganglia, melanophores, Schwann cells, autonomic ganglia, adrenal medulla cells, and some of the cartilages in the head region.

With this great variety of possibilities in their differentiation, the neural crest cells have been attractive objects for study, in culture as well as *in vivo*. It has been shown that their first interaction, at their migratory phase, is to repel each other. Niu and Twitty (1948) showed that even when only two neural crest cells from an amphibian embryo were placed in a capillary tube, in contact with each other, they immediately moved away from one another (Fig. 8.3). Larger

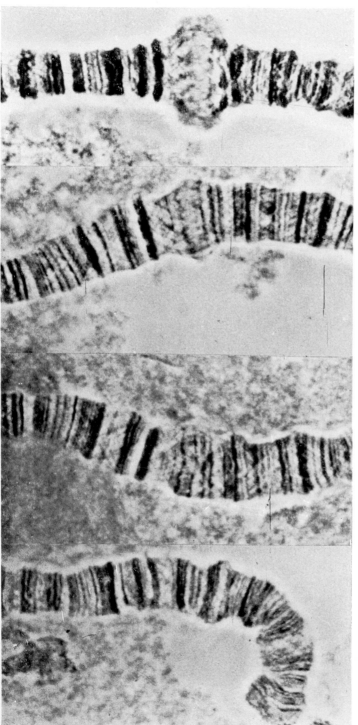

Plate 1.1. A sequence of photographs showing the development of a chromosome puff in the fly *Chironomus*. This particular puff appears in response to sugar treatment (*see* Beermann, 1973).

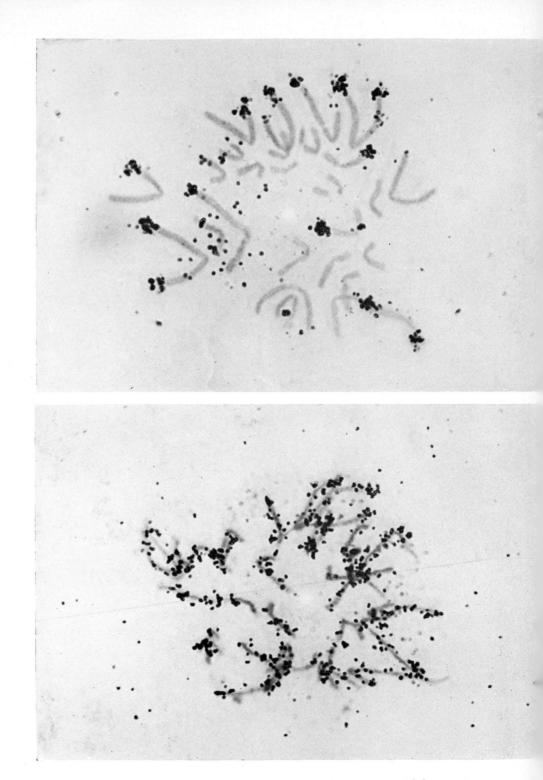

Plate 1.2. Autoradiographs showing the late-replicating regions of chromosomes (heavily-grained) in embryonic cells of *Rana pipiens* after colcemide treatment and labelling with tritiated thymidine.

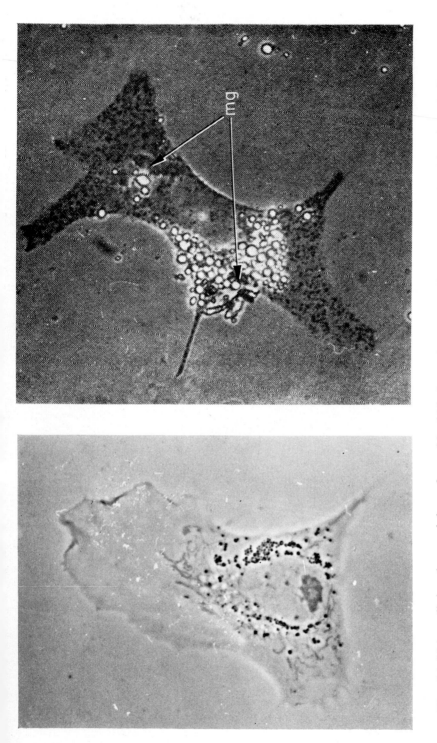

Plate 1.3. Living cells in culture and the surfaces with which they make contact with other cells. (a) A chick embryonic fibroblast, showing its extended, thin border, the ruffled membrane. (*From an original kindly donated by* S. Pegrem). (b) A neural crest cell from *Xenopus laevis*. (*From* Gabie and Andrew, 1969, with kind permission). *mg*: melanin granules.

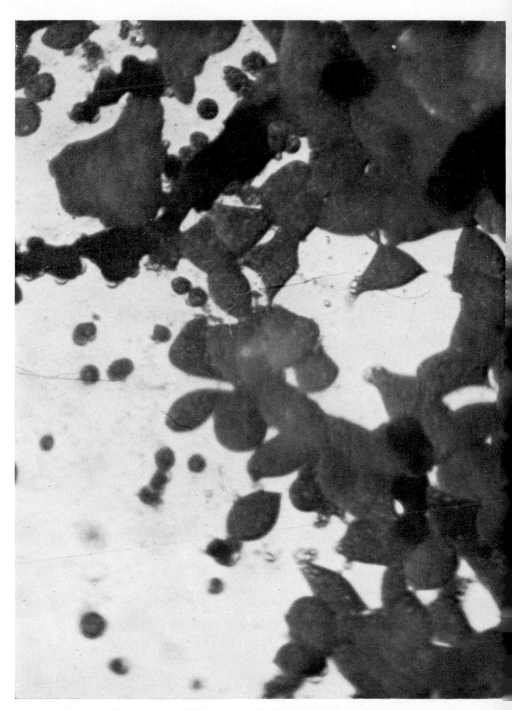

Plate 1.4. Gastrula cells of *Xenopus laevis,* in process of reaggregating. Note the pseudopodia (some like transparent bubbles at the edge of the cell) with which they make contact and explore the surfaces of other cells. (*Original kindly donated by* A. S. Curtis).

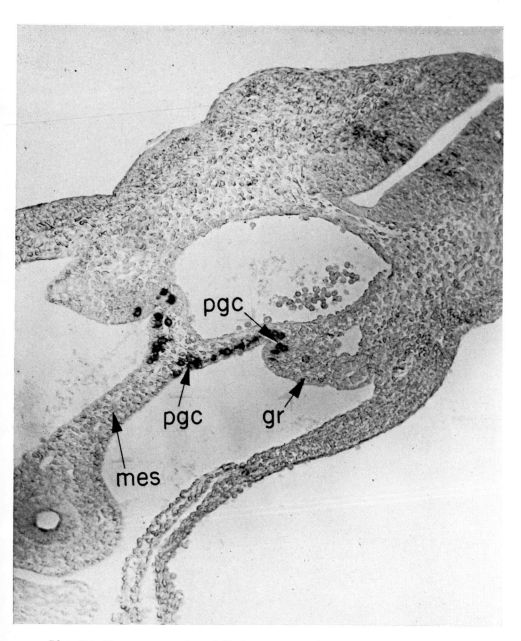

Plate 2.1. Transverse section of 10-day mouse embryo, showing primordial germ cells (darkly stained, for alkaline phosphatase) migrating up the central mesentery into the genital ridges. *Labelling*: *gr* = genital ridge; *pgc* = primordial germ cells; *mes* = dorsal mesentery of gut. (*From* Mintz (1957), by kind permission).

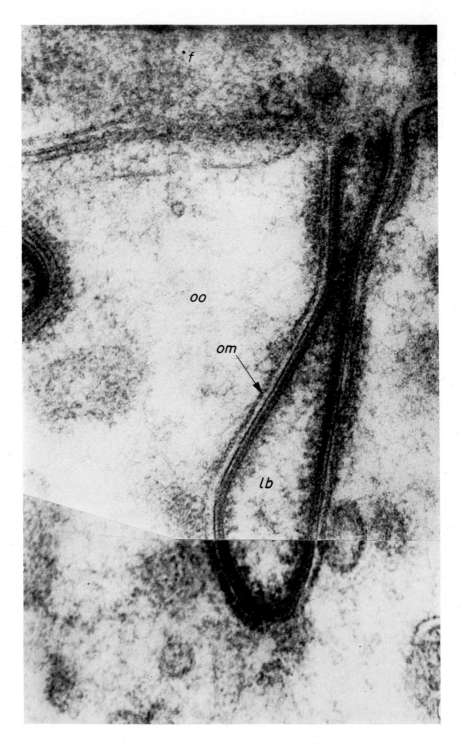

Plate 2.2. Electron micrograph (montage) showing a 'lining body' projecting from a follicle cell into the cytoplasm of the oocyte in the hen's ovary. (*Original kindly donated by* R. Bellairs.) *oo*=oocyte; *f*=follicle cell; *l.b.*=lining body; *om*=oocyte membrane.

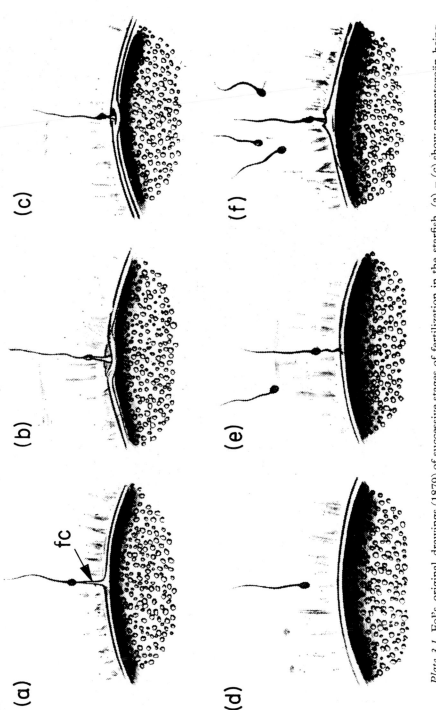

Plate 3.1. Fol's original drawings (1879) of successive stages of fertilization in the starfish. (a) – (c) show spermatozoön being drawn towards the egg by a fertilization cone (*fc*), which is extruded in response to the proximity of the spermatozoön, as shown in (e). There is no further response to supernumerary spermatozoa, once one has become attached as in (f).

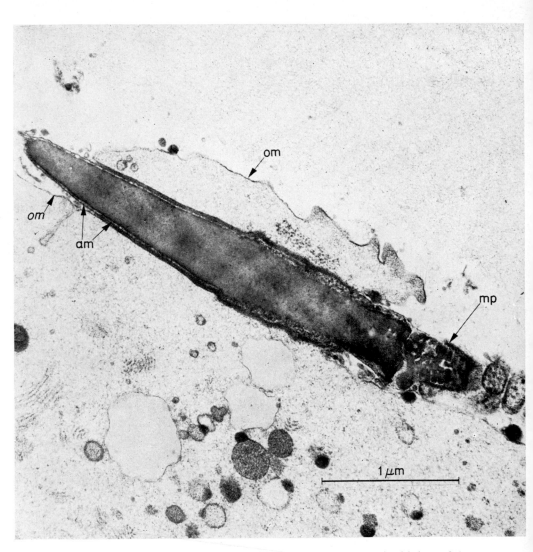

Plate 3.2. Electron micrograph showing the sperm head and midpiece of the rat in process of entering the oocyte, in a sideways-on position. *om* = oocyte membrane; *am* = acrosomal membrane; *mp* = midpiece. Magnification × 37 500. (*From an original kindly donated by* R. Presley.)

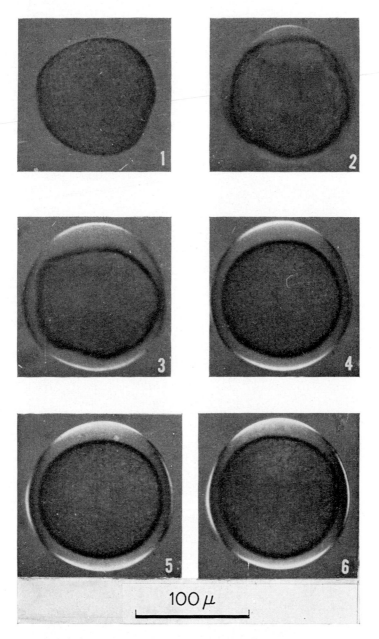

Plate 3.3. Sequence of photographs taken in polarized light of living sea-urchin (*Paracentrotus lividus*) egg during the elevation of the fertilization membrane. A zone of birefringence appears at the point of sperm entry (photo 2) and spreads outward from there. Birefringence also appears at the opposite pole of the egg (photo 3), and after the second polar body has been extruded (photo 5) the birefringence extends round the whole surface of the egg. (*From* Mitchison and Swann, 1952, with kind permission.)

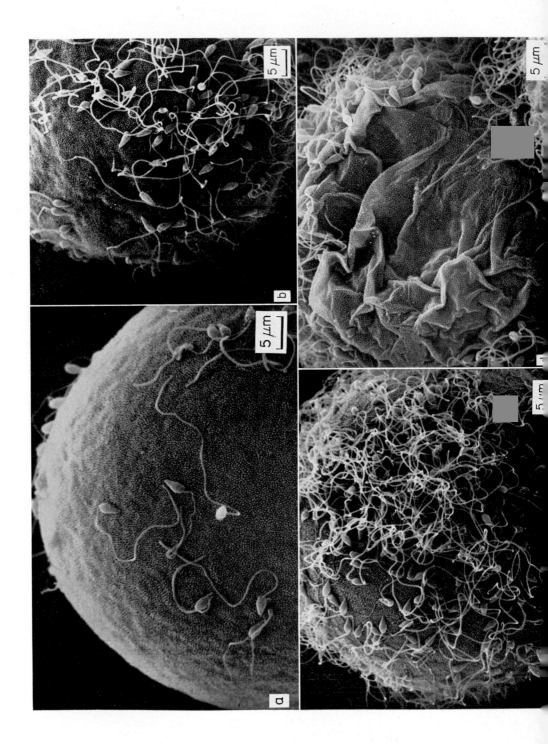

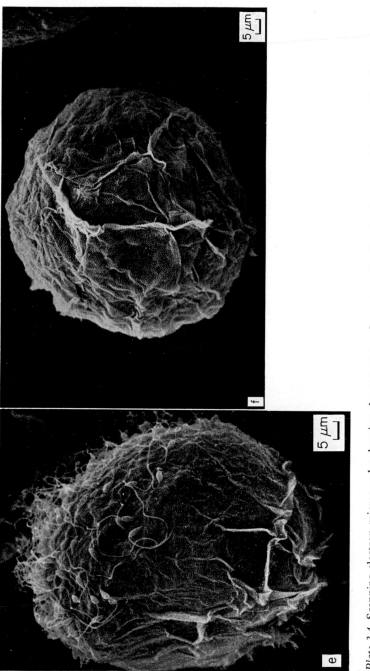

Plate 3.4. Scanning electron micrographs showing the sequence of events at fertilization in sea-urchin eggs. (a) a few sperm have reached the egg surface and some are penetrating it (above). (b) and (c) Many more sperm attach themselves to the egg, before folds of the fertilization membrane appear and cause the excess spermatozoa to be 'sloughed off' as shown sequentially in (d) to (f). (*From* Tegner and Epel, 1973 with kind permission).

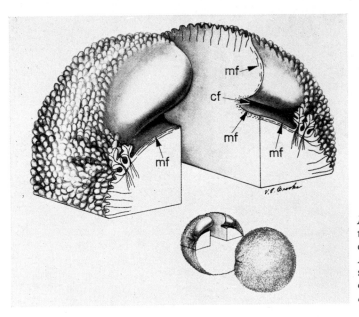

Plate 4.1. (a & b). The orientation of microfilaments in the egg of the invertebrate *Aequoria.* (a) Diagram showing sectors of the cleaving portion of the egg. *mf:* microfilaments; *cf:* cleavage furrow.

(b) Electron micrograph of the furrow region, showing microfilaments (*mf*) orientated parallel to the furrow. (*From* Szolossi, 1970, with kind permission.)

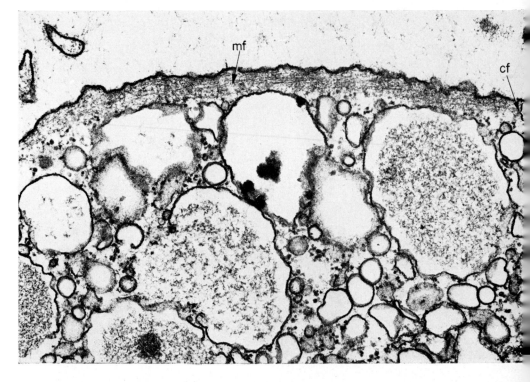

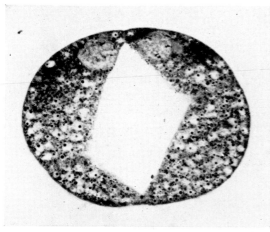

Plate 4.2. Section through the two-cell stage of cleavage in the snail *Lymnaea stagnalis*, showing the large spaces and very small zones of contact between the blastomeres. The cell nuclei are seen at top of picture. (*From an original kindly donated by* J. van den Biggelaar).

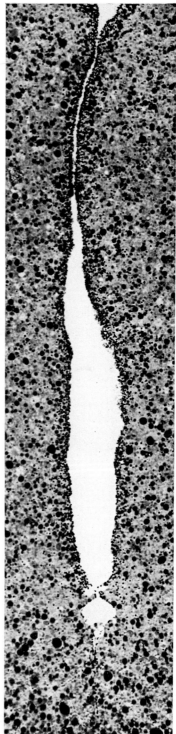

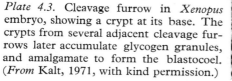

Plate 4.3. Cleavage furrow in *Xenopus* embryo, showing a crypt at its base. The crypts from several adjacent cleavage furrows later accumulate glycogen granules, and amalgamate to form the blastocoel. (*From* Kalt, 1971, with kind permission.)

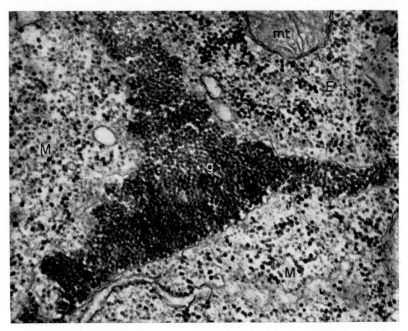

Plate 5.1. Electron micrograph showing the glycogen- and RNA- containing granules that lie between dorsal mesoderm and ectoderm cells in the gastrula of *Xenopus laevis*. Stain: Reynolds' lead citrate. Magnification × 56 000. *E:* ectoderm cell; *M:* mesoderm cell; *mt:* mitochondrion; *g:* granules. (*From an original kindly donated by* P. Hyatt.)

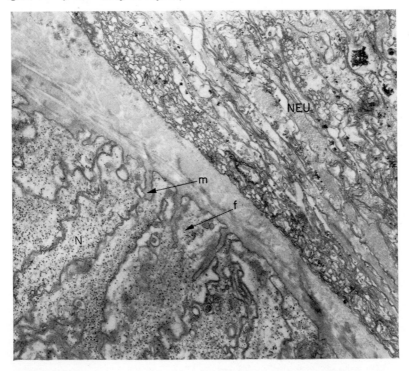

Plate 6.1. Electron micrograph of the notochord and adjacent neural tissue, section, (cf. Flood, 1970). (*From an original kindly donated by* P. R. Flood). *m:* membrane inpocketings; *f:* filaments (myotendinous junction?). N: notochord; NEU: neural tissue.

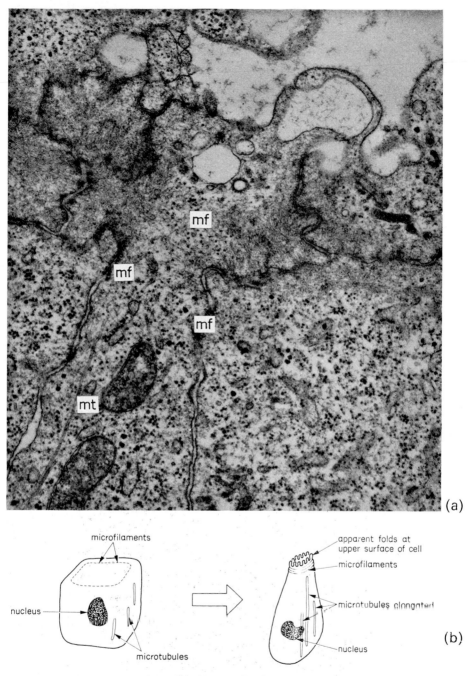

(a)

(b)

Plate 6.2. The roles of microfilaments and microtubules during neurulation, according to Schroeder (1973). (a) electron micrograph showing the upper ends of three adjacent neural tube cells of a chick embryo, with microfilaments (*mf*) cut transversely and microtubules (*mt*) orientated vertically in the cells. (b) Diagrams after Schroeder, to show the changes in shape of these components in a single cell during neurulation.

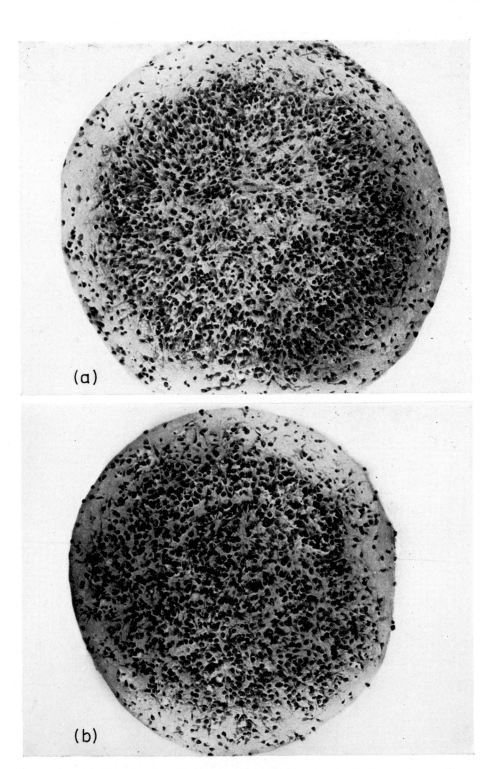

Plate 9.1. Photographs of reaggregated cerebellar cortex cells from (a) normal and (b) 'reeler' mutant mice. Note the much more regular and structured arrangement in (a). (*From* DeLong and Sidman, 1970, with kind permission.)

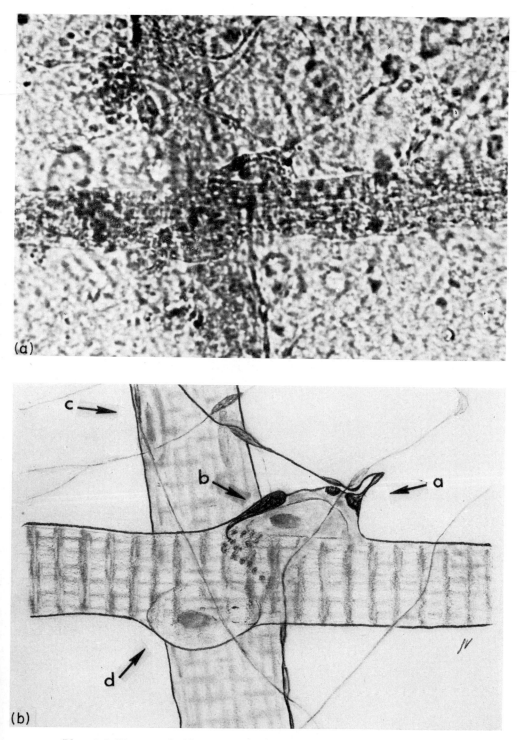

Plate 9.2. Photograph (a) and explanatory outline of it (b), showing muscle straps in culture, with four nuclei (a,b,c,d) bulging outwards towards the nerve endings which have made contact with them. (*From* Veneroni and Murray, 1969, with kind permission.)

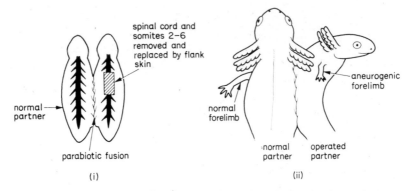

spinal cord and
somites 2–6
removed and
replaced by flank
skin

normal
partner

parabiotic fusion

(i)

normal
forelimb

aneurogenic
forelimb

normal
partner

operated
partner

(ii)

Plate 9.3. Aneurogenic limbs. (a) Method of producing aneurogenic limbs in
in *Ambystoma*, by removal of part of the nervous system and then parabiosis
with a normal embryo to promote survival. (i) Operation, at tailbud stage;
(ii) resultant, stunted limb in the larva. (*Redrawn after* Piatt, 1942).

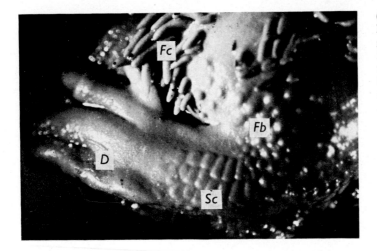

(b) Photographs of stunted
hind limbs of the chick,
grown aneurogenically by
isolating the limb bud onto
the chorioallantoic mem-
brane. (See Bradley, 1970).
(*Originals kindly provided
by* S. Bradley).

(i) External view, and (ii)
section of the limb bud. *Fb:*
feather buds; *Fc:* feather
cones; Sc: scales; D: digits;
C: cartilage. Arrow in (ii)
points to a malformed joint,
resulting from the lack of
movement in the limb.

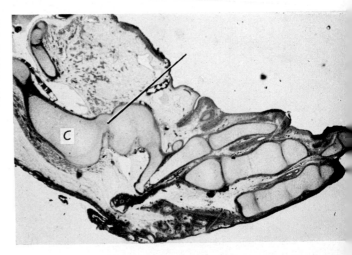

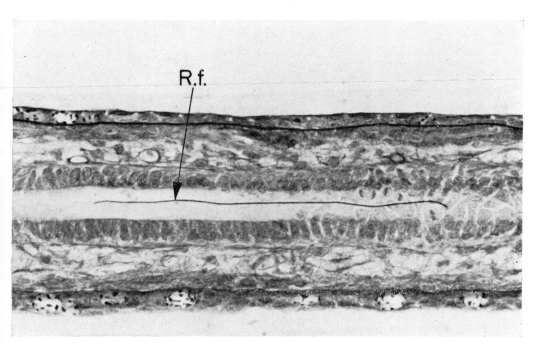

R.f.

Plate 9.4. Photograph of a longitudinal section through the tail end of the spinal cord in *Xenopus*, showing a Reissner's fibre (R.f.) running down the centre of the canal. (*From an original kindly donated by* R. Hauser.)

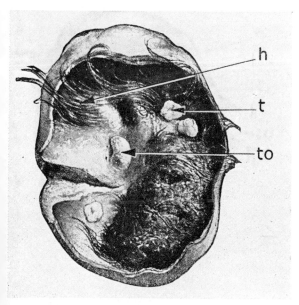

h

t

to

Plate 11.1. Photograph of a teratoma, removed from a human ovary. (*From* L. B. ＿ey, *Developmental Anatomy*, 6th. edition, with kind permission of the author and W. B. Saunders Co.) *h*: hair; *t*: tooth; *to*: tongue rudiment.

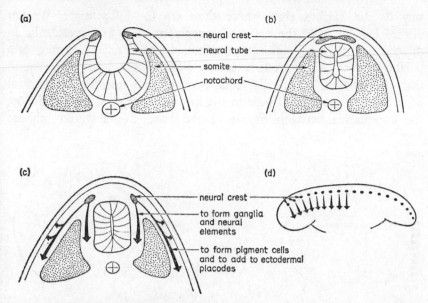

Fig. 8.2 The origins and interactions of neural crest cells. (a–c) Diagrammatic transverse sections of the axial tissues of a vertebrate embryo, during and after the end of neurulation. Neural crest cells (shaded) become detached from the edges of the neural tube, then form dorsolateral clusters from which individual cells migrate in the directions shown by arrows in (c) and (d), to form pigment cells and nerve ganglion cells. In (d) a lateral view of the embryo is shown, with the neural crest clusters segmentally arranged.

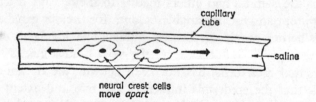

Fig. 8.3 Mutual repulsion by neural crest cells, shown when two cells are placed in a capillary tube (see Twitty and Niu, 1948).

groups of neural crest cells also showed an immediate tendency to disperse, when placed together in culture. There appeared to be a negative chemotaxis between them – though it was never shown whether or not any secretion of the cells was responsible for their mutual repulsion.

In order to reach specific destinations where they will differentiate, the neural crest cells must be able to interact, not only with each other but also with other tissues of the embryo so that they find their

151

F

way to the various sites where they are to differentiate. Weston (1963) has followed the paths of movement of neural crest cells in the chick by grafting radioactively-labelled neural tube and crest into unlabelled host embryos in a lateral position (Fig. 8.4). He found that the labelled crest cells tended to migrate towards the ventral edge of their own neural tube in the first instance, and then to follow the boundaries between tissues of the host. Some would migrate

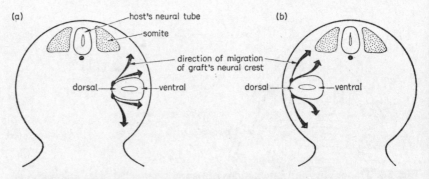

Fig. 8.4 Weston's experiments (1963) showing that neural crest cells migrate according to the orientation of the neural tube with which they are in contact. In (a), a radioactively-labelled neural tube with neural crest is grafted laterally into a host chick embryo, with its dorsal surface inwards; in (b) the graft is placed with its dorsal surface facing outwards. The arrows show the directions in which labelled neural crest cells migrated in each case.

lateral to the somites and others medial to them, thus reaching positions appropriate for melanoblasts and for nerve ganglia, respectively. Whether purely mechanical guidance is involved, or whether some more specific influence from the axial tissues guides the course of the neural crest cells, has not been shown. We do know, in amphibians, that the epidermis influences the rate and extent of migration of the melanoblasts. Dalton (1950) showed in a classic experiment that the difference in colouring between black and white axolotls was due to the inability of melanoblasts to migrate under the white epidermis, and not to any defect in the synthesis of melanin. He showed this by exchanging triangular pieces of skin between black and white animals, as shown in Fig. 8.5, and also by culturing melanoblasts of the two types, which both proved equally capable of forming melanin.

The adrenal medulla, to which the neural crest contributes, is part of the wider system of 'chromaffin' cells distributed through the autonomic nervous system and some other tissues of vertebrates. As

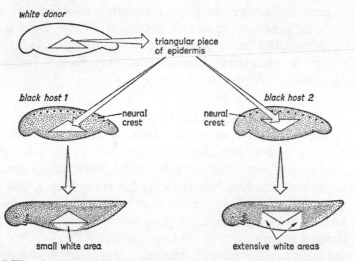

Fig. 8.5 The classic experiment of Dalton (1950) showing that the epidermis of white axolotls inhibits the migration of melanoblasts from the neural crest. A triangular piece of epidermis from a donor of the white strain was placed either with its apex towards the neural crest (left hand series of diagrams), or with the broad base of the triangle towards the neural crest (right hand series) on a host embryo of the black strain. Resultant larvae had white ventral areas indicating the inability of melanoblasts to cross the white skin, in the right-hand series. In the left-hand series, only the area immediately under the centre of the graft was white, and even here some melanoblasts had penetrated from areas lateral to this.

mentioned earlier (7.1.6) Pearse (1969, 1971) has suggested that all cells in the body which secrete polypeptide hormones originate from the neural crest. Using fluorescent staining methods, and also the technique of grafting between quail and chick embryos, whose cells are distinguishable histologically (see LeDouarin and Teillet, 1971), evidence was obtained that neural crest cells migrated to the gut region as well as to the adrenal medulla and chromaffin system, and gave rise to endocrine cells of the gut. Further work by Le-Douarin and Teillet (1973) indicates, however, that all the crest cells which migrate to these regions become ganglion cells rather than secretory cells. So more evidence is needed before Pearse's earlier dictum can be generally accepted.

There is no neural crest in invertebrate embryos, but there are neurosecretory cells associated with the cephalic ganglia in ascidians, cephalopod molluscs and insects. In insects these cells play a role in controlling the moulting and metamorphosis of the larva – a topic to which we shall give more attention in Chapter 10.

In vertebrates another category of structure to which the neural crest is thought to contribute are the teeth. These develop by a series of interactions between ectoderm and mesoderm, which have been studied in some detail in mammals. This work is reviewed in the next section.

8.2 Cellular interactions in the development of teeth

The structure of teeth varies in different vertebrates. We shall deal here only with mammalian teeth on which considerable experimental work has been done recently. It has been shown that their two components, ectoderm which forms the enamel layer, and mesoderm (possibly added to by neural crest) which forms dentine and pulp, are mutually dependent for their normal development. Either layer, isolated *in vitro*, cannot undergo normal differentiation, but if the other layer is put with it, it recovers its capacity to differentiate.

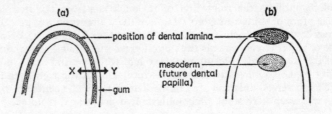
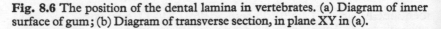

Fig. 8.6 The position of the dental lamina in vertebrates. (a) Diagram of inner surface of gum; (b) Diagram of transverse section, in plane XY in (a).

The first rudiment of future teeth in mammalian embryos is the dental lamina, a thickened band of ectoderm along the crest of the gum (Fig. 8.6). This develops apparently independently of the mesoderm at first. Next, a series of aggregates of mesoderm cells appears: the tooth papillae (Fig. 8.7a). It is not known whether these are 'induced' in any way by the dental lamina. But we do know that it is in response to the presence of each dental papilla that a downgrowth of cells occurs from the dental lamina and forms the rudiment of the enamel organ (Fig. 8.7b). If a tooth papilla is transplanted under another part of the oral ectoderm, or if it is placed in contact with uncommitted ectoderm *in vitro*, it will induce the formation of an enamel organ in the ectoderm. Conversely, if the mesoderm of the dental papilla is removed from an embryo or from an organ culture of the tooth rudiment, the enamel organ does not form.

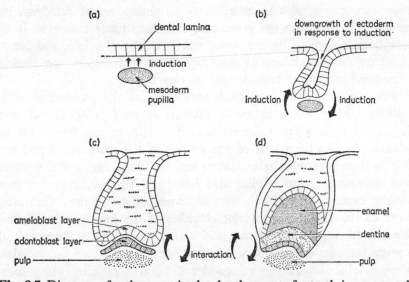

Fig. 8.7 Diagrams of early stages in the development of a tooth in a mammal, indicating the interactions (arrows) between ectodermal and mesodermal layers. (a) Mesoderm papilla induces dental lamina to form a downgrowth as in (b), and interactions between this downgrowth and the mesoderm result in the formation of an ameloblast layer and an odontoblast layer, as in (c). Further interactions control the laying down of enamel and dentine between these two layers, shown in (d).

The next stage in the differentiation of the tooth is that the ameloblast layer of the enamel organ begins to lay down enamel, while an odontoblast layer appears in the tooth papilla which lays down dentine (Fig. 8.7c, d). These two layers are mutually dependent and will not develop in isolation from each other. Glasstone (1936) was one of the first to observe that neither odontoblast layer or ameloblast layer could complete its differentiation alone *in vitro*, but that if both layers were combined, normal tooth tissues formed. Her findings have since been confirmed by other workers (Koch, 1967; Slavkin *et al.*, 1969a). The final shaping of the tooth *in vitro* also requires the presence of both layers. Kollar and Baird (1969) have shown that in tooth rudiments of mice, it is the mesoderm that determines whether the molar or the incisor tooth shape develops. They showed that incisor mesoderm with molar ectoderm produces an incisor tooth, while molar mesoderm with incisor ectoderm produces a molar tooth. Kollar and Baird also showed, however, that the age of the ectoderm governs the extent to which it can be influenced by the type of tooth mesoderm put with it. This is a general

155

point that also applies in other cases of 'induction' of ectoderm by mesoderm. The older the ectoderm, the more likely it is already to have been influenced by the mesoderm at its site of origin, and hence to be committed already to one type of differentiation and unable to respond to another mesodermal influence.

The effects of other, non-tooth mesoderm on the ectodermal tooth rudiment have also been tried. Slavkin *et al.* (1969a) found that dermal fibroblasts transplanted to the mouth region were able to orientate and to form part of the matrix of the tooth rudiment, but that in their presence the ectoderm was not then able to differentiate into an enamel organ. Kollar and Baird (1970) also isolated mouse incisor enamel organs with various types of mesoderm. Culturing these combinates in the anterior chamber of the eye, they found that the enamel organ was readily induced to form other structures if foreign mesoderm was put with it. It was induced to form bristles, when combined with snout mesoderm, or to keratinize, when combined with foot-plate mesoderm. Reciprocal combinations were also tried: foot-plate ectoderm combined with dental papilla mesoderm produced normal teeth, but snout ectoderm with dental papilla mesoderm produced bristles, indicating that this snout ectoderm was already 'committed' (or, as most embryologists say, 'determined') to form hairs, having already received some induction stimulus from its own mesoderm. This is an example of another general phenomenon, already discussed in Chapter 6, which is that differentiation proceeds craniocaudally in time sequence, in animal embryos. So at the time when snout ectoderm has already become 'determined', the ectoderm of hind extremities such as the foot may not yet be determined. Hence, caudal embryonic tissues tend to have their normal potentialities overruled by more cranial tissues, in transplant experiments. In the intact embryo too, events initiated in the head region tend to control events in more caudal regions, producing an apparent 'head dominance'. This is also seen in adults of some invertebrates.

The determination of individual regions in the tooth rudiments is evidently a gradual process, according to experimental work in which parts of it have been isolated to test their ability to reconstitute a whole tooth. Koch *et al.* (1970) showed that the proximal third of an incisor tooth rudiment differentiated more successfully than either the middle or distal thirds, and was able to reconstitute the whole tooth rudiment whereas the distal two thirds were rarely able to do so. In all these isolates, both mesoderm and ectoderm components were included.

It is not known what factor or influence passes from the dental papilla to the ectoderm, or what materials may be interchanged between the two layers later on. It has been shown that the early tooth rudiments interact successfully through a millipore filter. So the initial induction from mesoderm to ectoderm must be mediated by some factor in solution (Koch, 1967). At later stages, the interacting regions are zones of extracellular material, the beginnings of enamel and dentine, which separate the ameloblasts from the odontoblasts (cf. Fig. 8.7d). Slavkin et al. (1969) examined the ultrastructure of this extracellular material in rabbit incisor rudiments, and saw thin, hair-like pseudopodia (filopodia) extending from the mesenchyme through the pre-dentine, to attach to a PAS-positive (i.e. mucopolysaccharide) basement membrane of the ectoderm. In addition, fibrils were seen accumulating in the intercellular space, which may have been related to the deposition of dentine and enamel. In further investigation on the nature of the matrix material, Slavkin et al. (1970) found that it contained four methylated species of RNA, all of low molecular weight, besides PAS-positive material. They also showed that the matrix, put together with either the epithelial or the mesenchymal layer of the tooth rudiment, stimulated the differentiation of these cells. According to Slavkin and Bavetta (1970) the RNA component of the matrix must be present, for it to exert this morphogenetic stimulation. This provides a further example of the role of intercellular matrix materials in organogenesis (cf. Section 7.3.2). In this case, radioactive labelling experiments indicate that the matrix is derived from both ectodermal and mesodermal layers of the tooth rudiment. As we have seen, it also stimulates the morphogenesis of both these layers.

In birds, a horny beak develops instead of teeth. Here too, the mesoderm seems to determine the kind of differentiation that takes place in the ectodermal tissue. Hayashi (1965) showed that when oral mesoderm from the duck was combined with oral ectoderm from the chick, a duck-like beak with tooth-ridges developed. The reciprocal combination gave rise to a chick-like beak with no tooth ridges. So far it has not been tested whether oral mesoderm from a bird combined with ectoderm from a mammal will produce a beaked mammal: this would be a bizarre phenomenon indeed! Of course there is one famous and extraordinary mammal which has a beak: the duck-bill platypus. But the beak is not developed from exactly the same areas of cells as the teeth, which are present as well. At any rate, the oral ectoderm of this mammal is evidently capable of producing horny material for a beak.

When we come to consider the interactions between mesoderm and ectoderm cells that occur in the formation of more widespread skin appendages such as hairs, feathers and scales, we find that there is considerable interchangeability between them, according to the type of mesoderm that is combined with epidermis. One can, in fact, get mammalian cells to participate in forming feathers, and avian cells to form hairs, by recombination experiments *in vitro*. In all these cases, the mesoderm determines what kind of structure will form, unless the ectoderm with which it is combined is older and has already been committed to another path of differentiation.

8.3   Feathers, hairs and scales

All of these appendages develop from tracts of epidermis which follow definite patterns, characteristic of the species. Like the teeth, they depend for their initial outgrowth upon some stimulus from the underlying mesoderm (dermis), which forms an aggregate or papilla under the future feather, hair or scale rudiment. Shortly after the dermal aggregations have appeared there is an ingrowth of epidermis above each of them, not unlike the ingrowth which forms the enamel organ of the teeth. From the centre of this ingrowth (which is flatter and shallower in the case of a scale), the appendage then grows outwards as a hollow structure, whose core becomes filled gradually by the deposition of keratin, and by penetration of blood vessels and nerve endings into its basal region. Fig. 8.1 shows diagrammatically the morphology of developing hairs, feathers and scales for comparison.

A great deal of experimental work has been carried out by Sengel and his associates on developing feathers in chick embryos: this is reviewed by Sengel (1971). He has found that the presence or absence of feathers in regions of the body, as well as their arrangement in tracts and groups, and their individual morphogenesis, are all determined by the orientation and the origin of the mesoderm that lies beneath and forms the feather papillae. For instance, if mesoderm from one of the featherless 'brood patches' (apteria) in the ventral region is transplanted under the ectoderm of other regions of the body at early embryonic stages, a much reduced number of feathers is formed in these other regions. Conversely, if mesoderm from other regions is combined with apterium ectoderm, feathers do form in this ectoderm (Sengel, Dhouailly and Kieny, 1969). Sengel has also tried altering the orientation of the dermal layer *in situ* on the embryo, by rotating it 90° or 180°, and has found that the orientation of the feather tracts is altered accordingly. Rotation of the ectoderm has no such effect, provided it is done early enough, for it develops in accordance with the orientation of the mesoderm below it.

It will be remembered that the dermal mesoderm is derived from the dermatome layer of the somites (see Fig. 6.5). Hence, damage to the somites may affect feather development. Mauger (1970) found that X-raying of somites in the chick embryo resulted in 'apteria' arising in the dorsal regions.

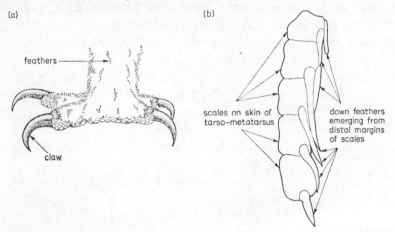

Fig. 8.8 Feather-tipped scales on the legs and feet of some birds. (a) Feathered leg and foot of the barn-owl (sketched from the illustration by George Rankin in Landsborough Thomson's book, 1910). (b) Detail of scales terminating in down feathers, on a young chicken's foot. (*Redrawn from a photograph by* Rawles, 1963).

Not all areas of the body in birds carry feathers: the legs and feet bear scales instead. In some larger birds and in some strains of domestic fowls, however, there is more feathering of the leg regions than in other birds and there are also composite, feather-tipped scales on the feet (Fig. 8.8). The existence of these raises an interesting question as to how flexible the potentialities for either feather or scale development are, in ectoderm cells of avian embryos. Rawles (1963) showed in beautifully-designed experiments on chick embryos

that leg mesoderm tends to induce scales to form, and dorsal mesoderm induces feathers, even when these areas of mesoderm, or the
ectoderm overlying them, are interchanged. Thus, dorsal ectoderm
can be induced by leg mesoderm to form scales instead of feathers:
conversely, leg ectoderm can be induced to form feathers if dorsal
mesoderm is put with it (Fig. 8.9). In both these types of operation
it is essential to test the ectoderm at a stage before it has been acted
on by the mesoderm of its own region, however. Rawles obtained
some cases in which the ectoderm was old enough to be in a border-
line condition: these produced feather-tipped scales. She found that
the ectoderm of feathered regions becomes determined earlier than
that of scale-clad regions: this is an example of the craniocaudal and
also dorso-ventral sequence of embryonic development in vertebrates.
It may also explain why, when Sengel and Patau (1969) made interchanges of feather and scale mesoderm between duck and chick
embryos, they found that feather development prevailed over that
of scales in the majority of cases.

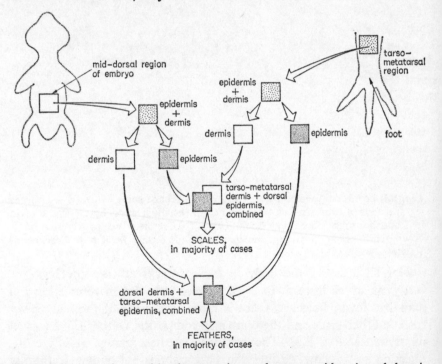

Fig. 8.9 Scheme of recombination experiments between epidermis and dermis
of feathered (back) regions and scaled (leg) regions in the chick embryo, and
their results (*Adapted from* Rawles, 1963).

Recently a surprising example of the ability of ectoderm from unusual sources to form feathers has been demonstrated by Coulombre and Coulombre (1971). They have shown that corneal tissue of the eye can be induced to form both feathers and scales, if either feather or scale mesoderm is introduced into the anterior chamber of the eye. In this case the cornea reverses a course of differentiation on which it was already started, and embarks on another course of differentiation instead. Clearly it cannot have been irrevocably determined as cornea.

We know very little about the cellular interactions involved in the formation of either reptilian scales or mammalian hairs, since both reptile and mammal embryos are difficult material for operations. Of mammalian hairs it can be said however, that their cells are capable of interacting with the cells of feather rudiments. When potential feather cells from a chick embryo are placed in culture with cells of a mouse hair rudiment, they may combine to form a chimaeric organ, which may be either a feather or a hair, according to the age of the embryos used and the degree to which the rudiments are already determined to form the organ normal for their own species. Thus, mouse skin tissue of up to 13 days' gestation can be induced to form feathers as well as hairs in any mixture of mouse and chick skin cells, but if the mouse embryo is 14-15 days or older, only hairs will form, whatever the combination with chick embryo cells (Garber, Kollar and Moscona, 1968). The ectoderm of each species evidently may show wider potentialities than are normally realized in that class of vertebrates. This is a point of general interest to students of evolution as well as to geneticists and developmental biologists. Douailly (1973) however re-tested the differentiation of combinates between mouse and chick dermis and epidermis, grown on the chorioallantoic membrane of the chick embryo. In her combinates, the epidermis nearly always formed the organ – hair or feather – appropriate to *its* origin and not to that of the dermis. Thus, chick dermis could induce mouse epidermis to form hairs, and mouse dermis could induce chick epidermis to form feathers. She concluded that the induction stimulus for hair or feather must be common to both species, but that the reaction to it is a genetic characteristic of the epidermis.

We know little yet, unfortunately, about the mechanisms by which ectoderm and mesoderm cells interact in these scale, hair or feather rudiments of vertebrate embryos. It is clear that at least two

161

steps must be involved: first, the stimulus to the ectoderm to start development by growing downwards, above each mesodermal papilla; and secondly, the more specific stimulus which will, if the ectoderm is not yet otherwise committed, induce it to form a structure appropriate to the region of the body and the species of animal from which the cells are derived.

8.4　Salivary glands

The salivary glands are organs whose secretory cells are entirely ectodermal, so that mesoderm does not contribute to their structure until they acquire connective tissue and vascular supplies. However, their development depends from the beginning on an interaction with the mesoderm lying below the future salivary epithelium. Grobstein and his associates (see Grobstein, 1953b) demonstrated this dependence of the salivary ectoderm upon some stimulus from underlying mesenchyme (Table 8.1) in organ cultures of salivary glands from mouse embryos. It was at first thought that with mesenchymes other than that normally associated with the salivary gland no differentiation of salivary ectoderm was possible. However, Cunha (1972) has pointed out that tissues of the submandibular salivary gland in the mouse respond to androgens, and he postulated that only androgen-sensitive mesenchymes could be expected to promote salivary differentiation. He therefore tested mesenchyme from the seminal vesicle, preputial gland and urogenital sinus of 13-19 day mouse embryos, culturing these with submandibular gland epithelium in the anterior chamber of the eye in male mice. When the mesenchyme was taken from 13-14 day embyros, i.e. during its androgen-sensitive period, it was able to promote the formation of a normal salivary gland, whereas mesoderm taken from other stages, not sensitive to androgens, produced atypical differentiation. This is a most interesting finding and needs further investigation. In the meantime, Lawson (1972) has shown that at least one other mesenchyme, that of the lung, which is *not* androgen-sensitive, will support salivary differentiation. She first established that either submandibular or parotid gland epithelium would differentiate successfully when their two mesenchymes were exchanged, but that the morphology of the gland corresponded to the source of the mesenchyme. Submandibular mesenchyme caused greater growth and the formation of more secretory crypts or acini than did parotid mesenchyme: this is a normal difference in the morphology of the two types of salivary

162

glands with their own mesenchymes. Submandibular gland mesenchyme also promoted an increase in amylase activity, when placed
with parotid ectoderm. When other kinds of mesenchyme were
combined with parotid ectoderm, Lawson found that lung mesenchyme allowed some development of the gland, but not so much as
its own mesenchyme. Stomach and pancreatic mesenchyme on the
other hand did not allow normal differentiation of the gland at all:
only limited histogenesis occurred. With all mesenchymes that
Lawson tested, it looked as if the epithelium took on a form
resembling the initial stages of the organ from which the mesenchyme
was derived (see Table 8.1).

Table 8.1 Effects of various mesenchymes on salivary gland ectoderm
(Data based on Lawson, 1972).

Epithelium	Mesenchyme	Result
Parotid gland	Parotid gland, in normal amounts.	Normal differentiation (control) Relatively few, loosely-packed acini. Amylase activity.
Parotid gland	Parotid gland, increased amount.	Differentiation and morphology as above, but more amylase activity.
Parotid gland	Submandibular gland.	Normal differentiation. Relatively more, closer-packed acini and greater amylase activity.
Parotid gland	Lung	Fairly normal differentiation. About one-fifth normal amylase activity.
Parotid gland	Stomach	Limited histogenesis only. Very much reduced amylase activity.
Parotid gland	Pancreatic	Limited histogenesis only. Very little amylase activity.
Submandibular gland	Submandibular gland	Normal differentiation (control) Large nos. of closely-packed acini. Normal amylase activity.
Submandibular gland	Parotid gland	Normal differentiation. Fewer, less closely-packed acini. Normal amylase activity.

Bernfield (1970) drew attention to the fact that a collagen matrix
is synthesized by mesenchyme cells placed in association with salivary
epithelium. This collagen is capable of crossing a filter placed between

the two tissues. However, collagenase treatment of salivary epithelium that has been in contact with mesenchyme does not arrest its development, though it does remove the matrix. On the other hand, a basement membrane of mucopolysaccharide that can be seen on the surface of the epithelial cells when they are viewed with the electron microscope, does appear essential for their differentiation. When it is removed by hyaluronidase treatment, the epithelial cells cease to differentiate (Bernfield and Wessels, 1970). It is possible that further ultrastructural studies may be able to tell us more about the mechanism by which the salivary mesenchyme normally interacts with its epithelium and facilitates the differentiation of the secretory cells. Spooner and Wessels (1972) have observed that microfilaments appear during histogenesis in the epithelial cells, and that if these filaments are disrupted by treating the cells with Cytochalasin B, the differentiation of the gland is inhibited. It seems possible therefore that cellular contractions play a part in the morphogenesis of the ectoderm of the salivary gland. Not all authors agree that Cytochalasin B affects microfilaments, however (cf. Section 4.2). Colchine on the other hand, which is found by some to disrupt microtubules in the cells but does not affect the microfilaments, does not block differentiation. Until we know more about the roles of microtubules and microfilaments, we cannot speculate further about the mechanism of the interaction between these two layers of the salivary gland.

8.5   Interactions in the skin of invertebrates

There is very little experimental evidence about cellular interactions in the skin and appendages of invertebrates. The group whose outer

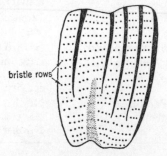

Fig. 8.10 The pattern of bristles on the integument of a leg segment in the bug, *Rhodnius* (*Adapted from* Locke, 1967). The first left femur of the fifth larval instar is shown with the integument flattened out to show the lower surface laterally.

covering or integument has come in for most study are the insects, on which Wigglesworth and his collaborators have worked for many years. Recent findings relate mainly to the pattern and arrangement of structures such as bristles, tracheae and integumentary ridges.

There is no direct evidence for cellular interactions during the formation of bristles and tracheae, but it has been argued from the regular spacing of these structures, (Fig. 8.10) that each bristle or trachea rudiment may exert an inhibitory influence, preventing cells in its immediate neighbourhood from forming another such structure. Wigglesworth (1940) noted that in the bug, *Rhodnius* there was always a minimum distance between each bristle, and he thought that this might be explained on the basis of some inhibitory influence which decreased with distance from the point of origin. Theories of this kind have often been invoked to explain the spacing of repeated structures in embryos: some of these have been mentioned already in Chapter 6. Wigglesworth's theory failed to answer one crucial question, however, which is, what initiates the formation of the first bristle at any particular place? We still do not know the answer to this.

The factors controlling the patterns of markings on the integument of insects have been studied experimentally by Locke (1967). He rotated pieces of integument, or grafted them to different levels along the craniocaudal axis (Fig 8.11), and obtained changes in the ridge patterns which were easiest to explain on the basis of a gradient-like influence spreading from cranial to caudal in each segment of the body. Thus, a graft placed alongside its original position produced no change in pattern, nor was there any change if the piece was grafted to an equivalent level in an adjacent segment (Fig. 8.11a, b). But if the craniocaudal level of the piece within one segment was changed, the bristle pattern was altered in ways that suggested a reorganization within a preexisting gradient of interaction between the tegumentary cells (Fig. 8.11c). Locke found similar evidence of gradient-like interactions between cells of each joint of the leg, when pieces of integument in these regions were transplanted or rotated as shown in Fig. 8.11.

This work on insects provides an example of how the existence of cellular interactions may be deduced in animals and in circumstances where it is impracticable to isolate the cells or to recombine different layers of tissue. It also exemplifies the fact that important interactions take place within the ectoderm alone, during organogenesis. We

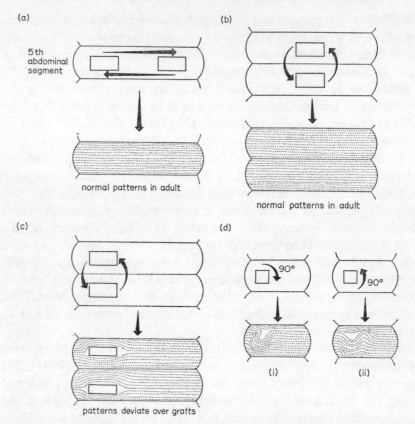

Fig. 8.11 Results of grafting pieces of integument to different positions and different segments, in insects (*Modified from* Locke, 1967). (a) Horizontal exchanges of pieces at same level in one segment, leads to normal bristle patterns. (b) Exchanges of pieces between similar levels in adjacent segments, also leads to a normal bristle pattern. (c) Exchanges of pieces to *different* levels of adjacent segments, cause deviation of the bristle rows. (d) 90° rotation of a piece causes characteristic deflections of the bristle rows.

shall see further examples of this later, when considering the development of the eye, in vertebrate embryos (see Section 8.8).

Before going on to those other examples of interaction within ectoderm, however, we must first consider a much larger type of appendage which, like the preceding appendages of vertebrates that we have discussed, consists of an outer layer of epidermis with a central core of mesoderm. This is the limb bud. In both vertebrates and invertebrates, the development of the limb involves complex cellular interactions. Most is known about these in vertebrates, so we shall consider them first, before saying a very few words about limb development in insects, the group of invertebrates on which most work has been done.

8.6   Cellular interactions in the vertebrate limb

The limb bud in tetrapod vertebrates develops from a lateral ridge of mesoderm which is covered by a layer of ectoderm. Fore and hind limb buds appear as swellings in this ridge, and are each covered by a cap of epidermis which remains thin except at the apex of the bud, where it thickens to form a ridge, known as the apical ectodermal ridge (Fig. 8.12). Early work carried out on amphibians by Harrison (1918) showed that the initial outgrowth of the limb bud was caused by some influence from the mesoderm. If mesoderm from the prospective limb region was transplanted to other parts of the body, it would form a swelling and induce the ectoderm to expand over it, forming a limb bud, with an apical ectodermal ridge too. It was later found, in bird embryos as well as in amphibians, that the mesoderm also determines whether the limb bud develops into a fore or a hind limb. Prospective leg mesoderm placed in the wing region of a chick embryo causes a leg to form here, and conversely, wing mesoderm placed in the leg region produces a wing (Saunders, 1972; Zwilling, 1972; two recent reviews).

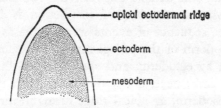
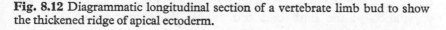

Fig. 8.12 Diagrammatic longitudinal section of a vertebrate limb bud to show the thickened ridge of apical ectoderm.

Subsequent work on chick embryos by Zwilling, Saunders and their collaborators (and dissentient rivals too!) has been concerned chiefly with the role of the apical ridge and its interactions with the mesoderm of the developing limb bud. The prolific work of Zwilling in this field is commemorated in a complete issue of the journal, *Developmental Biology* (Vol. 28, part 1, 1972). He and his colleagues established early on that if the ectoderm of the limb bud was removed, no digits formed in the resultant limb. It remained a matter of controversy for some time, however, whether this absence of digits was due to damage to the mesoderm (Amprino, 1965) or whether in a few cases digits *could* form in the absence of the ectoderm (Bell *et al.*, 1962). Saunders, Zwilling and others continued with numerous experiments in which the role of the apical ridge was investigated further. They rotated the ectodermal cap of the limb bud 90° or 180° and found that this altered the orientation of the digits correspondingly. They also grafted an additional apical ridge onto the limb bud and found that this resulted in duplications of the digits. Further light was thrown on the role of the apical ridge by a study of mutants in the chick. Polydactylous mutant chick embryos were found to have a wider apical ridge than normal, while 'wingless' mutants (which have reduced or deformed wings) developed an apical ridge at first, but this soon degenerated. Interestingly, it was then found that mesoderm grafted from wingless mutants into normal limb buds caused their apical ridge to regress and produced wingless chicks. So evidently the mesoderm of the mutant lacked some factor which is normally necessary to maintain the growth of the apical ridge. This was called the 'maintenance factor' and its discovery unleashed another heated controversy as to whether the mesoderm, or the ectoderm, was the main controller of limb development and of the number of digits that formed. Most people now agree that there must be reciprocal interactions between the two tissues, but we do not yet know what mechanism of interaction is involved, at the cellular level. The sequence of events seems to be as follows (cf. Fig. 8.13): first, mesoderm of the prospective limb region induces a bud to form, covered by ectoderm, and specifies that this shall be either a fore or a hind limb; next, under the influence of some factor from the limb bud mesoderm, an apical ridge forms in the ectoderm; then this ridge, maintained by factors from the mesoderm, controls the orientation of the limb bud and evokes the formation of digits. It should be remarked, however, that not all present investigators accept the somewhat complex interpretation of this chain of events that is shown in the figure, and recently Amprino and Ambrosi (1973) have put forward a simpler view of the interactions in the developing limb which is gaining some support.

Some, but not all of these relationships between mesoderm and ectoderm have been confirmed in the limb buds of amphibians and mammals. Tschumi (1957) observed an apical ridge in the frog, *Xenopus* and showed that if this ridge was rotated, the orientation of the limb and its digits were correspondingly altered, as in birds.

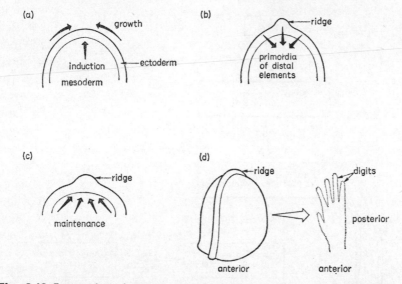

Fig. 8.13 Interactions between mesoderm and ectoderm in the developing limb bud. (a) Mesoderm induces formation of the apical ectodermal ridge; (b) the apical ridge controls the formation of distal structures, including digits, in the mesoderm. (c) The mesoderm provides a 'maintenance factor' essential for the survival of the apical ridge, and also specifies whether fore or hind limb shall form. (d) The orientation of the apical ridge controls the orientation of the limb and its digits.

He also found that removal of the ectodermal cap prevented digits from forming. In mammals, detailed histological and histochemical studies by Milaire (1965) show evidence of activity in the apical ridge, correlated with the development of the limb bud and its digits. Experimental work on limb buds in mammals like that of Saunders and Zwilling in the chick has not yet been possible, however.

The gradual shape changes in a growing limb bud are moulded mainly by a series of cell divisions and rearrangements within the mesoderm. These have been followed in some detail in the chick. Hornbruch and Wolpert (1970) observed a proximo-distal gradient in mitotic rate in the cells of chick limb buds, but they considered this insufficient to account for the changes in shape as the bud grew. Janners and Searls (1970) paid special attention to the mitotic rate in myoblasts and also to the increased rate of labelling with tritiated thymidine, indicating higher mitotic rates, in the cells of the future limb bud area as compared with the rest of the flank. This study also did not provide sufficient explanation for the shape changes in the limb. Ede and Law (1971) applied a computer programme system

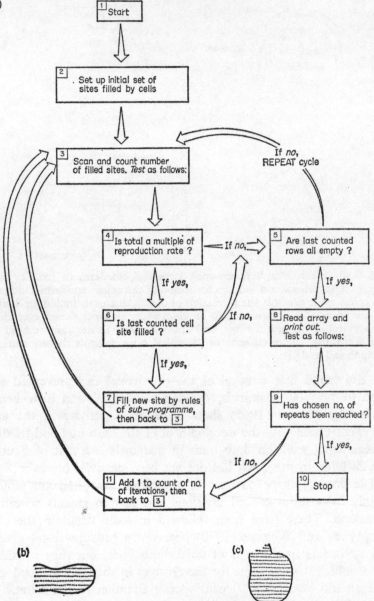

Fig. 8.14 (a) Main 'flow diagram' of a computer programme for limb development. (b) Print-out pattern obtained with this programme, giving the outline of a normal limb bud. (c) Print-out obtained from a programme involving less forward movement by the cells, giving the outline of a 'talpid' mutant limb bud. (*Adapted from* Ede and Law, 1969.)

to the study of the cell-interactions in the chick limb, comparing normal embryos with 'talpid' mutants in which the limb bud is shorter and broader. They were able to show that the stunted limb bud of the mutant could result from a failure of its mesoderm cells to migrate distalwards at the same rate as in normal limbs (Fig. 8.14). Ede and Agerbak (1968) had already shown that the talpid cells tended to aggregate into larger clumps than did those of normal limbs, after they had been disaggregated *in vitro*. These findings emphasize that besides the ectoderm/mesoderm interactions there are also important cell-interactions within the mesoderm of the limb, which mould its shape. One need not postulate more than the quite common activities of embryonic cells – i.e. mitosis, adhesion and forward shifting, but their orderly programming can result in a distinctive shape to the bud.

Cell interactions have also been studied, in mammals as well as in birds, at later stages of limb development when chondrogenesis begins. In the mouse, limb cartilage does not form if the ectoderm is removed prior to $10\frac{1}{2}$ days' gestation, according to experiments by Milaire and Mulnard (1968). They also found that in culture, fragments of spinal cord induce cartilage-formation in the limb, just as they do in the somites (Section 6.5). The somites, too, may be implicated in the control of limb development: Pinot (1970) reported better differentiation of limb bud grafts in the chick when somite tissue had been included with them.

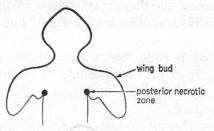

Fig. 8.15 Positions of the necrotic zones that mould external form in the wing buds of chick embryos.

Another type of cell-interaction which plays an important part in later limb morphogenesis is cell death. Saunders and co-workers (see Gasseling and Saunders, 1964) first drew attention to a special *posterior necrotic zone* of ectoderm cells, at the angle between the base of the wing and the flank of the chick embryo (Fig. 8.15). These cells even when they are transplanted to the anterior margin of the wing, still undergo necrosis at the time when they would have if left *in situ*. It has since been shown that there is also a *polarizing zone* of cells adjacent to the necrotic zone, and that when this is transplanted, it will alter the orientation of the limb bud, so that the zone still lies at the posterior margin (McCabe and Saunders, 1971). The polarizing zone is capable of inducing the cells adjacent to its new site to form a new necrotic zone. It appears then that the transplanted posterior necrotic zone in earlier experiments died either because it had already been acted on by the polarizing zone at its original site, or because part of the polarizing zone was transplanted with it, and later caused necrosis to occur. Between the digits of developing limbs there is also necrosis of ectoderm cells (Fig. 8.16), which helps to mould the form of the digits. In addition, some central mesoderm undergoes necrosis as the cartilage elements form (Dawd and Hinchliffe, 1971; Hinchliffe and Ede, 1968, 1973). This necrosis is more extensive in talpid mutants (Ede and Flint, 1972). Whether any products from the necrotic cells are essential in initiating differentiation in other parts of the limb, is not know. We may hope that some of the cellular interactions that take place in the limb will be elucidated by future ultrastructural work. Rajan and Hopkins (1970) and Kelley (1970) have already looked in detail at the central necrotic zone, but more observations are needed on the zone between ectoderm and mesoderm at the apex of the limb bud. Balinsky (1972) has observed a collagenous basement membrane on the inner surface of the apical ectoderm in both fore and hind limb buds of the toad *Bufo*. He has also seen processes from the mesenchyme cells extending into this membrane, which may provide means for chemical exchanges between the two layers. It should be possible by radioactive labelling to trace whether any substances are exchanged here during the development of the limb.

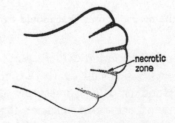

Fig. 8.16 Necrotic zones between the future digits, in a developing limb bud of a tetrapod vertebrate.

8.7 Limb development in insects

Experimental work on limb rudiments in insects is not in any way comparable to that on vertebrate limb buds. It provides an example of quite different problems in morphogenesis, however, and is therefore of some interest. The limb rudiments exist throughout larval development as minute, imaginal discs, just as do other organs such as the genital system (Section 7.2.5), the eyes and the antennae. These discs are so small and delicate that considerable skill is needed to remove or transplant them, let alone isolating parts of them to find out their prospective fates or to find evidence of cell-interactions. Nevertheless, some skilled work of this kind was done several years ago by Goldschmidt, Hannah and Piternick (1951). They managed to fuse pairs of leg discs in *Drosophila* and thus showed that these interacted in such a way as to produce nearly single legs. This indicates powers of adjustment and 'regulation' similar to those in vertebrates, and is somewhat surprising since the cells of insect embryos are much less able to adjust their course of differentiation to fit changing environments than are the cells of vertebrate embryos. The converse experiment, testing the ability of isolated parts of the leg disc to regulate towards forming a whole leg, was achieved by Schubiger (1971) and Bryant (1971). They injected portions of the leg disc into host larvae which were about to metamorphose. They found that the proximal half of the disc is able to form a complete leg, but an isolated distal half cannot reconstitute the proximal part: instead it tends to *reduplicate itself*. Ouwenweel (1971) has obtained similar results using transplantation methods. From all these results one may conclude that in the normal disc, the component cells exert mutual inhibitory actions which limit the number of potentialities that are realized in each area. All of the cells in a leg disc are closely related in terms of cell-lineage: Bryant (1970) and Bryant and Schneiderman (1969) showed that the limb discs in *Drosophila* develop as clones, from about 20 progenitor cells. Each clone forms a longitudinal strip of tissue on the leg or wing, so they must proliferate in a very orderly sequence in one dimension only. This makes it all the more remarkable that half-discs and double-discs in the experiments mentioned above were able to change their pattern of development and to regulate so well.

In the limb discs as in other imaginal discs, it has been shown that interactions with other tissues play a part in controlling their

173

development. For instance, the eversion of the discs at meta-morphosis is dependent on the hormone ecdysone, secreted by cells of the prothoracic gland (cf. Chapter 10). There may be interactions between grafted limb discs and host larval tissues, too, for repeated transplants of the discs into immature larvae causes 'transdetermintaion' to occur (cf. Table 10.2) so that some-thing other than a limb is eventually formed.

It can be seen from this brief account that some important types of cellular interaction have come to light from the studies on limb development in insects. They are not interactions unique to insects, however: regulation, cloning, and linear sequences of cell divisions occur in a number of other developing systems in both vertebrates and invertebrates.

8.8   Ectoderm/ectoderm interactions in the eye

The eye is composed mainly of nervous tissue and therefore develops entirely from ectoderm, with mesoderm contributing only to its outer protective coverings and its blood supply. We shall deal only with the eye of vertebrates, since the tissue and cell interactions which occur during the development of the eye have

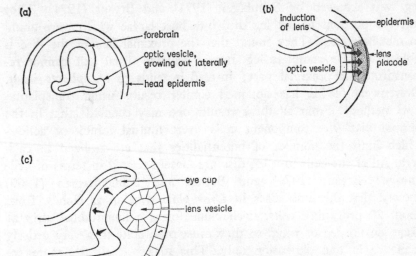

Fig. 8.17 Diagrams to show the outgrowth of the eye rudiment and its induction of the lens rudiment in vertebrate embryos. (a) Coronal section of head at level of forebrain with optic outgrowths beginning. (b) Enlarged view of optic vesicle coming into contact with the epidermis. (c) Invagination of the optic vesicle and formation of lens rudiment.

been studied in considerable detail in vertebrates, but very little yet in invertebrates.

The eye rudiment appears in vertebrate embryos as a ventro-lateral outgrowth from the forebrain. This optic vesicle makes contact with the inner surface of the head ectoderm, and interacts with it to induce the formation of a lens (Fig. 8.17). This is a very well-known example of embryonic induction, for it was demonstrated in amphibians long ago by Spemann (1903). He found that grafting the eye rudiment under the ectoderm in other parts of the body induced the formation of a lens in these abnormal sites, and that conversely, removal of the eye rudiment from its normal position before it had made contact with the surface ectoderm prevented a lens from forming. We shall mention here only a few much more recent findings, which relate to the nature of the interactions between the optic vesicle and the lens ectoderm.

Brahma (1959) investigated in *Xenopus* embryos the extent to which lens-formation is independent of contact with the eye cup. He found that ectoderm removed from over the optic vesicle at the tailbud stage was not yet capable of forming a lens in isolation, so the induction process was evidently not complete by that stage. On the other hand, Babcock (1963) removed the optic vesicle at the same stage and found that a lens could form in the ectoderm that was left *in situ*. So it seemed that other factors from tissues of the head region might be involved in lens induction. Adding this evidence to the considerable body of work on other species of amphibians (see Balinsky, 1970, for review) it has now been concluded that the head mesoderm, as well as the eye cup, takes part in inducing the lens to form.

There are several other interactions between the components of the developing eye, which have been revealed by experimental work on amphibians and birds. For instance, Twitty (1932) showed by grafts of the lens between different species of salamanders, that the size of the eye cup is affected by the size of the lens. A lens from a large species grafted into the eye cup of a smaller species causes the eye cup to grow faster than it would normally, and vice versa. Later, the cornea depends for its development on the presence of the lens, and the eyelids in their turn develop only if the cornea is present. All these interrelationships are summarized, with additional ones too, in Coulombre's diagram based on his work on the chick embryo (Fig. 8.18).

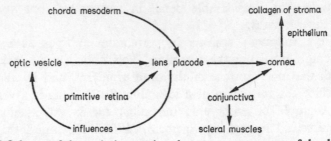

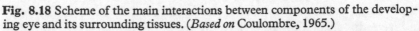

Fig. 8.18 Scheme of the main interactions between components of the developing eye and its surrounding tissues. (*Based on* Coulombre, 1965.)

During its development the component parts of the eye cup are able to undergo considerable regulation in their course of differentiation. In early stages the retina and the pigment layer may be interchanged, and a normal eye will still develop. More remarkable still, in adult urodele amphibians, parts of the retina and iris can differentiate and form a new lens, if the original one is removed. Classic experiments by Stone (1954) showed that a diffusible inhibitor is secreted into the anterior chamber from the normal lens, inhibiting lens development in any other parts of the eye. Removal of the lens causes loss of this inhibitor, so a new lens is then formed, usually from the dorsal iris. However, injection of fluid from the anterior chamber of a normal eye prevents lens regeneration in a lentectomized eye. Conversely, isolating the lens of a normal eye from the retina by inserting strips of mica, causes extra lenses to form because the inhibitor is not reaching the retina. Stone's crucial experiments are shown in Fig. 8.19.

All of these regenerative processes are accompanied by the appearance of lens proteins in the new lens, which are identical immunologically to the proteins of the normal lens (Braverman *et al.*, 1969). These proteins are not present in the normal retina or iris, but appear soon after lens regeneration has begun in either of these tissues (Campbell, 1965). The antigens have been localized by applying antibodies labelled with fluorescent tracers to sections of the eye (Coons and Kaplan, 1950).

The eye of the chick embryo has the same capacity as that of adult urodeles to regenerate a lens, from either neural retina, or pigmented layers. Coulombre and Coulombre (1970) found that this would also occur in organ cultures *in vitro*, but that under these conditions some neural retina must be present if the pigmented layer is to regenerate successfully. Surprisingly, they found that

176

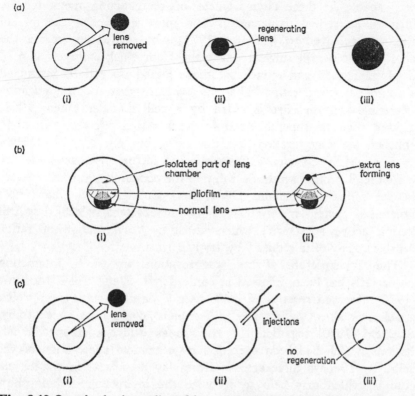

Fig. 8.19 Stone's classic studies of lens regeneration in urodele amphibians. (*Diagrams modified from* Stone, 1953.) (a) Normal regeneration: (i) lens is removed; (ii) regeneration begins from dorsal margin of iris; (ii) lens regeneration complete. (b) If part of the lens chamber is isolated by a film of mica, (i) an extra lens may be regenerated in this isolated region, (ii). (c) If, after removal of the lens as in (i), daily injections of aqueous humour from an eye containing a lens are given, as in (ii), lens regeneration is inhibited as in (iii).

neural retina from the eye of a mouse embryo was just as effective as that of the chick in stimulating lens regeneration by pigmented chick tissue. It has been known for some time that chick and mouse retina cells are capable of combining in culture and forming a chimaeric retina (Moscona, 1962). So it looks as if the 'signals' both for retinal differentiation and for lens regeneration are similar in the cells from these two widely different animals.

Perhaps the most remarkable degree of co-ordination between ectoderm cells of the eye is shown in experiments in which chick embryonic retina cells are disaggregated and then grown in culture, either *in vitro* or on the chorioallantoic membrane. The retina cells

177

are capable in these circumstances of reorganizing themselves to form intact retinal tissue again. The most spectacular example of this was achieved by Fujisawa (1971) who seeded an area of the chorioallantois with single retinal cells from chick embryos of 6-10 days' incubation and found that these joined up and reconstructed a complete neural retina. There is evidence that the reaggregation of retinal cells *in vitro* is aided by a cell exudate. Lilien (1968) showed that the supernatant from chick retina cell cultures would enhance the reaggregation of retina cells, but not of other tissues, so it evidently contains a tissue-specific factor. Cycloheximide, an inhibitor of protein synthesis, inhibits the accumulation of this factor in the retina supernatant. Daday and Creaser (1970) have now isolated a protein material from the supernatant, which they call 'retina active substance', which enhances reaggregation of retina cells. Its activity is reduced by trypsin treatment.

The ultrastructure of the reaggregation process in interacting retina cells has been followed in some detail. Sheffield and Moscona (1970) observed rosettes of cells after 7 days in culture, in which attachment sites, the *zonulae adhaerentes*, could be seen centrally. Sheffield (1970) regarded the zonulae as evidence of specific adhesions, which had been preceded by a random aggregation of the cells. By 10 days there are plexiform layers present in the retinal clumps, which may help to stabilize the intercellular relationships (Sheffield and Fischman, 1970).

There are many other examples in the developing nervous system of interactions between ectoderm cells, controlling the structure of neural tissues and the connections they make with end organs. This is too large a subject to include in the present chapter, so the next chapter will be devoted exclusively to it.

9

Interactions between cells of the early nervous system and other embryonic tissues

The cells of the nervous system are specialized for long-distance, rapid communication within the animal body. As these specializations are gradually acquired during their development (cf. Fig. 9.1), the cells become increasingly able to influence and to be influenced by other organs with which their cytoplasmic endings (dendrites and axons) make contact. We shall deal in this chapter with examples both of the influence of the nervous system on the development of end-organs with which it connects, and of the reciprocal influence of some end-organs on the developing nervous system. All the experimental evidence that we shall review about these events comes from vertebrates, since much less is known about equivalent events in invertebrates, except at post-embryonic stages which are beyond the scope of this account.

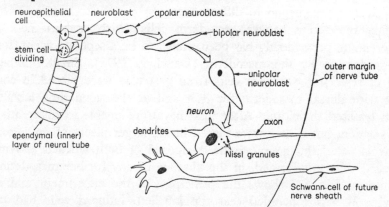

Fig. 9.1 Diagrammatic representation of the development of nerve cells, seen as if in a transverse section of the neural tube.

179

Besides interactions with other organs, there are also important interactions between component cells within the nervous system. These begin at early stages in the development of the neural tube and they control both its early morphology and the later details of its histological structure. We will consider first some examples of these internal interactions within the developing nervous system.

9.1 Cellular interactions within the central nervous system

9.1.1 *Interactions between induced ectoderm cells*

It used to be said that neural induction consisted of two events: the stimulus passing from the mesoderm to the ectoderm, which was called 'evocation', and a further process of organization within the ectoderm in response to continued contact with the mesoderm, which was called 'individuation' (Waddington, 1956). These two terms are now seldom used, but it has often been confirmed that for the ectoderm to respond adequately to the induction by the mesoderm, there must be an adequate further process of interaction between the induced ectoderm cells as a group. Experiments on cells of amphibian embryos in culture (Jones and Elsdale, 1963; Deuchar, 1970a, b) show that neural differentiation is initiated successfully in induced ectoderm only if a minimum of 30 to 40 ectoderm cells are present. For the differentiation to be maintained and for a viable neural tube to form, at least 100 ectoderm cells must be present initially. Fewer than 100 cells tend to degenerate (i.e. to round up, fall apart and die) after a few days in culture, even if they had at first begun to differentiate.

This interaction between ectoderm cells which enables them to differentiate permanently has been shown to be a separate event from the induction by the mesoderm (Deuchar, 1971). Small numbers (10 or fewer) of ectoderm cells from gastrulae were placed in contact with dorsal mesoderm for 3 h and at the same time labelled with tritiated thymidine. After this they were unable to differentiate in isolation, but if they were placed inside larger pieces of uninduced gastrula ectoderm, neural tissue composed of both labelled and unlabelled cells was formed, in the absence of any further mesodermal induction. Fig. 9.2 shows the principles of the experiment and its results. It was concluded that the labelled, induced cells had co-opted some of the unlabelled, uninduced ectoderm and together with it became organized into neural tissue. It is not yet known

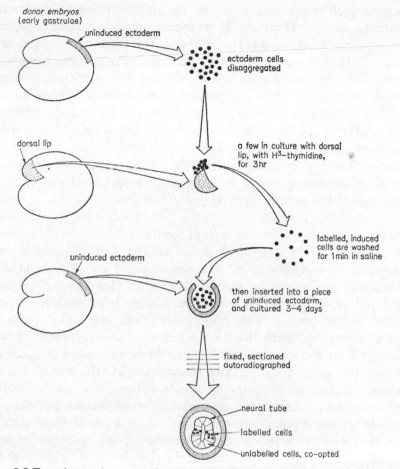

donor embryos
(early gastrulae)

uninduced ectoderm

ectoderm cells
disaggregated

dorsal lip

a few in culture with dorsal
lip, with H³-thymidine,
for 3hr

labelled, induced
cells are washed
for 1min in saline

uninduced ectoderm

then inserted into a piece
of uninduced ectoderm,
and cultured 3–4 days

fixed, sectioned
autoradiographed

neural tube

labelled cells

unlabelled cells, co-opted

Fig. 9.2 Experiment demonstrating the co-option of further ectoderm by a few induced cells, to form neural tissue. (Deuchar, 1971.)

by what mechanism this co-option of uninduced cells occurred. There are probably many other instances in development where uncommitted cells may be influenced by contact with other cells of their own kind which have received some stimulus initiating differentiation. For instance, the segmentation of somites (cf. Chapter 5) may be controlled in this way after some 'evocation' process has acted on the most anterior cells of the series.

We have no reason to suppose that immediately after neural induction the ectoderm cells have already acquired specializations for communication such as are found later in developing neurons. Nerve axons, dendrites and their connections at synapses do not appear until much later stages in the differentiation of the central nervous system. However, in a recent electron microscope study Hayes and Roberts (1973) have observed structures which look like the beginnings of synapses between neurites, appearing very soon after the neural tube has closed. So from this stage onwards it is possible that cells in the embryonic nervous system interact and co-ordinate its development by signals similar to nerve impulses, passing from cell to cell across structures like synapses. Until other workers have confirmed the existence of synapse-like structures in the neural tube of other vertebrates, however, one cannot assume that all developing neural cells have these special means of inter-communication at such early stages of embryonic life.

9.1.2 *Interactions between cells of the brain*

The shape of the developing brain is, like that of other organs, moulded by movements and rearrangements of its cells. Its internal structure, too, results from these cellular activities, as well as from the differentiation of the cells into neurons and neuroglia (nerve cells, and supporting cells, respectively). The cell bodies form the grey matter and their long cytoplasmic processes (dendrites and axons) form the white matter, in both brain and spinal cord. In the brain, the three main regions, forebrain, midbrain and hindbrain, appear at first as swellings externally. These arise partly because of differential mitotic rates and partly by migrations of cells outwards to form thicker layers of grey matter in these three regions. In the cerebral cortex of the forebrain region, successively more layers of cells migrate out and form complex groupings. Fig. 9.3 summarizes the chief phases of mitosis and migration that occur during the early development of the brain in vertebrates.

The only mechanism we know of that might explain the complex groupings of cells in the embryonic brain is that they may have differential adhesive affinities. There is in fact evidence of this, from studies of brain cells in culture, though one cannot be certain that the cells show the same properties *in vivo*. Adler (1970) was able to show that dissociated cells obtained from the neural tube of three to six-day old chick embryos gradually changed their pattern of reaggregation with advancing age of the embryo. There was an increasing tendency for cells which were still synthesizing DNA (i.e. still dividing and not yet differentiated into neurons) to aggregate separately from those that had ceased to synthesize DNA.

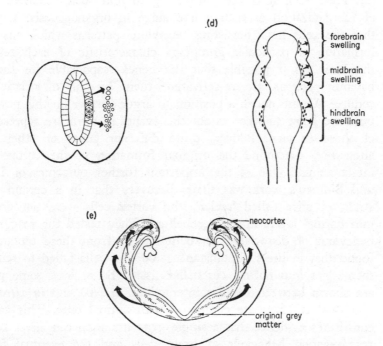

Fig. 9.3 Summary diagrams of the cell-interactions and migrations that take place during development of the brain in vertebrates. (a) Induction of neural plate by the underlying mesoderm cells; (b) Adhesive affinities between neural plate cells controlling neurulation; (c) Mitosis of ependymal layer cells, and outward migration of neuroblasts to form the mantle layer (future grey matter). (d) Phases of mitosis and cell migration cause a series of swellings to appear, and three main protuberances persist as the forebrain, midbrain and hindbrain. (e) Secondary migrations of cells from the original mantle layer form outer layers of grey matter, the cortex of the brain. (a), (b), (c) and (e) show transverse sections; (d) is a horizontal longitudinal section.

Later, regional specificities appeared: thus, optic lobe cells adhered to each other in preference to cells from any other region of the brain.

183

G

Some very interesting observations have been made on the ability of cells from the cerebral cortex of mouse embryos to reaggregate and to form normal brain tissue *in vitro*, after they have been completely dissociated experimentally (DeLong and Sidman, 1970). At quite late stages of development, shortly before birth, regions of the isocortex of both the hippocampal lobe and the cerebellum can be dissociated into single cells *in vitro*. These cells are then able to reaggregate again and to form a remarkably normal cortex (cf. Plate 9.1). It is very surprising to find such extensive powers of reorganization at such a late stage in organogenesis. It implies that the cells have persisting behaviour patterns which ensure the formation of particular groupings characteristic of each region of the brain. Is it possible that the genes responsible for these behaviour patterns are re-activated, then, by the dissociation procedure? We are not in a position to argue further on this possibility, until we know to what extent the events *in vitro* are representative of those *in vivo*. Perhaps quite different genes or other factors altogether, determine the original formation of the cortex in the intact animal. One of the important further outcomes of DeLong and Sidman's work was their discovery that in a certain mutant strain of mice called 'reeler', the cortex cells were not organized into normal layers in the cerebellum. They tested the reaggregation behaviour of dissociated cerebellum cells from these mutants, and found that in these circumstances too, the cells failed to reorganize themselves into normal cerebellar tissue. So at least some parallels are shown between cellular interactions *in vivo* and *in vitro*, when abnormal embryos are compared with normal ones. This case also constitutes evidence that a single gene mutation can affect both the reaggregation behaviour of brain cells and the eventual form of an area of the brain. Associated with the failing in cellular interaction and their inability to arrange themselves into the normal layers of cerebellar tissue, the function of this region is impaired and the animals cannot walk properly but tend to turn in circles: hence the name 'reeler'. A change in cell behaviour is linked with a change in animal behaviour.

It is possible that among the signals by which the cells of the prenatal cortex are able to inform and react to each other during their reaggregation, are patterns of electrical impulses such as occur in the fully-developed brain. Specific patterns of activity have been shown to arise in isolates of developing brains and spinal cords of mammals (see Crain, 1966). Corner and Bot (1969) found that isolates of brain tissue from the chick embryo 2-3 days before hatching (a stage equivalent to that at which DeLong and Sidman's experiments were carried out on the mouse) showed a definite, cyclic rhythm of electrical activity resembling the 'activated sleep' rhythm in an intact animal's brain.

9.2 Migrations during the formation of the nervous system

9.2.1 *Migrations of whole cells*
With the possible exception of germ cells, more migration occurs among cells of the nervous system than in any other tissue. We have already seen an extreme example of this, in the neural crest cells which become detached from the edges of the neural plate and migrate out to give rise to a variety of peripheral tissues including non-nervous ones (Section 8.1). A few more points may be added here about the extent of their migrations.

Among the nervous elements derived from the neural crest are the parasympathetic ganglia of the autonomic nervous system. These eventually lie peripherally, on the organs they supply, and to get to these positions neural crest cells may undergo more than one migration. Andrew (1970, 1971) has traced the origin of the parasympathetic ganglia of the gut wall of the chick embryo, and has found that they are derived from neural crest cells of the vagus ganglion in the neck, which have migrated secondarily to several regions of the gut. Her work has since been supplemented by the grafts of neural crest cells from quail embryos into chick embryos carried out by Le Douarin and Teillet (1973), which were referred to briefly in the last Chapter (Section 8.1). Since quail cells are easily distinguishable from those of the chick, these workers were able to confirm that the grafted neural crest cells contributed to the parasympathetic ganglia of the gut.

It was stressed in Chapter 5 that as a result of the cell migrations in the gastrula, possibilities of cellular interactions in early embryonic development are greatly increased. The same is true in the case of neural crest cells. Those which migrate laterally in the head region, interact with several areas (placodes) of surface ectoderm cells and combine with them to form the cranial ganglia. This interaction evidently involves mutual inhibitory processes as well as stimulatory ones, for it has been shown in chick embryos that removal of the neural crest will cause the placodal cells of the cranial ganglia to proliferate more than usual and thus to form normal-sized ganglia on their own (Hamburger, 1962). In contrast to this, we may cite an instance in which neural crest cells evidently develop independently of the ectodermal component of an organ and do not interact with it. Deol (1970) has observed that the dominant gene Wv in mice, which affects neural crest derivatives and causes a spotted coat colour owing to inadequate pigment-formation by melanoblasts, affects the inner ear also. The acoustic ganglion and the saccular portion of the inner ear, which are derived from the neural crest, are abnormal, but the vestibular region which is derived from placodal cells is unaffected. So in this case the ectodermal placode cells do not develop any extra material to compensate for the inadequacies of the neural crest cells, and there is no evidence that they have propensities which are normally curtailed by their association with the neural crest cells.

One could point to many other examples of cell migration in the nervous system (see Rakic and Sidman, 1969, for a description of cell migrations in the human fetal brain and for further references). From their first formation, the neuroblasts migrate outwards to form the mantle layer (Fig. 9.3), then groups of them collect to form dorsal and ventral horns of grey matter in the primitive brain and the spinal cord. During the later development of the brain, as we have already indicated, the layers of the cortex in both cerebrum and cerebellum are formed by a series of migrations of groups of cells from the original grey matter. Unfortunately little is known yet about what cellular interactions may control these migrations, so that they cannot be discussed in detail here. There is still a wide-open and fascinating field here for further investigation. The most hopeful line of attack so far has been that of DeLong and Sidman, studying cell behaviour in mutants that affect brain function. Not many of these 'psychomotor' mutants have yet been recognized, however.

9.2.2 *Migration of nerve processes*
A feature unique to nerve cells is their formation of long outgrowths, the axons or nerve fibres, which conduct impulses to other nerve cells or to end-organs. During embryonic development some of these axons grow beyond the confines of the central nervous system and find their way to their appropriate destinations.

Despite extensive studies both *in vivo* and *in vitro*, it is still not known how the tips of the axons are directed towards the correct end-organ. Experiments on nerve regeneration and on embryonic nerves grown *in vitro* have produced controversies as to whether some chemical stimulus, or some kind of guidance by mechanical contact, controls the direction in which the nerve fibres migrate. Schwann cells (the neural crest cells which form the nerve sheath) have been found to attract and guide the growth of axons, both when the axons are regenerating *in vivo* and when they are developing in tissue culture. Collagen fibrils will also orientate nerve axons, which tend to grow along them in culture (Abercrombie and Lubinska: see Curtis, 1962b). A protein extract from salivary glands, known as 'nerve growth factor' (Levi-Montalcini, 1958) has been shown to stimulate outgrowth of sensory nerve fibres both *in vivo* and *in vitro*, but it does not appear to influence the direction of their growth. At the stage when the first nerve fibres are growing towards their end-organs in the embryo, however, there are as yet no Schwann cells near the tips of the axons, very little connective tissue or collagen, and no circulating factors from salivary glands. So there is no obvious external agent, either mechanical or chemical, to guide the axons in the early embryo.

It was eventually suggested by some investigators that each nerve cell may have some 'chemical marker' giving it an affinity for a certain neuron which would be the next one in the pathway, and at the end of the series, an affinity for a certain end-organ. This idea was formulated as the 'neuronal specificity theory' and appeared to be supported by experiments on the eye of amphibians, initiated by Sperry (1943, 1944). But as we shall see later, more recent experiments by Gaze and his collaborators have shown that this theory requires some modification (see Section 9.4).

To mention one other mechanism by which nerves may be guided to their end-organs, it is possible that they send electrical signals down the growing axons. Lamont and Vernon (1967) grew pieces of spinal cord and of dorsal root ganglia in culture, and found that they sent out active nerve processes along which pulsations could be seen to pass. It has been noted in earlier paragraphs that there are electrical discharges between early embryonic cells, and that synapses may already exist between cells of the neural tube.

Once a nerve has reached its end-organ, it can make normal junctions with it *in vitro* as well as *in vivo*. Veneroni and Murray

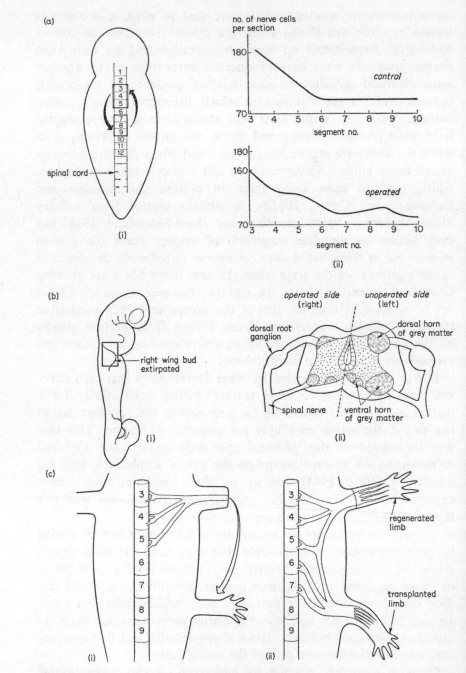

(a)

(i)

spinal cord

no. of nerve cells
per section

180

control

70
3   4   5   6   7   8   9   10
segment no.

(ii)

180
160

operated

70
3   4   5   6   7   8   9   10
segment no.

(b)

(i)

right wing bud
extirpated

operated side
(right)

unoperated side
(left)

dorsal root
ganglion

dorsal horn
of grey matter

spinal nerve

ventral horn
of grey matter

(ii)

(c)

(i)

3
4
5
6
7
8
9

3
4
5
6
7
8
9

regenerated
limb

transplanted
limb

(ii)

Fig. 9.4 A few of the classic transplantation experiments carried out by Detwiler and Hamburger in their studies on the development of the nervous system

(1969) and Shimada *et al.* (1969) have observed the formation of normal neuromuscular junctions in cultured tissues of the chick embryo. Cholinesterase activity may be detected in the early neuro-muscular junctions in intact newt embryos by histochemical methods (Kocher-Becker, Drews and Drews 1970) and also with the electron microscope (Sabatini *et al.*, 1963). Though this was not suggested by these authors, one wonders whether the enzyme activity may perhaps precede the formation of junctions and even act as a chemical signal to attract the nerve to the muscle. In their work quoted above, Veneroni and Murray observed that the exploring tips of the nerve fibres appeared to stimulate certain nuclei and the membrane of the muscle bundle (myotube) to bulge towards them, helping to anchor them in position (Plate 9.2). So it seems that there could be both mechanical and chemical interaction at this stage.

9.3 Interactions between nerve cells and end-organ cells

9.3.1 *Influences of end-organs on the nervous system*
It has been known since the classic experiments of Detwiler and Hamburger in the 1930s that the presence or absence of developing limb buds in amphibians and birds affects the size of the ventral horns of grey matter in the spinal cord. These horns are enlarged in the segments of the cord opposite the limbs (cervical and lumbar regions) and contain more motor neurons than in other regions. Detwiler and his colleagues carried out a series of transplantations and extirpations, both of segments of the spinal cord and of the limb buds (see Fig. 9.4) and found that there was always an en-

---

(see Detwiler, 1936). (a) Exchanges of brachial segments (3, 4, 5) of spinal cord with segments 7, 8 and 9, in *Ambystoma*, and resultant changes in the numbers of nerve cells maturing in these segments. (i) Operation, at tailbud stage; (ii) results, expressed graphically. (b) Extirpation of wing bud from right side of a chick embryo, and resultant reduction in size of dorsal and ventral horns of grey matter in the spinal cord, and of the dorsal root ganglia, on that side. (i) The operation, on a 3–4 day chick embryo; (ii) results seen in a transverse section of the spinal cord at a later stage. (c) Transplantation of forelimb in *Ambystoma* to the level of segments 7, 8 & 9, and the resultant pattern of re-innervation of both transplanted and regenerated forelimbs. (i) The operation; (ii) Nerves of segments 6 and 7 are attracted to innervate the transplanted limb: so is one of its normal nerves from segment 5. Meanwhile a forelimb has regenerated from the stump, and is innervated by segments 3, 4 and 5. So the nerve of segment 5 has 'divided its loyalties' between its proper segment and its proper limb.

largement of the ventral horn, in any region of the cord which came to lie level with the developing limb buds (see Detwiler, 1936 for review). Conversely, removal of the limb buds, or transfer of cervical and lumbar cord segments to non-limb levels could at certain stages prevent the usual enlargement of the ventral horns in these segments from occurring. Similar findings were reported later in chick embryos by Hamburger and Levi-Montalcini (1949). They also noted that in regions of the cord other than at limb levels, there was a large-scale death of neuroblasts. Cell death also occurred at limb levels but was not on such a large scale as in other regions. Evidently the numbers of neuroblasts that form throughout the central nervous system are in excess of the numbers of neurons eventually required, and this is another example of a morphogenetic process involving cell death. In this case, the cells are not so rigidly 'programmed' to die as are those of organs such as the limbs and the palate, however (cf. Chapters 7 and 8) since they respond to environmental influences. There is evidently some communication between the cells of the spinal cord and those of the limb buds, which results in a curtailment of cell death in the cord. The obvious route for this communication would seem to be via the axons of either motor or sensory spinal nerves, which are beginning to develop at this stage.

A quantitative study of the relationship between presence or absence of a limb bud and the numbers of neuroblasts that degenerate, was undertaken in the frog *Xenopus* by Hughes and his associates. They found that both sensory and motor neuroblasts were affected. Hughes and Tschumi (1958, 1960) did experiments in which they either removed the limb buds surgically, or suppressed their development by treatment with nitrogen mustard. They found that this caused more neuroblasts than normal to degenerate, both in the dorsal root ganglia and in the ventral horns opposite the absent limbs, and that this led to a reduced number of neurons in these regions compared with control, unoperated animals. Hughes and Lewis (1961) described the degeneration of neuroblasts in more detail, pointing out that as already noted in the chick by Hamburger and Levi-Montalcini, there was large-scale death of neuroblasts throughout the cord, and that the same was true in the spinal ganglia. Prestige (1967) did careful counts and estimated that normally, at least one third of the motor neuroblasts degenerated in the trunk region. This confirmed that the limb ablation experiments had merely enhanced a process that occurs in normal development of the spinal cord. Prestige also found that the reaction to limb ablation was more rapid, in lumbar segments, if the ablation was carried out at a late stage in maturation of the neuroblasts. Ablations at earlier stages did not show any effects until after a delay period. It seemed that the neuroblasts had to reach a certain stage of differentiation before the process either of neuron-formation or of degeneration could begin. It is an interesting problem, by what mechanism certain neuroblasts are selected to die, while others survive. So far we have no information as to how this cell-selection is made, either in normal or in operated animals.

9.3.2 Influence of the nervous system on the development of end organs

There are many examples known in biology and medicine of an organ failing to develop, or atrophying later, because of lack of a functional nerve supply. This often happens to muscles, which waste away if their motor nerves are permanently damaged or have failed to develop properly. It is also true of sense organs in the absence of sensory nerves. Taste buds, for instance, degenerate if they are denervated either during their development or at a later stage in the adult. After ten days, in adult rats, the denervated buds have completely disappeared, but as soon as their nerve has regenerated, they start to develop again. Iwayama and Nada (1969) noted that there was an increase in acid phosphatase activity in the degenerating cells. Zalewski (1970) showed however that even when undergoing chromatolysis, the first stage in their demolition, the nerve fibres are capable of maintaining the taste buds. Later, when the nerve fibres become engulfed by dark cells, probably phagocytes (Farbman, 1969), they are no longer capable of maintaining the taste buds.

Having just seen in the previous section that limb buds have effects on the development of the nervous system, we must now point out that the limbs are also affected in their development by presence or absence of a nerve supply. Piatt (1942) was able in urodeles to produce entirely 'aneurogenic' limbs which had never had a nerve supply throughout their development. This was done by an experiment in which embryos with a large section of their neural plate removed were grafted in parabiosis with normal embryos, to enable them to survive. Limbs did develop, in these nerveless parabionts, but they were small and distorted and, of course, insensitive and immobile (cf. Plate 9.3a). Small and distorted limbs also result if limb buds of chick embryos are grown in isolation from the nervous system in organ culture, either *in vitro* or on the chorioallantoic membrane (Plate 9.3b). The joints of these limbs do not form properly, since they are moulded partly by the movements of the limb that normally occur when it is innervated (Fell and Canti, 1934; Hunt, 1932; Bradley, 1970).

Not only the development, but also the regeneration of limbs which occurs in amphibians is affected by the presence or absence of nerves. Singer (1952) has collected substantial evidence showing that there is a quantitative relationship between the total volume of axonal material supplying a limb and the success of its regeneration. Diverting extra nerves to the limb stump increases the rate of its regeneration, or even in some cases results in a double regenerate. Injections of homogenates of nerve tissue have also been shown to accelerate regeneration (Lebowitz and Singer, 1970). and extracts of spinal cord appear to have the same effect (Deck, 1971). Conversely, reduction or removal of the nerves that normally supply a limb stump delays its regeneration if the blastema has not yet formed at the time of operation (Schotté and Harland, 1943). It would seem from these findings that some material may be exuded from the tips of the nerve fibres which has a stimulatory effect on limb regeneration. So far, no such material has been identified, however. Presumably it should be possible to isolate similar material from active homogenates of nerve tissue which have been used successfully in experiments.

In the tail, regeneration can occur independently of a nerve supply. A tail stump of *Xenopus* isolated *in vitro* undergoes normal regeneration, for instance (Hauser and Lehmann, 1961; Weber, 1962). There are some special cells in the nervous system of amphibians which do influence tail regeneration, however. Hauser (1969, 1972) has shown that in *Xenopus* the success of tail regeneration is affected by processes from cells of the *subcommissural organ* in the roof of the midbrain. This organ occurs in fish as well as amphibians (Olsson, 1955) and is thought possibly to be neurosecretory. Fibres, known as Reissner's fibres, extend from its surface cells down the spinal canal right to its caudal end (Plate 9.4). The integrity of these fibres is essential for a normal tail regenerate to form. Damage to the subcommissural organ, or interruption of

Reissner's fibres along the cord, causes the tail regenerate to be smaller and to develop more slowly than normal. It is still not known if any material, neurosecretory or otherwise, is conveyed to the tail via Reissner's fibres, and further study is needed before we can understand their function.

A part of the brain that is known to be neurosecretory is the hypothalamus. Its cells play a major role in the co-ordination of activities of endocrine glands, and they also affect the development of at least some of these glands. Much less is known about the function of cells of the hypothalamus in embryos than in adults, but Goos (1969) has identified some of the functions connected with development of the thyroid gland in *Xenopus*. He has noted cells in the pre-optic region of the hypothalamus which stain with pseudoisocyanine (PIC cells: see Fig. 9.5) and he has deduced that these cells secrete a thyroid-releasing factor, causing the thyroid gland to release its hormone into the bloodstream. The evidence for this is that treatment of *Xenopus* larvae with phenylthiourea (PTU) causes the PIC cells to become de-granulated, as if releasing their secretion in an attempt to re-awaken the activity of the thyroid gland. The PTU treatment also causes cells of the anterior pituitary gland which secrete thyroid-stimulating hormone to become active. They cause hypertrophy of the thyroid, in an ineffective attempt to compensate for the inhibitory effects of PTU. Further interactions were noted by Goos between these three sets of cells in the pituitary, hypothalamus and thyroid. One was that the deficiency in the thyroid caused by PTU treatment of larvae resulted in incomplete differentiation of the developing hypothalamus: so these two organs seem to interact reciprocally. Another interaction was between the cells within the hypothalamus: if the pre-optic area was removed, new PIC cells differentiated in a region just behind this. It seems as if the cells in this region had received some 'signal' informing them that the normal PIC cells were no longer present, and enabling them to realize their own potentiality to form PIC cells, which must normally be suppressed by some interaction with the pre-optic area.

Many more interactions of endocrine organs with the hypothalamic region of the brain are known, in adult mammals. These are beyond the scope of this book, but those interested may obtain further information from Austin and Short (1972). We shall discuss

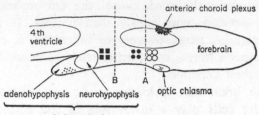

Fig. 9.5 Diagram of the positions of cells producing thyroid-releasing factor seen in a longitudinal section of the brain, in *Xenopus*. (*Modified from* Goos, 1969.) oo: cells in pre-optic nucleus (original TRF cells); ● ● : cells that appear after removal of the preoptic nucleus by transection of the brain at A; ■ ■: cells that appear after complete removal of the hypothalamus by transection at B. ....: cells in ventral, central adenohypophysis which secrete thyroid-stimulating hormone in response to TRF.

other activities of endocrine cells in development, in the next chapter.

9.4 Specificity in interneuronal and in nerve/end-organ relationships

It is impossible to trace in detail the development of all the connections made by neurons with each other and with end-organs throughout embryonic development. Histological work, however careful, does not reveal every single nerve axon and dendrite, nor does it show for certain which of the pathways are functional. Experiments on the behaviour of whole animals after operations on the nerves or their end-organs are the simplest way to test functions of nerve connections *in vivo*, but these are sometimes difficult to interpret because only a very limited range of behavioural responses can be distinguished in lower vertebrates. Amphibia have been used most often for work of this kind, because they are free-swimming and therefore accessible for experimentation at stages when the nervous system is still immature and the nerve connections have not yet all been made. The two types of neuronal connection whose investigation has led to the most instructive (though also controversial) findings are sensory connections to the skin, and connections between the retina of the eye and the optic tectum of the midbrain. We will take first the connections to the skin.

Sperry and Miner (1949) carried out experiments to test whether any regional specificity existed in the sensory connections to the

194

skin of the head in newts, and came to the conclusion that each region was reinnervated by its original nerve fibres, if these were cut and allowed to regenerate. Later, Miner (1956) rotated strips of lateral skin in tadpoles of *Rana* through 180°, replacing the strips so that the original belly skin now lay dorsally, and the back skin lay ventrally (see Fig. 9.6). She found that the adult frogs showed

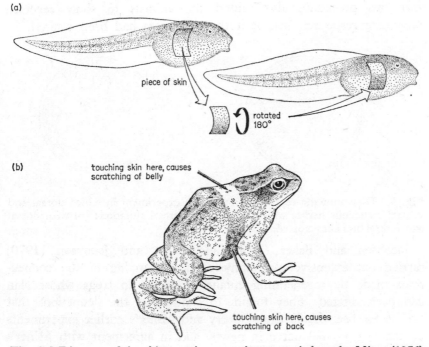

(a)

piece of skin

rotated 180°

(b)

touching skin here, causes scratching of belly

touching skin here, causes scratching of back

Fig. 9.6 Diagram of the skin-rotation experiment carried out by Miner (1956). (a) Operation, on tadpoles of *Rana*; (b) resultant frog and its responses to touch in the rotated areas.

reversed responses to these areas of the skin, and scratched their backs if the rotated skin now on the belly was touched, or scratched their bellies if the rotated skin now on the back was touched. Miner now suggested that the skin in some way specified the type of message carried along whatever sensory nerves supplied it and that regenerating nerves did *not* necessarily find their way back to make connections with the correct area of skin, when this was in an abnormal position. To say that they *did*, would imply that certain nerve fibres were in some way attracted to specific regions of the skin. Pursuing this interpretation further, it would mean also that the discrimination between sensations from back and belly is due to the particular nerve fibres along which the sensory impulses are carried, rather than to any different quality conferred on the impulses by cells in different regions of the skin. This last deduction was supported by experiments described by Szèkely (1971), in which he crossed the dorsal and ventral sensory nerves so that each was connected to the opposite area of skin (Fig. 9.7). It was found that this procedure also caused the animals to show reversed scratching responses, just as if the skin areas had been rotated.

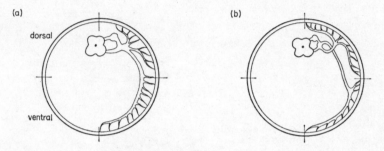

Fig. 9.7 Diagrammatic representation of the experiment in which dorsal and ventral cutaneous nerves were crossed. (a) Normal situation; (b) with dorsal and ventral branches crossed.

Jacobson and Baker (1968) and Baker and Jacobson (1970) carried out extensive electrophysiological mapping of the connections made by regenerating cutaneous nerves in frogs whose skin had been rotated. They found, in contrast to the deductions that had so far been made from Sperry and Miner's earlier experiments on the fifth cranial nerve in newts, and in agreement with Miner's later deductions from her skin rotations, that there was *no* selective outgrowth of nerve fibres to their correct skin regions: on the contrary, whichever nerves were nearest supplied the rotated skin. These investigators also observed the behaviour of the animals in rather more detail than had been done before and found, unexpectedly, that at first the animals showed *normal* responses when the rotated areas were touched: only later did they show the reversed response, as if it had been acquired gradually. This suggests that the skin receptors in the reversed piece *do* send some specific signal along whatever sensory nerves connect with them, which *is* eventually interpreted as 'dorsal' or 'ventral' according to the type of skin and not its new position, and according to the type of signal, not the type of nerve along which it is carried. Jacobson and Baker found no evidence of consistent differences in the patterns of impulses passing along different cutaneous nerves, however.

There are other examples, often quoted in accounts of the development of nerve connections (e.g. Szèkely, 1966; Gaze, 1970), which show that end-organs convey sensory impulses which are recognized by the animal as originating from that organ, despite an abnormal route along which the impulse may come. To give two of these examples: an extra leg grafted to the thorax in a chick embryo will, if pinched after hatching, cause the chicken to limp on its normal leg of that side (Fig. 9.8a). An extra eye grafted onto a newt tadpole will result in blinking of the normal eye if this graft

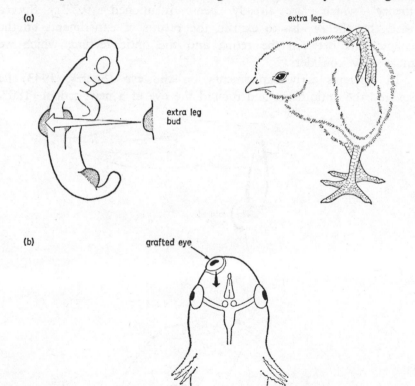

Fig. 9.8 Some experiments that demonstrated organ-specific sensory stimuli. (see Szèkely, 1966). (a) Grafting of an extra leg bud onto the dorsal thorax of a chick embryo, gives a leg with sensory but no motor nerves. When this leg is pinched, the hatched chick limps on its normal leg of that side. (b) If an extra eye is grafted onto the head of a newt larva and establishes sensory connections with the brain, stimulation of this eye causes blinking of the host's normal eye on that side.

is touched (Fig. 9.8b). The sensory messages received from these oddly-placed organs appear to have been recognized as 'leg' or 'eye' by the central nervous system, even though they were carried along the nearest sensory nerves and not the normal ones. These sensory impulses were evidently capable of activating the motor pathways to the host's normal leg or eye muscles.

We have now reached a rather different interpretation of the kinds of experiments carried out by Sperry and Miner, which contrasts with their original view that some chemical specificity caused the nerves to find their correct connections in the skin, despite its altered orientation. This view fitted with the 'neuronal specificity theory' which has already been mentioned (9.2.2). Sperry used this theory also to explain the results of experiments on the connections between the retina and the optic tectum, which we must now consider.

In Sperry's early experiments on the eye (Sperry, 1943) he severed the optic nerve and rotated the eye of a newt tadpole 180°,

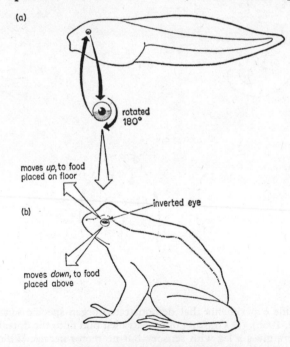

Fig. 9.9 Eye rotation experiments in the frog *Rana* (cf. Sperry, 1944). Severing of the optic nerve and rotation of the eye 180° before replacing it, as shown in (a), causes reversed visual reactions in the resultant frog, shown in (b). Sperry's original experiments were on newts (see text).

then replaced the eye and allowed the optic nerve fibres to regenerate (Fig. 9.9). He then tested the newt's visual responses and
found that these were upside-down. When food was held above it,
the newt dived down to get it, and if food was placed on the bottom
of the tank, it looked for it above. So Sperry deduced that fibres
from each part of the retina had found their way back to the proper
areas of the optic tectum in the brain, despite their rotated position, i.e. that there was some kind of specificity which guided them
to the correct neurons in the tectum. However, it has since been
shown that this interpretation was unsatisfactory. Recent electrophysiological techniques have made it possible to map in detail the
precise points on the optic tectum to which each point on the retina
connects, both in normal animals and in those with rotated eyes.

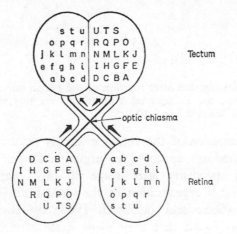

Fig. 9.10 Map of the retinotectal connections in *Xenopus*. (*Modified from*
Jacobson, 1968 and 1970.) The letters of the alphabet designate points on the
retina which, when stimulated, activate neurons in the correspondingly-lettered
region of the optic tectum. The mapping was achieved by means of micro-
electrodes. In *Xenopus*, nearly all the optic nerve fibres cross at the optic
chiasma and run to the tectum of the opposite side.

Gaze, Jacobson and their collaborators have made maps of this kind
for the retinotectal projections in *Xenopus* (see Fig. 9.10). From
these and from experiments similar to Sperry's, they have collected
a number of findings which argue against the idea that any cell-to-
cell specificity exists in retinotectal connections. Their main findings are set out below.

First, Jacobson (1965) followed electrophysiologically the re-
connections that were made as optic fibres regenerated to the

tectum, after rotation of the eye in the tadpole. He found that if the rotation had been done before stage 30 (late tailbud stage), the connections were exactly as in a normal eye, i.e. there was no evidence of changes due to rotation. If the eye was rotated after stage 30, however, there was a reversal of the retinotectal map (Fig. 9.11) and the animal often showed reversed visual responses. So the

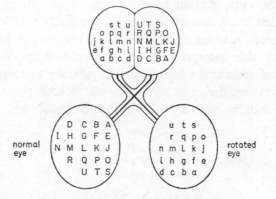

Fig. 9.11 Map showing that after rotation of the eye on one side, at late larva stages in *Xenopus*, the retinotectal connections on this side are reversed Lettering conventions as in Fig. 9.10.

regional differentiation of the retina evidently occurred at stage 30. So far, these findings appeared to argue in favour of there being some neuronal specificity after stage 30 in *Xenopus*. However, when the course of regeneration of the optic fibres was followed at several stages (Gaze and Jacobson, 1963) by electrophysiological mapping, it was found that they made quite random, disorganized connections at first, which only gradually assumed a regular pattern in the tectum. So if there was any neuronal specificity, this did not show at first, but may eventually have been acquired as they formed and re-formed several successive connections during development. Sperry himself came round to this viewpoint, after similar findings from mapping work on the tectum of *Rana* (Sperry, 1965). More recently, too, both electrophysiological and histological studies on the development of retinotectal connections in *Xenopus* larvae (Gaze, Chung and Keating, 1972; Lazar, 1973) have shown that there is a gradual restriction and modification of the synapses made by each optic fibre with cells in the tectum: the first connections that they make are not their definitive ones, for the tectal cells are also acquiring new dendritic branches as they develop.

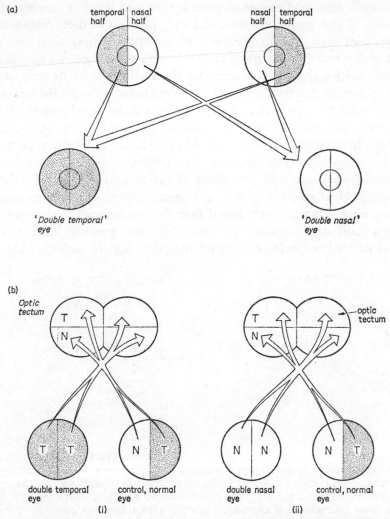

Fig. 9.12 'Compound' eyes and their retinotectal connections in *Xenopus* (see Gaze, Jacobson and Szèkely, 1965). (a) The method of forming a 'compound' eye, by recombining either two nasal or two temporal halves after bisecting two eyes. (b) The compound eyes are found to connect with the whole of the tectum, just as does the normal eye. (i) Double-temporal eye compared with normal eye; (ii) double-nasal eye compared with normal eye.

A further type of experiment to test the specificity of retino-tectal connections was carried out on *Xenopus* by Gaze, Jacobson and Szèkely (1965). Attardi and Sperry (1963) had found that a nasal or a temporal half of the retina of the goldfish, if the other half had been removed, would connect with only its proper half of the optic tectum. Gaze *et al.* tried joining *two* nasal or *two* temporal halves of the eye in *Xenopus,* and they found that these 'compound' eyes then connected to the whole of the optic tectum, with no signs of preference for the nasal or the temporal area of it (Fig. 9.12). Some critics argued, however, that the tectum might be only semi-developed if it were connected to a double-nasal or double-temporal eye, since the retina is known to influence the development of the tectum in some animals (cf. Schmatolla, 1972). To answer this point, Straznicky, Gaze and Keating (1971), uncrossed the optic nerve roots some time after inserting the compound eye on one side. So the abnormal eye now sent fibres to the tectum which had hitherto developed in conjuction with a normal, control eye and so should be normal. Again, it was found that the regenerated fibres from the compound eye colonized the whole of this 'control' optic tectum and showed no preference for one or other side of it (Fig. 9.13). So

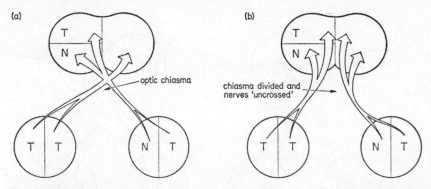

Fig. 9.13 The 'uncrossing' experiment of Straznicky, Gaze and Keating (1971)· In (a), a double-temporal eye is first allowed to connect with the developing tectum of the opposite side, via the optic chiasma. Later, in (b), the optic chiasma is divided, and the nerve fibres made to re-connect with the tectum of the *same* side instead of crossing. Thus the normal eye now connects with the possibly 'abnormal' tectum resulting from association with a double-temporal eye at first. Since it connects to all parts of this tectum, it is argued that the tectum is *not* deficient or abnormal.

this finding contradicted the earlier indications from Attardi and Sperry's work on goldfish that connections would be made with only half of the optic tectum in such cases. There is still a controversy about the degree of specificity between different areas of the retina and those of the tectum in *Xenopus,* however, for Straznicky (1973) has now found that removal of parts of the tectum in young frogs causes defects of vision in corresponding areas of the retina.

It looks as if these retinal cells have been unable to find appropriate regions of tectum with which to make connections, when fairly large areas of the tectum are absent. Possibly there is some degree of *regional* specificity in the retinotectal connections, enough to make certain connections preferred, so that if an excess of optic fibres are competing for smaller areas than normal, those that originate from the appropriate regions of the retina will complete their connections first. The exact sequence of connections made in these partial tecta will no doubt be followed in further work.

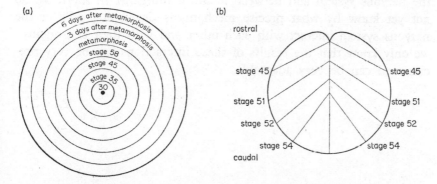

Fig. 9.14 The patterns of sequential maturation of cells in retina and tectum of *Xenopus*, seen from labelling experiments with tritiated thymidine. (a) In the retina, concentric rings of cells mature, in sequence from the centre to the periphery. Only the extreme central zone has mature cells by the larval stage 30 when the retinotectal connections appear to become 'determined'. (b) A quite different pattern is seen in the sequence of maturation of cells in the tectum. (*Modified from* Straznicky and Gaze, 1971, and Feldman and Gaze, 1972.)

Recently the course of development of neurons in the retina and tectum of *Xenopus* has been followed by Straznicky and Gaze (1971, 1972) to see if there is any evidence for a point-to-point correspondence in the two organs. Their first rather surprising finding was that at stage 30 (when, according to studies of behaviour after rotation of the eye, the regions of the retina seem to have become specified) only a very small central area of definitive neurons have formed in the retina. For a long period after this, neurons are added in concentric rings, proceeding peripherally (cf. Fig. 9.14a). When the pattern of development of neurons in the tectum is followed, however, no such concentric sequence is seen, but instead the sequence is from rostroventral to caudomedial (Fig. 9.14b). There seems no way of relating these very different patterns of

203

development in retina and tectum to the final maps of retinotectal projections (cf. Fig. 9.10). So Gaze and Keating (1972) now feel that the neuronal specificity theory requires some modifications, and have suggested that there is a more flexible control of nerve conections in embryonic development, which they call 'systems matching'. Since there is no clear-cut experimental support for this theory yet, we have to await further ideas and evidence, before it is possible to conclude what kind of control in the vertebrate embryo governs the cell-to-cell connections that are made, both within the nervous system and between it and end-organs. In short, we do not yet know by what precise mechanisms developing cells of the nervous system interact with each other and with their end-organs: we only know that the results of these interactions can be very precise, and can be very adaptable.

10 Long-distance cellular interactions mediated by hormones

We saw in Chapter 1 that there is a special type of cellular interaction that may occur over long distances within the animal organism, which is mediated by hormones. Hormones are often referred to in simple terms as 'chemical messengers'. They may be defined, very broadly, as distinct, usually steroid or polypeptide molecules which are synthesized within specialized secretory cells and released into the bloodstream, where they circulate and produce specific effects on target organs. They are destroyed in the body by metabolic processes, so their action lasts only as long as they continue to be released into the bloodstream by the cells which secrete them. Most of the hormone-secreting cells are located in special endocrine glands, which do not become functional until late in embryonic or larval life. So only a limited number of events in development – those which occur at late stages – can be controlled by endogenous hormones. Among these are the general rate of growth of the body, the maturation of the reproductive system, and the process of metamorphosis into an adult organism that occurs in some groups of animals, most notably the Amphibia and the insects.

It is impossible here to review the wealth of literature on hormone action which may be relevant to our understanding of the interactions between endocrine cells and other tissues during the

Table 10.1 Hormones that affect developmental processes in vertebrates.

Source	Hormone	Main Effects
Pituitary gland	Growth hormone	Enhanced cell-proliferation and protein synthesis.
	Thyrotropic hormone(s)	Growth enhanced and differentiation in the thyroid gland.
	Gonadotropic hormones	Enhanced proliferation of stem cells, growth of ovarian follicles, secretory activity in interstitial cells.
	Prolactin	Milk (casein) secretion in mammary gland.
Thyroid gland	Thyroxine and related subs.	Increased metabolic rate and protein synthesis. In amphibians, precipitates metamorphic changes which include: regression of tail, growth of limbs, ossification, collagen synthesis, maturation of neurons (but degeneration of Mauthner neurons), maturation of liver enzymes.
Ovary	Oestrogens	Appearance of female secondary sexual characters: proliferation and secretion in epithelial cells of oviduct (birds) or uterus (mammals); initial changes in mammary glands.
(Corpus luteum)	Progesterone	Further proliferation of uterine epithelium and maintenance of pregnancy (mammals); further differentiation of mammary glands.
Testis	Testosterone	Development of male gonoducts, genitalia and secondary sexual features.
	Unknown factor	Degeneration of female ducts.
Metanephric kidney	Erythropoietin	Proliferation of red blood stem cells and synthesis of haemoglobin.
Thymus	Thymosine	Proliferation of lymphocytes.
Pancreas	Insulin	Essential for normal development of appendages. Initiates differentiation in the mammary gland.
Adrenal gland	Cortisol	Essential for normal development of many organs at late stages. Stimulates late stages of differentiation in the mammary gland.

development of animal embryos. Below, we will consider only a few examples of hormonal effects on development that have been well worked out in the past (cf. Table 10.1), and we shall then go on to a more general consideration of the mechanisms by which hormones may influence cell differentiation. It is a relief, at last, to be able in this chapter to say that in many cases we know and have isolated the substance responsible for these long-distance cellular interactions: this is far more than can be said for most of the types of cellular interaction mentioned so far in this book. What remains a puzzle, however, is why hormones affect some tissues and not others, and why, on the other hand, thyroid hormone is able to affect a variety of tissues at metamorphosis in amphibians in very different ways. These are topics which will need further discussion below. It may be necessary to argue that every tissue has specialized hormone receptors at its cell membranes and that these act selectively, enabling the cell to respond to some hormones but not others, and different cells to respond in different ways to the same hormone.

10.1 Processes of cell proliferation and growth
Most of our knowledge about hormone effects on cell proliferation comes from studies on post-embryonic tissues, many of which are still capable of proliferation throughout life. There are certain well-known examples in vertebrates, which have been reviewed by Turkington (1971), of tissues that normally proliferate continuously with an innate rhythm of activity, but which change this rhythm in response to hormone treatment. There are other tissues in the adult which do not proliferate continuously, but which will do so in response to certain hormones. All of these cases are worth some attention here, since as soon as the relevant hormones are being produced, in the late embryo or the larva, it is likely that responses in these tissues could start to occur.

Of the continuously proliferating tissues, the gastric lining (mucosa) is one: this shows a diurnal rhythm of mitotic activity, highest at night. The same is true of the epidermal layer of the skin. Both these tissues respond to 'stress' hormones by a fall in mitotic rate which may last several hours. Adrenocorticotrophic hormone (ACTH) has the maximum effect on gastric mucosa, and the epidermis responds to adrenalin. Another important substance which decreases mitotic rate in epidermis is the epidermal extract named 'chalone' (Bullough and Laurence, 1964), about which more will be said in Chapter 11. A further example of a continuously proliferating tissue in the adult is the lymphopoietic system; this responds to the hormone thymosine from the thymus gland, which stimulates mitosis of the lymphoid cells.

The well-known 'growth hormone', or *somatotrophin*, secreted by cells of the pituitary gland does not appear to have any distinct effects on embryonic organs, since most of these have completed their main phase of growth before this hormone is circulating. It has no growth-promoting effects when tried on early mammalian embryos grown *in vitro*, either (Bonsor, 1972). Its effects are more readily seen in individual tissues grown in culture. The type of cell that does respond readily to growth hormone and is therefore used in assays of the hormone's activity, is the chondrocyte. Some hours after the administration of growth hormone, there is a wave of DNA synthesis in the chondrocytes of the ribs of young rats (Daughaday and Reader, 1966). This effect is not shown by chondrocytes *in vitro*, however, unless a factor from the serum which is believed to mediate the action of the hormone is present. This mediator has become called the 'thymidine factor' (Van Wyk *et al.*, 1969), since it enhances the DNA synthesis in the chondrocytes, which is measured by their uptake of radioactive thymidine. As we shall see later (Section 10.5), there are a number of intermediaries between a hormone and events in the cell, many of which are believed to operate in the cell membrane; but this serum factor is a different kind of intermediary, external to the cells.

Another hormone which has general effects on growth, because it increases metabolic rate and therefore tends to increase the uptake of nutrients into the cells, is thyroxine. Thyroxine deficiency during late embryonic life shows particularly in mammals, whose thyroid gland is active during the latter half of gestation. They are born underweight if the thyroid has not been active up to normal levels, and if the mother's supply of the hormone has not been able to compensate for this.

Turkington (loc. cit. 1971) also mentions several adult tissues which do not undergo continuous proliferation, but show a burst of mitotic activity under hormonal stimulation. Among these are the endometrial lining of the uterus in mammals, which proliferates in response to oestrogens; the adrenal cortex which will proliferate after treatment with ACTH; the prostate gland which responds to testosterone and the liver which regenerates after part of it has been removed. It is not certain that the agent which controls liver regeneration is a hormone: there has been conflicting evidence from experiments in which animals subjected to partial removal of the liver were grafted in parabiosis with normal animals and, in some cases, the mitotic rate in the normal liver increased. Serum from these partially hepatectomized animals has also in some cases been found to stimulate mitosis in liver tissue, both *in vivo* and in organ cultures. This work has been done mainly on adult animals, and is reviewed by Goss (1964).

Another adult organ which responds in its development and growth to hormone treatment is the mammary gland (see Section 10.3). As is the case with the other organs and tissues mentioned so far, there are not yet any proven effects of hormones on the mammary gland at embryonic stages. This is due to a dearth of experimental work, rather than to negative results: it is very probable that the hormones secreted by endocrine glands of the embryo late in its development do affect the proliferation of the tissues that we have mentioned above.

Let us now turn to some of the hormonal responses that are definitely known to occur during developmental stages.

10.2 Metamorphosis

One of the most spectacular hormone-controlled events in animal development, and one which involves many different tissues, is the process of metamorphosis (Fig. 10.1). We know most about its hormonal control in amphibians. In the following account, amphibian metamorphosis will therefore be dealt with first, then brief reference will be made to the metamorphosis of insects, which is also controlled by changes in hormone production but which shows some contrasts to the process in amphibians.

10.2.1 *Amphibians*

In amphibians, it has been known for a long time that the hormone or hormones produced by the thyroid gland at peak concentrations towards the end of larval life initiate(s) metamorphosis. Thyroxine is effective in inducing metamorphosis, when simply fed to the tadpoles by introducing it into the aquarium water. Normally thyroid hormone is secreted into the bloodstream in response to thyrotrophic hormone, which is released from the pituitary gland

209

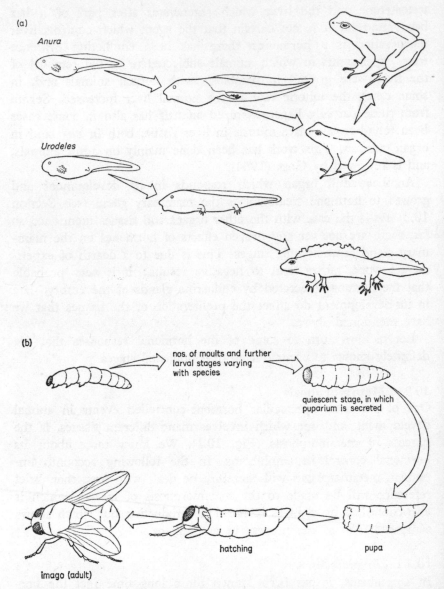

(a)

Anura

Urodeles

(b)

nos. of moults and further
larval stages varying
with species

quiescent stage, in which
puparium is secreted

hatching

pupa

imago (adult)

Fig. 10.1 Summary sketches of the process of metamorphosis. (a) in amphibians, (b) in insects of the 'holometabolous' type.

in response to yet another hormone, called thyroid-releasing factor, produced by the hypothalamus. Thyrotrophic hormone is secreted by specialized cells distributed throughout the anterior lobe of the pituitary gland. These cells become active in late larval life, just before the activity of secretory follicles in the thyroid becomes evident (Saxén *et al.*, 1957).

One of the most enigmatic features of thyroxine's action is that it has such different effects on different tissues. In the larval tail, it causes wholesale destruction of tissue and regression of growth to occur, so that the tail is eventually resorbed into the body. This regression will also occur when an isolated tail tip is treated with thyroxine *in vitro* (Weber, 1967). Some time ago Schwind (1933) showed that the regression response was specific to tail tissue only, and that if an eye was grafted onto the tail, the eye remained intact throughout the period of resorption of the tail. In limb buds, the very reverse of the tail's response to thyroxine occurs: they undergo rapid growth and differentiation. Finer points of contrast between different tissues are seen when the effects of thyroxine on different areas of the skin and of the brain are compared. The skin of the trunk lays down more collagen at metamorphosis, while the tail skin, by contrast, undergoes a destruction of its collagen layer. This distinction is maintained, even when crystals of thyroxine are implanted under the skin at two closely adjacent spots at the border between trunk and tail (Gross, 1964). In the brain, implanted crystals of thyroxine cause a rapid maturation of all neurons, *except* for the two giant Mauthner's neurons in the hindbrain, which degenerate in response to thyroxine (Weiss and Rosetti, 1957). These giant neurons function in the rapid swimming and the 'escape' reaction of the larva, and they fall out of use at metamorphosis (Fig. 10.2).

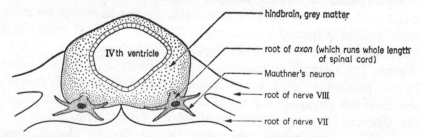
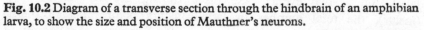

Fig. 10.2 Diagram of a transverse section through the hindbrain of an amphibian larva, to show the size and position of Mauthner's neurons.

Thyroxine also initiates a number of specific biochemical changes in larval organs. For instance, in the retina of the eye there is a change in the type of pigment synthesized. Instead of the larval type, porphyropsin, the adult-form rhodopsin begins to be formed

211

(Fig. 10.3). This change, which involves switching *off* two synthetic pathways, can also be brought about in retina tissue cultivated *in vitro* in the presence of thyroxine (Ohtsu, Naito and Wilt, 1964). In the liver, enzymes concerned with urea synthesis are brought into action by thyroxine at metamorphosis (Cohen, 1970; Tata, 1971). One of the chief enzymes studied, carbamyl phosphate synthetase, is a good indicator of these changes in function in the liver.

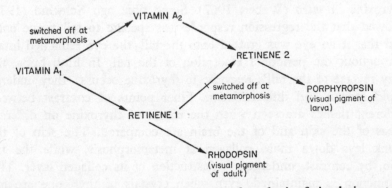

Fig. 10.3 Scheme of the alternative pathways of synthesis of visual pigment in amphibians. (*Based on* Ohtsu, Naito and Wilt, 1964.)

10.2.2   Insects

When we now turn to the insects, which are the largest class of invertebrates in which metamorphosis and its control have been studied experimentally, we find that the hormonal events that initiate metamorphosis are very different from those in amphibians. Metamorphosis in insects involves a change from a situation in which each moult produces a new larva, to one in which no more larval cuticle is formed but instead a different, imaginal cuticle and new, imaginal organs such as legs, wings, antennae and genitalia appear. The changed situation is brought about by, first, a decline in the production of a larval hormone called 'juvenile hormone', and then the *disappearance* of an endocrine gland (Fig. 10.4) called the thoracic (or prothoracic) gland which causes moulting in the larva. These two events take place during the last larval stage (instar), the total number of larval instars before metamorphosis being characteristic of each species.   If, however, the supply of juvenile hormone is maintained experimentally, and active thoracic glands are implanted, the larval moults can be prolonged and metamorphosis delayed indefinitely. The gland responsible for secreting juvenile hormone is the corpus allatum near the brain (Fig. 10.4;

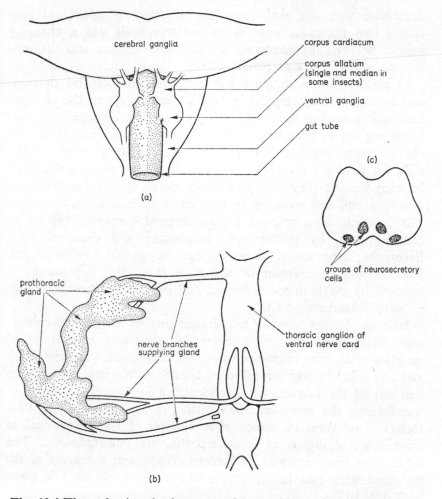

Fig. 10.4 The endocrine glands concerned in moulting and metamorphosis in insects. (a) Dorsal view of the brain and associated structures in a moth. (b) The prothoracic gland, seen on one side of the body only, with its nerve supply. (c) Transverse section of the brain in moths, showing the position of the neurosecretory cells. (*Modified from* Balinsky, 1970.) Compare with Fig. 2.14.

cf. also Fig. 2.14) which forms part of a ring gland in Diptera. This gland is also essential for the development of the gonads at metamorphosis (see Section 10.3). In the bug *Rhodnius*, which has been used for much experimental work by Wigglesworth and his associates (see Wigglesworth, 1954), the corpora allata cease to produce juvenile hormone at the fifth larval instar, and the thoracic glands

213

degenerate after the next moult has occurred. Wigglesworth concluded that the main stimulus to metamorphosis was a changing balance between the secretions of the corpus allatum and the prothoracic gland (see also Novak, 1967).

In insects just as in amphibians, different tissues and different regions of the body respond in very different ways to the changed hormonal environment at metamorphosis. The epidermis of different regions in both larva and adult secretes a cuticle of characteristic colour pattern. This pattern is an innate property of the cells, for if pieces of epidermis are grafted elsewhere on the body between moults, they still produce colouring typical for their site of origin, and will show up in contrast to their new surroundings. The imaginal discs respond to the changed hormonal balance at metamorphosis by proliferating, evaginating and eventually differentiating into wings, antennae, eyes or genital system of the adult. The final eversion of the discs that form appendages is initiated by the hormone ecdysone, and can also be brought about *in vitro* (Mandaron, 1970).

It is not known whether the strange process of 'transdetermination' which can be brought about by successive transplants of imaginal discs into immature larvae in *Drosophila* (Sections 7.2.5 and 8.7) is in any way controlled by hormones. We have no evidence that any of the hormones so far identified in insects is capable of transforming the course of differentiation in an imaginal disc. What factors cause them to change after a series of transplants and to form different organs at metamorphosis, are still unknown. The subject has been reviewed by Hadorn (1965), and a scheme of the transformations that he and his colleagues have obtained is given in Table 10.2.

In insects just as in vertebrates, there is evidence that the

Table 10.2 Directions of 'transdeterminations' in imaginal discs of *Drosophila*

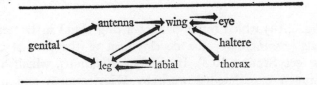

Note: the extent to which transdeterminations occur varies in different *Drosophila* species. Some mutations also cause transdeterminations.

hormone-production of some endocrine glands is under the control of a higher centre. In vertebrates, this is the pituitary gland, which in its turn is activated by neurosecretory cells in the hypothalamus of the brain (cf. Chapter 9). A midline region of the brain of insects, called the *pars intercerebralis* (Figs. 10.4 and 2.14) contains neurosecretory cells which produce a hormone controlling the functions of the prothoracic glands (Wigglesworth, 1940). The chemistry of this hormone has been investigated by Williams (1967), but the extent of its effects is still not fully known. There are also other neurosecretory cells in the brain of insects, but a point of major difference between these and vertebrate neurosecretory cells is that they often extend right to the peripheral organs, instead of discharging their products into the bloodstream (Maddrell, 1967). For example in aphids and cockroaches, axons from neurosecretory cells run all the way to the heart and gut, and in the bug *Rhodnius* they run to the epidermis and to the vertical tergosternal muscles. In animals as small as insects it is of course easily possible for cells similar to neurons to have processes long enough to make direct contact with target organs, and this is probably a much more rapid and efficient method of communication than via the blood (haemolymph), whose circulation is incomplete and sluggish owing to the absence of blood vessels and the very weak propulsive force of the primitive heart. Possibly the secretions of the cells that make direct contact with target organs should not strictly be classified as hormones. Those secretions that control the thoracic glands do travel via the blood, however, so may be regarded as fully-evolved hormones.

Hormonal intercommunication between cells is effective only in animals which either are small enough, or have efficient enough circulatory systems, for the target cells to be reached rapidly. In the vertebrates, the autonomic nervous system often helps to implement the effects of releases of hormones and to speed up the responses in target organs. The hypothalamus and the pituitary gland in vertebrates are at the centre of control, not only of the endocrine system but also of several sympathetic and parasympathetic nervous pathways which accelerate responses in many effector organs of the body. Some of the kinds of event that are under both hormonal and nervous control are shown in Fig. 10.5: these also occur in late embryos, as soon as the pituitary gland and the hypothalamus are functional. In this figure too, the hormonal controls in insects that have been mentioned above are set out for comparison in (b).

H

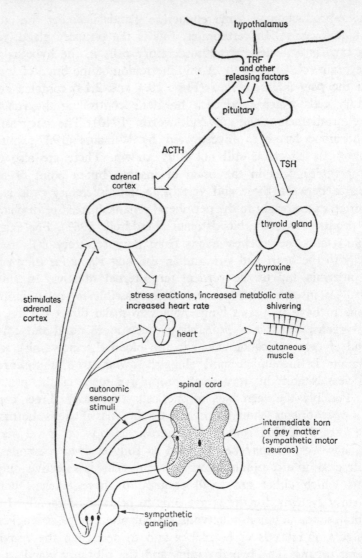

Fig. 10.5 (a) Some hormonal and neural interactions in vertebrates: simplified.

10.3 Hormonal effects on the development of the reproductive system

10.3.1 *Gonads and gonoducts*

We have seen in Chapter 7 examples of interactions between gonads and their respective ducts during development. It was shown that in mammals, the testis causes development of male ducts

216

and at the same time causes the female duct rudiments to regress, but that the ovary appears to exert no positive effect on the gonoducts. The contrasting situation in insects, where gonoducts seem

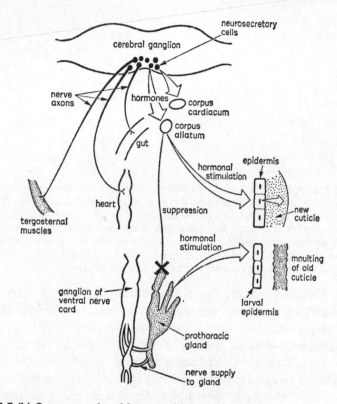

Fig. 10.5 (b) Some neural and hormonal interactions in insects, simplified.

to influence the development of the gonads rather than *vice versa*, was also described. Some of these effects could not be classed strictly as hormonal, however, since they occurred only when the organs were in close proximity, and passage of a stimulatory substance via the bloodstream was not implicated. In Jost's testis grafting experiments in rats, only the ducts of that side were affected, while those of the opposite side of the body remained female. In further experiments, Jost (1948) did demonstrate an effect due to the hormone testosterone from the testis, however. Genitalia in the foetus were masculinized on *both* sides of the body by the testosterone from a graft on one side only.

217

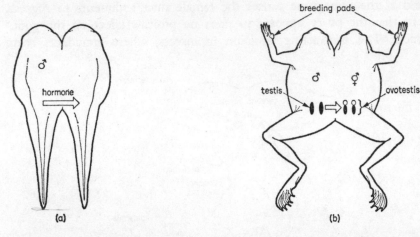

Fig. 10.6 Principles of Witschi's parabiosis experiments to demonstrate the influence of sex hormones on gonadal development in amphibians. (See Witschi, 1938.) (a) Pairs of newly-hatched larvae are grafted in parabiosis. In some of these pairs, one partner will have the male sex chromosome constitution and the other the female constitution. (b) At metamorphosis, the genetic females of such pairs are found to have a gonad which is partly transformed into a testis, and they develop secondary sexual characters like a male, too.

Other hormonal effects from the developing gonads have been demonstrated by parabiosis experiments in lower vertebrates. Witschi (1938) carried out some classic parabiosis work on amphibian larvae (Fig. 10.6) and showed that a genetically female partner could undergo masculinization, including partial transformation of the gonad into an ovotestis, when she was grafted in parabiosis with a male, but that the male was never feminized. So it seemed that here, as in mammals, there was no positive effect from the female gonad although there was from that of the male. This does not mean that the testis is never able to become feminized by oestrogenic hormones however, for in *Xenopus* Gallien (1953) and Chang and Witschi (1956) have found that adding oestrogenic hormones to the aquarium water causes 100% conversion of males into females, which later lay eggs.

In all vertebrates studied so far, it has been found that the gonads are susceptible to transformation by either oestrogenic or androgenic hormones, provided that these are administered before they have gone far with differentiation under the control of innate genetic factors (originating from the sex chromosomes). We have not very much evidence on birds and mammals because of the difficulty of doing experiments on these at early enough stages *in vivo*, but studies *in vitro* on the gonads of chick and mammal embryos (see Price and Ortiz, 1965; Hamilton and Teng, 1965) have shown that it is possible to initiate sex reversal in the gonad at early stages by treatment with the relevant hormones. Among mammals, the marsupials have proved useful for studies of hormonal effects on the differentiation of the genital system, because they are born when this is still immature. The classic experiments of Burns (1950) in which he administered androgens and oestrogens to newborn opossums showed that these could influence the gonads towards either the male or the female type. The gonoducts then developed correspondingly, under the influence of the transformed gonads.

In birds, both male and female gonadal hormones have been found to influence the development of gonoducts, genitalia and secondary sexual characters (see Romanoff, 1960 for review). The effects of the hormones have been tested in organ cultures *in vitro*. Much of this work has been done at Wolff's laboratories in France. There is some controversy as to what factors cause the female (Mullerian) ducts to regress in male birds, since both androgens and oestrogens have been found to cause them to regress *in vitro*. Table 10.3 summarizes these findings. Earlier, Wolff and Wolff (1951) showed that X-irradiation of the ovaries in the duck, leading to a decline in the output of ovarian hormones, caused the gonoducts to become progressively more like those of the male in their development. The syrinx (avian larynx) also became enlarged as in the male, and the genital tubercle took on a male form. So, unlike the situation in mammals, it looks as if in birds the ovary does have a positive effect on gonoducts. To show that this is truly an effect of ovarian hormones, one would have to measure the oestrogen-production of the ovary during its development and also to show that its effects on the gonoducts are normally mediated through the bloodstream.

It has to be emphasized that there is still some doubt as to whether the gonads of vertebrate embryos produce hormones that are chemically identical to those of adult gonads. So far, the main tests have been of their biological effects, compared with those of purified hormones, *in vivo* and *in vitro*. It is difficult to get enough gonadal material from embryos, to be able to extract sufficient hormone to characterize chemically.

We know very much less about the roles of hormones in the development of the genital system in invertebrates than we do in

219

Table 10.3 Effects of androgens and oestrogens on gonoducts of birds, grown *in vitro*. (Based on data reviewed by Hamilton and Teng, 1965.)

Agent	Male, Müllerian duct	Female, Müllerian duct
Control medium, no hormones.	No swelling. Fragmentation at 8½ days.	Swelling at cloacal region, then regression starting at 6 days.
Oestradiol-17β, 1μg ml⁻¹	Both doses same effect as in female at 8 days: no effect if explanted at 8½ days.	Immense swelling at cloacal region: delay of regression.
Oestradiol-17β >30μg ml⁻¹		Inhibits swelling and accelerates regression.
Testosterone <1μg ml⁻¹	Stimulated at 8 days: inhibited if explanted from 8½ days on. Accelerates regression	Inhibits swelling at cloaca and accelerates regression.
Testosterone, 10-100 μg ml⁻¹		
Androsterone, <30μg ml⁻¹	Stimulated at 8 days: regresses if explanted later. Accelerates regression	Inhibits swelling at cloaca and accelerates regression.
Androsterone, >30 μg ml⁻¹		

vertebrates. There are, however, a few well-studied examples which have been reviewed by Charniaux-Cotton (1965). She herself has worked on Crustacea, and she showed that in two groups, the amphipods and the isopods, the ovary of the female controls the development of female secondary sexual characters. These characters are absent if the ovary is removed, but reappear if another ovary is grafted in its place later. In the male, there is a pair of 'androgenic glands' near the distal ends of the gonoducts, and these are responsible for the development of male secondary sexual characters. They will also cause masculinization of an ovary if this is grafted into the male. Removal of the androgenic glands allows female characteristics to develop. Charniaux-Cotton concluded that it is the presence or absence of an androgenic hormone secreted by these glands that determines the sex as male or female in these animals: in other words, the testis plays only a subsidiary role, and the ovary has a positive role only in the female; it cannot overrule the androgenic glands of the male.

Relexans (1973) has found a contrasting situation to this, in the hermaphrodite oligohaete worm, *Eisenia foetida*. (This, like the common earthworm *Lumbricus*, has male and female organs developing simultaneously in different segments of the body.) Relaxans found that gonads of *either* sex undergo sex-reversal if they are transplanted into the segment containing gonads of the opposite sex. So there are evidently factors from both male and female sex organs of the host, capable of reversing the sexual differentiation of a graft. It is not clear yet whether these factors are hormones, nor whether they arise from the gonads or from other tissues of that segment.

We know a little about the control of sexual development in molluscs, from organ-culture work (see Gomot, 1970). Those molluscs which undergo male development first, and then transform into females later in life, are of particular interest. For instance, in the prosobranch mollusc *Calyptraea*, the development of the male penis depends upon some factor produced by the optic tentacle. Even a penis which has regressed after the animal has become a female, may be induced to grow again when cultured together with a tentacle from a mature male. In contrast, culturing a normal penis either in haemolymph from a female, or together with central nerve ganglia from a female, causes it to regress. It is believed that this regression is due to a hormone secreted by the nerve ganglion cells into the haemolymph. Another, short-lived factor circulating in the haemolymph of the female causes genital tract rudiments to develop into female ducts, in organ culture. This factor is also evidently a hormone produced by the nerve ganglia, for culture of the tract rudiments together with ganglia from a male which is transforming into a female also causes them to develop into female ducts. Finally, the gonads of *Calyptraea* are themselves influenced by factors in the haemolymph. Gonads cultured in haemolymph of a mature male will become testes, and if cultured in haemolymph from a female they will become ovaries.

In insects, the initial development of the genital system which occurs at metamorphosis is dependent on some hormonal control from the corpus allatum: if this is removed, no genital system forms (Davey, 1965). There is also at least one testicular hormone which controls the development of secondary sexual characters (see the review by Charniaux-Cotton, 1965). The secretion of this hormone is under the control of the neurosecretory cells of the brain (Fig. 10.4). It is interesting, and again reminiscent of the situation in the hypothalamus in vertebrates, that these neurosecretory cells of insects have different patterns of activity in males and females.

10.3.2   The oviduct of birds

In his review already mentioned, Turkington (1971) quotes the effects of hormones on the function of the oviduct in the hen, as an example of cells producing easily-recognizable proteins in response to a precisely-known hormonal stimulus. These effects are seen in post-embryonic stages, as are the effects on mammary glands that we shall be considering later, but they are developmental events and so come within the scope of the present account.

The immature oviduct of the chick is stimulated to grow and to acquire secretory cells in its lining, under the influence of oestrogens produced by the ovary (Brant and Nalbandov, 1956; Ljungkvist, 1967). Particular secretory cells called tubular gland cells produce two specific proteins, lysozyme and ovalbumin. A second type of cell, the goblet cell, also develops under the influence of oestrogens, but its special secretion, the protein avidin, is secreted only in response to the hormone progesterone. These proteins have been located in the cells in histological sections, by immunofluorescent techniques (see Kohler *et al.*, 1968).

Other changes, including enhanced DNA and RNA synthesis, are induced in the oviduct of the hen by further administration of oestrogens. There is a parallel in the response of the mammalian uterine endometrium to oestrogens (cf. Section 10.1), but no special 'marker' proteins have been singled out for study there as they have in the avian oviduct. The oviducal proteins of birds are important because they contribute to the albumen and the shell, which form round the egg as it passes down the duct to the exterior.

10.3.3   The mammary gland

Turkington (1971, loc. cit.) has done a great deal of work on the biochemical changes that take place in the mammary gland as it develops under the influence of a succession of hormones. We will pick out just a few salient points from those that he stresses, which give further insight into the ways in which hormones influence cellular differentiation.

Most of Turkington's studies were carried out on organ cultures *in vitro*, but the normality of development was assessed by comparison with the morphology and histology of the gland *in vivo* at equivalent stages. He found that in order to achieve normal growth and morphogenesis, followed by synthesis of milk proteins, at least three hormones were needed in succession. Insulin administered first, caused a proliferation of the epithelial cells. A growth factor extracted from the salivary glands of mice, called 'epithelial growth factor' or 'EGF', also stimulated proliferation, but the number of cells dividing could only be maintained if insulin was present too. A further effect of insulin, besides stimulation of DNA synthesis and of mitosis, is to activate the synthesis of histones. Oestrogens inhibit the proliferation of the mammary gland, but progesterone on the other hand stimulates its further development into branched ducts. Finally, the production of milk proteins requires the presence of insulin, hydrocortisone (cortisol, as it is now called) from the adrenal cortex and prolactin (a hormone from the posterior lobe of the pituitary gland).

The sequence of hormones required for each step in mammary gland differentiation is set out in Fig. 10.7. It is of particular interest that in male rats, the responses to these hormones are blocked by testosterone (Kratchowil, 1971). Turkington has also followed the rates of DNA and RNA synthesis in the presence of these hormones, and one of his interesting general conclusions from this work is that a phase of cell division is necessary, prior to any major step in cell differentiation. During the division process, he believes, there is an opportunity for new genes to become activated, and perhaps for some genes to be suppressed. This is a plausible suggestion, but there is no decisive evidence for it yet: only the circumstantial evidence that once cells start to differentiate, they usually cease to divide, and that responses to inductor tissues and hormones often show first as an increase in mitotic rate.

10.4 Hormones governing distinct steps in differentiation in other tissues

Continuing this account of some of hormonal effects on late developmental processes in animals, we will now consider two more examples in which a very clear-cut biochemical response is shown by the target cells on which a hormone acts (Turkington, 1971). These two further examples are the retina in birds, and the erythropoietic cells of the bone marrow in mammals. In the retina of the chick embryo, the enzyme glutamine synthetase is detectable from about 6 days' incubation, and it shows a marked increase in activity at 16 days. It is thought that this rise in activity is due to the action of the hormone hydroxycortisone from the adrenal cortex, because administration of hydroxycorticosteroids either to the chick embryo

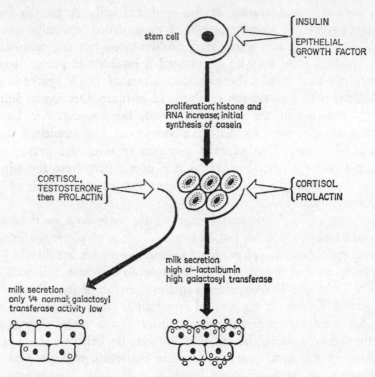

Fig. 10.7 Scheme of the actions of hormones on the differentiation of mammary gland cells. (*Based on* Turkington, 1971.)

in ovo, or to cultures of retina tissue, causes the rise in enzyme activity to appear earlier than usual (Moscona and Piddington, 1966, 1967). It is not certain, however, that the adrenal cortex of the normal chick embryo would already be producing hydroxycortisone at this stage: it remains possible that the rise in glutamine synthetase activity in the retina is normally independent of hormone action.

The bone marrow in developing mammals provides another example of cells producing a specific protein in response to the action of a specific hormone. Erythropoietin, produced by cells of the metanephric kidney, stimulates the stem cells of the erythropoietic line to synthesize haemoglobin and to differentiate into erythrocytes (Cole and Paul, 1966).

If we now include the avian oviduct and the mammalian mammary gland in this series of examples, we can set out for each of these four systems a precise scheme of the cellular interactions that take place (Table 10.4). In each case, one particular, known cell type is producing a known chemical secretion which circulates in the bloodstream to reach and interact with known target cells. These cells then give a specific response, producing one or more specific proteins.

Table 10.4 Proteins secreted in response to hormone action on cells

Source of hormone	Hormone	Target cells	Proteins produced
Ovarian cells	Oestrogen	Oviduct (birds)	Lysozyme
		Tubular gland cells (birds)	Ovalbumin
Corpus luteum	Progesterone	Goblet cells of oviduct (birds)	Avidin
Pancreas, β-cells	Insulin		
Adrenal cortex	Cortisol	Mammary gland	Casein
Pituitary	Prolactin		
Adrenal cortex	Cortisol	Retina cell (birds)	Glutamine synthetase
Metanephric kidney	Erythropoietin	Erythrocytes of bone marrow	Haemoglobin
Pituitary gland	Thyroid-stimulating hormone	Thyroid gland	Thyroglobulin

We must now consider more carefully the mechanisms by which hormones act on target cells and the mechanisms by which these cells respond to them.

10.5 The mechanisms of intercellular communication via hormones

Theories about the actions of hormones and other inducers of differentiation on embryonic cells have been completely revolutionized by the discovery of the cyclase system and the central role of the mononucleotide *cyclic adenosine monophosphate* (cyclic AMP, or cAMP) in cellular metabolism. The very wide range of functions carried out by means of cAMP in animal and plant cells has been reviewed by Bitensky and Gorman (1973). They have described this nucleotide as 'ubiquitous and indispensable in intercellular communication'. Among the general kinds of activity in which it participates is the mobilization of substrates by cells, so that a variety of proteins can then be synthesized. It is also responsible for special functions of certain cells – for instance it participates in neural excitation, and in the responses of pigment cells to hormones. The release of certain hormones from secretory cells is also initiated by cyclic AMP. These are sufficient examples to show that cyclic AMP is implicated both in hormone secretion and in the responses of target cells to hormones: moreover, all processes of cellular differentiation which involve the production of new proteins, structural or enzymatic, must depend at some stage on cAMP. Table 10.5 lists some of the events in which cyclic AMP has been shown to be involved. It has also been claimed that agents which increase the production of cAMP in mammalian cells in culture, inhibit the movement of these cells and hence speed up the onset of their differentiation (Johnson *et al.*, 1971). The corresponding guanosine mononucleotide, cGMP, is thought to take part in reactions that are antagonistic to those of cAMP. Thus, it tends to increase cell movements, particularly chemotactic ones (Estensen *et al.*, 1973).

Table 10.5   Summary of events controlled by cyclic AMP

Type of event	*Examples*
Substrate mobilization } Substrate utilization	Action of glucagon in liver
Melanocyte regulation	MSH activates adenyl cyclase in the melanocytes
Hormone release	TRF activates adenyl cyclase in TSH cells of pituitary gland
Neural excitation	Rhodopsin regulates cyclase activity in visual cells of retina
Regulation of water balance Cell movement	cAMP initiates aggregation in slime molds cAMP increases may inhibit movement in embryonic tissue cells
Synaptic transmission	cAMP appears: its role is uncertain
Smooth muscle contraction	Prostaglandins stimulate accumulation of cAMP in many tissues
Gene expression	cAMP added to cultures of bacteria and phages alters gene expression

Nearly all of the evidence about the role of cAMP comes from studies of cells in an advanced stage of differentiation, and there is a great need for work on its role in embryonic tissues. The only major morphogenetic role it has been shown to play so far is in the aggregation of slime moulds. Konijn *et al.* (1967) showed that addition of a purified extract from the bacterium *E. coli* which they identified as cAMP caused cultures of slime mould cells to aggregate by an apparent chemotaxis towards the substance, in the way that they normally do in response to the substance 'acrasin' produced by the cells themselves. The aggregation results in the formation of a fruiting body. Barkley (1969) has now concluded from chromatographic and spectrophotometric studies on 'acrasin' that it is identical to 3'-5 cAMP. It is tempting to wonder whether the aggregation behaviour shown by cells of animal embryos could also be a response to cAMP secreted into the medium by some cells. However, this one substance could hardly account for the *selective* adhesions shown by cells of like tissue type, as described in earlier chapters.

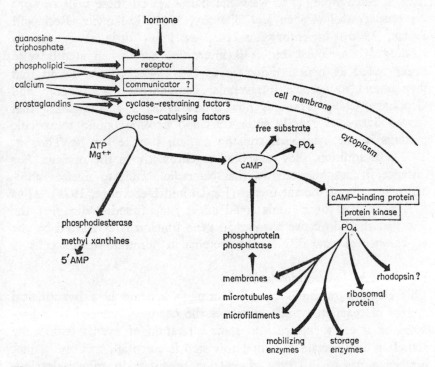

Fig. 10.8 Possible mechanisms of response of a cell to hormone action, involving cyclic AMP. (Based on the scheme put forward by Bitensky and Gorman, 1972.)

Bitensky and Gorman (loc. cit.) gave a diagrammatic summary of the processes that appear to be involved in the responses of cells to hormones: their diagram is reproduced in simplified form in Fig. 10.8. It assumes that a certain degree of differentiation has already occurred in the cell; for instance, it already has a specific receptor for the hormone, and it already has a variety of enzymes

227

and structural elements in its cytoplasm. Since hormones exert their effects late in development, this model of a partly-differentiated cell seems fair. Let us now consider some of the items suggested in the diagram, and see if there are examples of these in developing tissues.

10.5.1 *The hormone receptor*

This is thought of as a site where the hormone is firmly bound, either at the cell membrane or intracellularly. There is evidence that liver cells bind the hormone glucagon, and that adrenal cells bind ACTH, since attempts to wash the hormones off these cells *in vitro* are unsuccessful (Cohen and Bitensky, 1969; Lefkowitz, Roth and Pastan, 1970). In embryonic cells, we know little about surface binding of hormones, but it is interesting that when sterols were being tested as neural inductors in the 1940s, it was found that they bound to specific intracellular components (Waddington and Goodhart, 1949). There is now much evidence accumulating that steroid hormones readily enter cells and become bound to specific proteins in the cytoplasm and the nucleus (see review by Thomas, 1973). In addition, they may bind to chromatin in the nucleus. For instance, it has been shown that oestradiol binds to nuclear chromatin in cells of the rat uterus (Jensen and DeSombre, 1973). Thus it is clear that the steroids could affect gene function directly: the fact that they do cause changes in gene funcion is evidenced by the appearance of new dRNA and proteins in hormone-treated cells.

10.5.2 *The 'communicator'*

This element in Bitensky and Gorman's scheme is a hypothetical system of reactions which translates the change produced by a hormone at a cell's surface into some alteration of events within the cell. It is less certain now that this step is essential, and one cannot find examples of this type of reaction precisely, in animal development. The nearest analogy from developmental processes is perhaps the 'thymidine factor' which mediates the effects of growth hormone to chondrocytes (cf. Section 10.1). But this factor circulates in the serum and is not a part of the cell membrane as depicted in Fig. 10.8.

10.5.3 *The cyclase system*

This is present in all tissue cells that have been studied, and since cAMP plays a part in so many essential metabolic processes in the cell, one may assume that all viable cells in embryos as well as adults possess a cyclase system. It would be an enormous task for anyone to investigate whether or not cAMP is detectable in all kinds of embryonic cells, but a useful line of attack might be to introduce radioactively-labelled cAMP into some of the large blastomeres of early cleavage stages and to follow its fate. It should at least be possible to see if there are specific binding sites for cAMP in the cell, as suggested in the diagram. As noted in Chapter 6, cyclic AMP has been detected in the neural tube cells of the chick embryo.

10.5.4 *Phosphodiesterase and phosphoprotein phosphatase*

These two enzymes interact with cAMP, influencing its availability in the cell. As it happens, we do know that both enzymes are present in yolky eggs of vertebrates, where they play a part in breaking down the proteins of yolk. So their presence in all early embryonic cells which contain yolk may be assumed. Since it is rather unlikely that two such essential enzymes in energy metabolism would be lost and then reacquired later, one may also assume that they are probably present in all vertebrate cells during development.

10.5.5 *Special cytoplasmic components: membranes, microtubules etc.*

We have already seen that microtubules and microfilaments are present in embryonic cells from cleavage stages onwards, and that these appear to play a role in the changes in cell shape that occur, for instance, in cleavage and in neurulation (cf. Chapters 4 and 6). New membrane structures are a regular feature of differentiating cells, when these are viewed with the electron microscope. To take two widely-differing examples: the notochord of Amphibia, and the ommatidia (visual units) of the eye in insects, show a great increase in membranous cytoplasmic structures as they differentiate (Waddington, 1962). The other items on Bitensky and Gorman's list are all categories of substance that are normally present in all cells, except for rhodopsin which occurs only in the cells of the retina. Miller, Gorman and Bitensky (1971) have shown that cyclase activity in the membranes of the rod cells of the retina is regulated by the photochemical state of rhodopsin.

10.6 **Problems for future research**

Looking back at the examples of cellular interaction via hormones that have been given in this chapter, it can be seen that to explain all these cases by reference to some universal cAMP-mediated response in the target cells leaves a large number of interesting *differences* in hormone action still unaccounted for. Why, for instance, does the same hormone, thyroxine, initiate breakdown processes in some skin cells of the amphibian larva, and synthetic processes in other, apparently identical skin cells? Why does it accelerate maturation in most neurons, but degeneration in Mauthner's neurons? There must be very different factors in these cells controlling their cytoplasmic and nuclear responses to the hormone, but we know nothing about these as yet. There is also the problem of the tissues that respond to a variety of different hormones in a controlled sequence: for instance, the mammary gland epithelium and the epithelium of the oviduct. It is not necessarily different cell types that respond to each hormone, in these cases. Must one argue, then, that every cell possesses a variety of hormone-binding sites at its surface, which are active at different times? Again, we can only admit ignorance about this as yet.

Finally, there is the central problem for the developmental biologists, of how hormones and other extracellular molecules are able to bring about differentiation processes that require gene action. Can they transmit their effects through the cytoplasm and the nuclear membrane of the genome of the cell immediately, or must there first be a period of DNA synthesis and mitoses, to expose the genome to the cytoplasm with its new activators and/or repressors? This was implied in Turkington's suggestions about the need for mitoses in the mammary gland before differentiation could begin in response to hormones. We have seen above, however, that some hormones can act directly on the genome. Besides the evidence reviewed by Jensen and DeSombre (loc. cit.) on oestradiol, O'Malley and co-workers (1971) have produced evidence that progesterone acts on the genome of cells of the oviduct in mammals.

In invertebrates too, hormones may act on the genome directly, though not necessarily always in such a way as to promote the synthesis of new proteins. Novak (1967) suggested that the juvenile hormone of insects acts on DNA in the epidermal cells, causing it to replicate and at the same time *inhibiting* the translation of any new genetic information, so that the larval stage of differentiation persists. Unfortunately this idea was not extended to explain the reason for the changes that *do* occur at metamorphosis: we are left to

230

assume that, unless inhibited by the juvenile hormone, the DNA of the insect epidermis would transcribe new genetic information every time it replicated after a moult.

In conclusion, it is worth stressing again that despite several problems that still need investigation, the known effects of hormones in animal tissues provide the most clear-cut examples of how cells may communicate from a distance. The mechanism can be explained with some degree of precision at the cellular level, since we know which cells produce a given hormone, we know the structure of the hormone in many cases, and in the examples given in Sections 10.3 and 10.4, we see a precise response in the cell because it produces easily-detectable new proteins. The big gap in our knowledge is the lack of evidence for events as precise as these in embryonic tissues. It is understandable that the pioneering work on these events should have been done first on post-embryonic material, which is more abundant and does not always require ultramicromethods for detection of specific molecules in it. But it is important that these findings should be shown definitely to apply to developing tissues too. After all, it is possible that an entirely new hypothesis may be needed to explain hormone effects in embryos: we do not know yet.

11

Protective interactions between developing cells and tissues

We have now seen many examples of the fact that all animals have a series of controlling processes built into the mechanisms of their development and growth. In this book we are concentrating on the controls that are mediated by varieties of cellular interactions, promoting the development of organs and tissues along their normal path. Thus, the individual blastomeres in a cleaving egg, the groups of cells that form germ layers in a gastrula, the layers of tissue within each organ, and finally certain specialized cells within the nervous system and endocrine organs, carry on a series of interactions that are essential for normal development and maintenance of the organism.

Some of its built-in control processes also enable the organism to adapt to losses of tissue or to damage, and may therefore be regarded as *protecting* it against adverse circumstances of several kinds. In this chapter we shall select a few examples which emphasize the protective functions of cellular interactions in animal development. A few of these will refer back to early embryonic stages, but in addition, there are more specialized protective mechanisms that arise late in development and extend into post-embryonic life: for instance, immune reactions and the process of regeneration and repair of lost or damaged tissues. Finally, the control of normal and abnormal growth in embryonic tissues and tumours must also be considered briefly.

11.1 Protective interactions in early development

11.1.1 *Elimination of abnormal cells*
Individual cells of an embryo may become abnormal because of malfunctions or losses of genes. Chromosomal abnormalities may

232

arise at any cell division, though they are much rarer during mitotic divisions of somatic cells than during gametogenesis when meiotic divisions, pairing and crossing-over of chromosomes occur, with their concomitant risks that the chromosome pairs may fail either to rejoin or to move apart correctly. Gene mutations may occur at any time, but the normal mutation rate for most genes whose mutation rate has been measured is low, of the order of 1/20 000 genes. If a very early embryonic cell undergoes a mutation which results in defective metabolism and causes it to die, its continued presence and the possible spread of lytic enzymes from it to other cells could seriously impede the development of the whole embryo. One very rarely sees cleaving embryos of vertebrates carrying one or two dead cells and yet remaining normal as a whole. In Roux's famous pricking experiment on two-cell stage frog embryos (Roux, 1888), the one dead cell impeded the development of the other one. In the mouse, if one or two abnormal cells appear during early cleavage the embryo is not likely to survive. Clearly, then, it would be an advantage if these embryos had some mechanism of removing their abnormal cells. Unfortunately there is no such mechanism in higher organisms until the circulatory system has begun to develop and macrophages carried in the blood or other body fluids can then engulf the dead cells. So if early embryonic cells of vertebrates become abnormal, the chances are that they will jeopardize the survival of the whole embryo. There is a little evidence that this happens in human embryos too, though these have not of course been studied experimentally in the same way as other mammal embryos. Singh and Carr (1968) examined 168 human embryos that had died and aborted at early stages, and found that 73 of these had chromosomal abnormalities in some of their cells. Sixteen of the remaining 95 embryos had gross abnormalities of the organs, which included types of abnormality that in other cases have been found to result either from gene mutations or from abnormalities of the chromosomes. So it was concluded that chromosomal and gene abnormalities in a few cells were the original cause of death in these embyros. As more becomes known about the genetics of both vertebrate and invertebrate animals, it may be possible to confirm more examples of the elimination of genetic abnormality either through death of the whole organism, or at later stages through death and subsequent phagocytosis of the abnormal cells.

Besides these examples in which the cells that die are clearly abnormal, there are several instances, as we have seen in Chapters 7, 8 and 9, in which cell death occurs as a normal part of development. (E.g. in the epidermis of the inner edge of the palate and of the posterior margin of the wing bud, and in the motor and sensory neuroblasts in the spinal cord and ganglia.) In these cases, we have no evidence yet as to whether the cells that die carry any genetic or chromosomal abnormalities: all we know is that something in their genetic make-up determines that these cells shall die at a very precise stage in development. In all of these cases the dead cells are eventually eliminated by phagocytes of the surrounding tissues. In the interdigital zones of the developing limb of the rat, where cell death occurs on a large scale as the digits form (Ballard and Holt, 1968), these phagocytes, rich in acid phosphatase enzymes, can be seen engulfing the dead cells. A similar process of phagocytosis is seen at the posterior necrotic zone of the wing in the chick (cf. Fallon and Saunders, 1968). Vertebrates are not exclusive in this phenomenon: Whitten (1969) has pointed out that in insects also, there is some cell death in the imaginal discs that give rise to limbs.

The widespread occurrence of cell death in normal ontogeny was first stressed by Glücksmann (1951), and the general problem of how cells gradually senesce and die has been reviewed by Strehler (1962) who dealt not only with the redundant cells in normal ontogeny but also with the stable population of cells that survive into adult life. Once cells have died, we can never know if their genes were abnormal unless some obvious change has been produced in them or in their DNA which is not a part of the process of death. It is only if they 'sicken' or senesce gradually, that abnormalities other than chromosomal ones might be detectable before their death. Often one cannot recognize the doomed cells of an embryo until they are in an advanced state of necrosis, however. So there is still much speculation about the extent to which abnormalities may be 'weeded out' by means of cell death in embryonic organs. In the nervous system of vertebrates, and in the ovaries of mammals before birth, for instance, there is such large-scale loss of cells by death that one feels sure that some process of genetic selection must be involved. Is it a group of descendants of one particular abnormal stem cell – i.e. the *clone* derived from it – that dies? This is a difficult point to investigate, and we have no clear evidence along these lines as yet.

11.1.2 *Cell affinities and cloning, seen as protective mechanisms*
We saw in Chapters 5 and 6 that there is a sequence of cell and tissue movements and adhesions during gastrulation and neurulation which ensures that certain groups of cells will come into contact and be enabled to interact. The interaction in its turn ensures the normality of the next stages of differentiation in those cells. The importance of these early cell and tissue movements is evident from cases where there has been abnormal gastrulation and this results in failures of differentiation. For instance, if the mesoderm of an amphibian gastrula is prevented from invaginating under the ectoderm, by placing the embryo in hypertonic saline, the dorsal ectoderm fails to form a nervous system (Holtfreter, 1933). In hybrids which fail to gastrulate, as we saw in Chapter 5, development is arrested. In this latter instance, it will be remembered, the reaggregation behaviour of the hybrid cells was also abnormal (Johnson, 1969), implying that the abnormal gastrulation was at least partly due to abnormal affinities between the cells.

There are two 'protective' results of the cell affinities in early embryos. One, which we have just stressed, is that the normal cell affinities ensure normal tissue movements, normal interactions and therefore normal subsequent differentiation. Conversely, it can also be argued that the inability of hybrid or other genetically abnormal cells to adhere and to gastrulate will ensure that no embryos can develop further from these, and so eliminates them from the population. It is true of all vertebrates and also of some invertebrates that without normal gastrulation, no nervous system can form and little further differentiation of other tissues or organs can take place, so the embryo is likely to die. Even if a partially developed individual does reach larval stages, it is unlikely to survive for long. To take an example from invertebrates: lack of gastrulation in some molluscs would result in failure of the gut to invaginate and to make contact with the future shell gland: hence the larva would have no shell, besides its other deficiencies, and would be most unlikely to survive the hazards of predators and of desiccation.

As was suggested in Chapter 5, one possible mechanism by which cells of early embryos recognize others of their own tissue type is that they may carry at their surface molecules with distinct structural configurations, perhaps analogous to antigens. These surface molecules must, like every other molecule of the cell, be products of the activity of its own genes, or of the genes in the stem cell from which it was derived. A clone of cells, originating from one stem cell by mitosis, could inherit a set of structural elements and of antigen-like molecules passed on to them from the stem cell. This would mean that as mitosis and growth proceeded in any one organ or tissue, any clone of cells in it would be able to continue to adhere because they continued to carry identical surface molecules responsible for adhesion. So it would not always be necessary for each daughter cell of a division to synthesize its own specific molecules afresh, or for there to be any lag period before it was able to adhere to other cells of its clone.

Many tissues and organs in both vertebrates and invertebrates are built up from clones of cells during embryonic development. As we saw in Chapter 8 (Section 8.7), the limbs of insects are made up of twenty clones of cells, each forming a longitudinal strip of tissue. In the frog, *Xenopus*, each skin gland is a clone derived from one cell (McGarry and Vanable, 1969). In the mouse embryo, experiments in which cells of different genotypes are mixed (Mintz, 1967, 1970; Moore and Mintz, 1972) indicate that a high degree of cloning occurs in several tissues. If morulae from different strains carrying marker genes (e.g. for coat colour) are fused, the resultant mice often show patches or even stripes of differently coloured fur (cf. Fig. 4.9). Mintz argued that each stripe or patch could be a clone arising from a single neural crest cell. The distribution of cells of different genotypes has since been traced in several embryonic tissues of chimaeric mice, and the findings are consistent with the idea that each organ is derived from one or a very few progenitor cells. This evidence is still controversial, for it involves several assumptions about the times at which the cells and organs differentiate. The uncertainties have been discussed by McLaren (1972). If the evidence is found to stand, however, it confirms that cells derived from the same original stem cell tend to adhere together and to have the highest adhesive affinity. If this is because they carry special structural molecules at their surfaces which have affinities similar to those of antigens and antibodies, then this mechanism of interaction of the cells may be compared with some aspects of the immune system. The cell-recognition processes which govern adhesions of embryonic cells may in fact be the forerunners of the protective immune mechanisms which develop much later in embryonic life. We will now go on to consider these.

11.2 Cell-recognition and the immune system

In vertebrate embryos, a specialized line of cells is produced late in development which becomes responsible for recognizing the antigens present throughout the embryo, and of reacting to any abnormal, 'foreign' antigens by producing complementary molecules called 'antibodies' which combine with the antigens. This specialized population of cells develops in the thymus gland or some equivalent organ, and from there colonizes the spleen and the lymph nodes, and circulates in the blood and lymph. In birds there is a diverticulum of the hind gut, called the bursa of Fabricius, where some of these cells are produced. When any of the circulating cells of this lymphocyte system encounter a foreign antigen, produced perhaps by an infective mirco-organism or a cell of some other individual of the same species, the lymphocytes undergo some response, probably chemical, which they carry back to cells in the spleen and the lymph nodes. These spleen and lymph node cells then produce an antibody specific to that antigen. They also proliferate rapidly, forming clones of cells all carrying the same antibody. The antibody is either liberated into the blood serum (serum antibody) or carried in the blood on special lymphocytes (cell-bound antibody), and it eventually reaches the site where the foreign antigen is present. When the antibody combines with the antigen, this neutralizes any toxic effect that it may have had. If the antigen is carried by a foreign cell, this cell also becomes 'opsonized' as a result of events associated with the antigen/antibody reaction: i.e. it is immobilized, its cell membrane is weakened, and it becomes more easily digested and engulfed by phagocytes.

The immune mechanism is a most important cellular interaction which, like the controls exerted by hormones, extends into post-embryonic life. It protects the organism against external infective agents. It is most highly developed in mammals, but exists in similar form in birds and amphibians and, as far as evidence goes, also in reptiles and fish (Burnet, 1968). Evidence for competence to mount immune reactions is obtained by testing the ability of individuals to produce serum antibodies in response to injections of antigens or of foreign micro-organisms, and also by skin grafts to test for 'transplantation immunity' – i.e. the ability to reject foreign cells. A skin graft from any other individual will be rejected, unless the individual has a very closely similar genetic constitution. Identical twins, and animals from highly inbred strains, will tolerate each other's skin grafts.

A special example of a foreign tissue which is tolerated despite being genetically different from the host organism, is the implanted mammal embryo in the uterus of the pregnant female. As it has half its genes from its father, this embryo is far from genetically identical to the mother, and carries transplantation antigens which, if any of its cells are injected into the female's blood or transplanted to other parts of the body experimentally, produce an antibody reaction and result in rejection of the transplant. There has been much discussion of the reasons why the fetus is not rejected like a foreign graft (see Edwards, 1970 for a brief review). Current evidence indicates that lymphocytes from the maternal blood have some difficulty in penetrating the endometrial layer of the uterine wall, though they can easily pass through its outer, myometrial layer. So antibody-carrying cells may not be able to reach the fetal tissues effectively when it is implanted in the uterus, though they can when it is transplanted to less 'privileged' sites. This suggestion was supported by experiments in which skin grafts placed in the uterus of the rabbit were shown to survive much longer than they would at other sites in the body (McLean and Scothorne, 1972). It has also been shown that when maternal antibodies are allowed direct access to rat embryos by growing them *in vitro* in serum which contains antibodies to the outer membranes of the yolk sac (New and Brent, 1972), they seriously inhibit growth of the embryo and cause its eventual death.

There are many other groups of animals, besides the mammals, in which viviparity occurs (i.e. the embryos develop within the female and the young are born alive). Some snakes are viviparous (Bellairs, 1951); so are some fishes, and some of the invertebrates, notoriously *Peripatus edwardsii*, a primitive worm-like arthropod: In all these cases, there must be some mechanisms for ensuring that no rejection reactions occur between the cells of the maternal immune or phagocytic system and the cells of the embryonic tissues. We do not yet know what these mechanisms are, and it will be a fascinating subject for future research.

Invertebrates, although they do not have organs as highly specialized as the thymus and the spleen of vertebrates, do have a system of circulating cells capable of recognizing foreign materials which are then engulfed by phagocytes. The body fluids of some invertebrates have also been found to contain substances which act in a similar

238

way to agglutinins and cause clumping of mammalian red blood cells (see Burnet, 1968).

An important corollary to the recognition of foreign cells and proteins by the immune system is that it should also be able to recognize and *not* to react against the organism's own cells and antigens. In early embryos, at the stages when the cells depend for their movement and adhesions on mutual recognition mechanisms, the immune system has not yet formed. This is why experimental embryologists are able to graft tissues from other individuals or even other species into host embryos, without their being rejected. If the grafts are small, however, and are left *in situ* until stages when the host's immune system is functional, they may eventually be rejected. Despite their long association with the embryo during the time that the immune system is developing, the grafts are evidently recognized as foreign. Some time ago, however, Billingham, Brent and Medawar (1953) showed that if foreign lymphoid cells were injected into newborn rats and mice, whose immune systems were still immature, this rendered the animals 'tolerant' to skin grafts from the donors of the foreign cells. In this case it seemed that exposure of the foreign cells to the immune system while it was still developing allowed them to be accepted as 'self' and not as foreign, and that subsequently other, skin cells from the same donor were also treated as 'self'. This provided evidence that the early immune cells of a developing animal register as 'self' all those cells and antigens that they meet before they are mature. It thus provided an explanation of why they do not attack any cells of the animal's own tissue, except in some autoimmune diseases. In these diseases it is thought that new antigens are liberated from the affected tissues which were not previously in circulation, so the immune system reacts against them because it has not encountered them before.

Recent advances in medical research on organ transplants, and other work in which it has now been found possible in adults to induce tolerance to foreign tissues, has thrown doubt on earlier ideas as to how each individual embryo acquires tolerance to its own cells. The 'clonal selection theory' of Burnet (see his book, 1969) no longer seems adequate, and Bretscher and Cohn (1968, 1970) have suggested a more complex mechanism by which self-non-self discrimination could arise in embryonic development. They envisage that antigen-sensitive cells (e.g. lymphocytes) in any organism may either be paralysed, or be induced to form antibody, according to whether the cell's surface receptors each bind only one, or two or more antigen determinants. On this theory, low doses of antigen, which lead to the binding of only one antigen determinant at each receptor on the sensitive cell, might be expected to paralyse the cell. This might be the situation in a developing embryo, where the new cell types in each tissue were appearing only gradually and releasing only small amounts of antigen into the bloodstream. This might then preclude the production of antibody to these new cell types. But there is not enough evidence yet to substantiate this theory, and the problem of self-recognition in animal cells remains a matter for controversy and much further research.

To ensure that the developing immune system is exposed to all the antigens that will be formed within its own organism and learns to recognize these as 'self', the timing of its own maturity must not be in advance of the complete differentiation of all cell types in embryonic organs. Ideally, no new cellular antigens should appear in the embryonic organs after the immune system's development is complete and it is ready to destroy all unfamiliar cells and molecules. This nicety of timing could perhaps have evolved by some process of 'natural selection', for any new cells appearing too late would be eliminated. When we look at the time of onset of immune reactions in some of the animals that have been studied, we see that it occurs late enough for it to be likely that all cell differentiation in other organs is complete by then. In amphibians, for instance, cellular immune reactions seem to be weak or absent during larval life but to mature after metamorphosis, when all the new adult tissues have formed. Orfila and DeParis (1970) found that skin grafts from other individuals were tolerated by larvae of the salamander *Pleurodeles waltlii* but were rejected soon after metamorphosis. (This shows, incidentally, that the graft cells had *not* been learnt as 'self' by the immature lymphocytes, even though at least some of their antigens must have entered the circulation of the host.)

A detailed study of the maturation of the immune system in larvae of the frog, *Xenopus* has been carried out by Manning (1971) and Horton (1969). They have shown that the onset of rejection of skin grafts coincides with the first appearance of lymphocytes in the thymus gland. If the thymus gland is removed before this time, the skin graft rejection reaction is reduced or absent. Cells morphologically similar to lymphocytes are found to be in circulation before they are seen in the thymus gland, and their presence in the gland is thought to indicate the last step in development of their specificity against different antigens. The competence of the lymphocytes to act on foreign cells is most conveniently tested in tissue culture. When tests of this type were applied to lymphocytes of *Xenopus*, Kidder, Ruben and Stevens (1973) showed that they could produce antibodies which agglutinated foreign red blood cells, by mid-larval stages (stage 50) when the forelimb buds had appeared. This stage coincides with the maturation of the thymus and other lymphoid organs also. But it does not necessarily mean that all cellular immune reactions are already operating *in vivo*, for there are other controlling factors which affect the efficiency of the immune responses in the intact animal. One is the accessibility of the foreign antigens to the cells of the immune system, as we saw in the case of implanted mammal embryos, and another is the ability of the cells to proliferate in sufficient numbers to gain the upper hand in any combat with foreign material. In tissue culture, both of these factors can be controlled by the experimenter.

Although we have not sufficient evidence about this yet, it is possible that hormones either from the pituitary gland or other sources may control the onset of maturity in the immune system. At the time of metamorphosis in amphibians, the endocrine system also has reached full maturity. If the maturation of the lymphocytes, as well as their proliferation, depended on some hormone similar to thymosine (cf. Chapter 10), this would mean that the immune system could not become mature until a late stage in development when most organs, including the endocrine ones, were fully differentiated and functional. This would ensure that there was no interference with normal development and differentiation by 'non-self' reactions. When considered in this light, it is seen that hormonal controls may also have a protective aspect: they could ensure that a number of maturation processes occur at appropriate times, and that correlated events are suitably synchronized. It could well be this system of hormonal controls which, among its many other effects, protects the embryo from destruction by its own major protective system, the immune system.

11.3 The protective interactions of regeneration and repair

It is well known that whereas some of the lower vertebrates can regenerate whole limbs and other large organs, and invertebrates can sometimes regenerate a whole new body from a small part, the higher vertebrates such as birds and mammals have much more restricted powers of regeneration. Generally speaking, the smaller and simpler the animal, the more extensive are its regenerative powers. In the following account we shall discuss only a few features of regeneration and repair processes in vertebrates, with emphasis on the cellular interactions that are involved in these protective events.

11.3.1 *Regeneration*

As we saw in Chapter 9, the nerve supply influences the success of growth and regeneration, particularly in limbs of amphibians. Denervated limbs do not regenerate well, whereas diverting extra nerves to a limb stump may increase the normal rate of regeneration, or even induce a duplication of the regenerated limb. Pituitary hormones also play a role in controlling regeneration. Hypophysectomy (removal of the anterior pituitary lobe) retards limb regeneration, but this may be restored to normal by pituitary implants or by administration of growth hormone. These apparently important interactions between pituitary cells, nerve endings and regeneration blastemal cells are not essential however, for an isolated limb blastema will start regeneration *in vitro* (Stocum, 1968). It can then be made to continue its regeneration by transplanting it, even to a foreign site such as the tail. There have been some claims that a limb blastema in this position is induced to form a tail, by interaction with the tissues of the tail stump, but these cases have not always been very convincing. The regeneration of the tail itself is influenced by the subcommissural organ in the brain, as we saw in Chapter 9, but this interaction is not essential either, for tail stumps will also regenerate when isolated *in vitro* (Hauser and Lehmann, 1962).

Within the regenerating limb, cellular interactions between mesoderm and ectoderm govern its normal development just as they do in ontogeny. Michael and Faber (1971) found that removal of the ectoderm from a forelimb blastema of the axolotl made it unable to form digits, just as the denuded embryonic limb bud fails to form digits (cf. Chapter 8).

The initial sequence of cellular activities in regeneration is similar to that in wound-healing, and its main purpose is threefold: to seal the cut edge of the stump of the organ, to remove dead cells and to mobilize raw materials to the site ready for use by the newly-differentiating blastema cells. Needham (1952) pointed out that the essential first step is wound closure, which is achieved very rapidly either

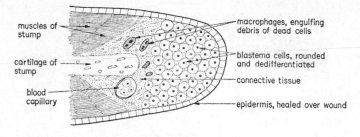

muscles of stump

macrophages, engulfing debris of dead cells

cartilage of stump

blastema cells, rounded and dedifferentiated

connective tissue

blood capillary

epidermis, healed over wound

Fig. 11.1 Diagram of a longitudinal section of a regenerating appendage in an amphibian, to show the blastema.

by the stretching of surrounding tissues if the gap is small, or by a blood clot. Later, epidermal cells proliferate and migrate under the scab formed by this clot. After dead cells have been removed by phagocytosis, remarkable changes occur in some of the internal cells. These undergo a process of *de-differentiation,* reversing the sequence of changes that they went through during their initial differentiation, and they form a blastema of rounded cells which are capable of reconstituting the new organ and all its component tissues (Fig. 11.1). This is a remarkable phenomenon and implies that the cells must have received some information from the wounded area 'instructing' them to de-differentiate and then to start a new process of gene activity leading to their proliferation, redifferentiation and further growth. We do not know what cells, or what stimuli initiate these elaborate events of regeneration. In smaller-scale repair processes such as wound-healing, however, it is thought that the loss of a mitotic inhibitor may be involved (see Section 11.3.2).

11.3.2 *Wound healing*

The activities of the epidermal and dermal cells during healing of a skin wounds in mammals have been described by Ross (1969). We will summarize his account briefly here (see also Fig. 11.2). As in regeneration, the first step is a flow of blood to the wound area. This blood clots and thus closes the wound; the clot then loses fluid and forms a hard, protective scab within a few hours. There is then a leakage of blood serum and antibodies into the wound area, which causes it to become inflamed. It has been found that the dying tissues produce some substance which causes the nearby blood capillaries to leak – perhaps because it loosens the adhesions between their endothelial cells. White blood cells then enter the wound area too: first the

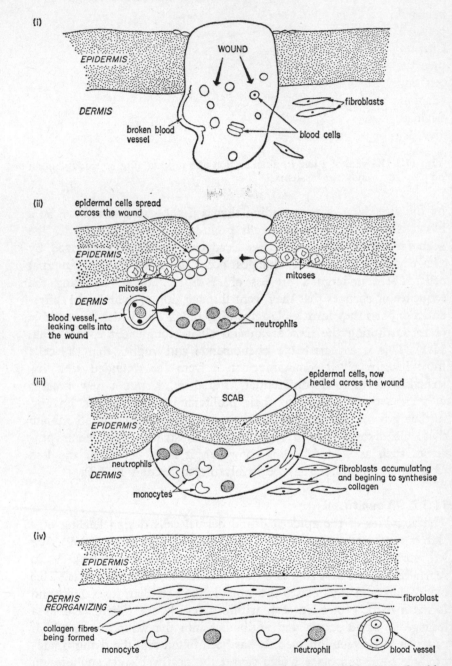

Fig. 11.2 Schematic drawings showing steps in the process of healing of a skin wound. (*Modified from* Ross, 1969.) (i) At the time of wounding; (ii) one day later; (iii) two days later; (iv) seven days later (reduced magnification).

neutrophils emerge from the blood vessel walls by pushing their way between the endothelial cells (by diapedesis, like the germ cells: cf. Chapter 2). These neutrophils engulf bacteria by phagocytosis, and then themselves break down rapidly, liberating enzymes which attack cell debris, making it easily ingested by other phagocytes. Within 12 hours another type of white cells, the monocytes, appear. These originate in the bone marrow, and when they reach the wound site they acquire rough endoplasmic reticulum in their cytoplasm, indicating that they are synthesizing new proteins. They function for a long time as macrophages.

When all the above events, which constitute the inflammatory response, begin to subside, fibroblasts appear from the dermal layer and lay down collagen to form the scar tissue. Ross has followed in detail the production of this collagen within the endoplasmic reticulum of the cells and its passage from there out into the extra-cellular spaces. It is gradually rearranged and remodelled within the next few weeks, so that the scar tissue becomes more like the original skin. During this time too, the epidermal cells have proliferated and have moved across under the scab: Ross points out that to do this they probably rearrange their desmosomal attachments between the cells. Another unusual feature of their behaviour is that they engulf strands of serum protein or fibrin that lie in their path.

Ross concludes that there must be some interaction between the epidermal and the dermal cells which ensures that both layers complete their healing simultaneously. He suggests that one factor could be the activity of epidermal collagenase, which helps to break down and remould the collagen in the dermis. Croft and Tarin (1970) observed that the epidermal cells seemed to attach to dermal cells as they moved across the wound area. There must also be interactions within the epidermal layer, which cause the cells to stop proliferating and migrating once they have met over the wound area. Some limit must be set to the migration of the fibroblasts too, so that they do not either over-populate a wound or spread into surrounding areas and lay down excessive collagen there. Possibly both types of cell stop moving on making contact with others of their kind, because of the phenomenon of 'contact inhibition' (cf. Chapter 1) described first in fibroblasts by Abercrombie and Heaysman (1954).

It is still not clear what initiates these several steps in wound healing. One suggestion has been that the damaged tissues produce a 'wound hormone' which stimulates the activity of the fibroblasts and epidermal cells. A similar hormone, it was thought, might trigger the regenerative events in lower vertebrates. The only experimental evidence advanced recently for the existence of any such wound hormone, however, is the claim by Joseph and Dyson (1970) that in rabbits, abdominal wounding enhanced the regeneration of epidermal tissue of the ear. They did not mention whether there was any reciprocal effect from the wounded ear tissue on the healing of the abdominal wound. The 'wound hormone' theory has fallen out of favour generally, since no one has yet isolated any substance from wounded tissues which has the properties ascribed to this hormone. There is now much more support for a converse type of theory, which suggests that normal adult tissues are restrained in their growth by mitotic inhibitors called 'chalones' (Bullough and Laurence, 1960). The name is derived from the Greek word $\chi\alpha\lambda\grave{\alpha}\omega$ meaning 'to make slack', in contrast to 'hormone' which comes from the Greek $o\rho\mu\grave{\alpha}\omega$ meaning 'to arouse'. According to Bullough's theory, wounding causes a loss of chalone from the damaged tissues and thus allows an increase in their mitotic activity. He suggests that each tissue produces its own specific chalone, which acts only on its own cells and not on those of any other tissue or organ. Stimulated by these ideas, a number of workers have now isolated aqueous extracts from various tissues and have found that these have chalone-like properties, inhibiting mitosis in their own tissue but not in others. Bullough has also extended his theory to explain the phenomenon of compensatory growth, which we must now discuss in more detail, since it is another example of protective interactions between cells in the developing organism.

11.4 Cellular interactions in compensatory growth

Compensatory growth is a phenomenon which occurs in adult as well as in developing animals. It has been studied most in adult mammals, birds and amphibians. In all of these forms, when part or the whole of an organ is absent, either from natural causes or as a result of experimental surgery, the rest of that organ, or its pair on the opposite side of the body, will enlarge to make good the deficiency. For example, loss of one kidney causes the other one to enlarge, and loss of liver tissue causes the rest of the liver to proliferate. The term compensatory growth may also be applied to cases where an undamaged organ undergoes hypertrophy (i.e. overgrowth) when it is unusually overworked: a good example is the enlargement of the heart in some individuals with circulatory defects that impose an extra burden on the heart muscle. These types of unusual growth are protective, in that they tend to restore the normal functioning of the organism and to enable it to survive.

Theories about the cellular interactions that control compensatory growth have been many and unsatisfactory. Present views are influenced by Bullough's 'chalone' theory mentioned above. He envisages that a specific chalone in each organ is at equilibrium concentration with some of the same chalone circulating in the blood (cf. Fig. 11.3). He reasons that if part of the organ is lost, the equilibrium

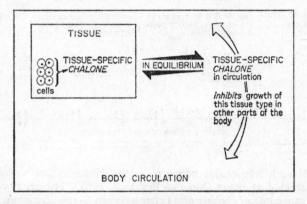

Fig. 11.3 Diagram to show a possible mechanism of control of growth in individual tissues by means of tissue-specific chalones. As the tissue grows, the concentration of chalone is thought to build up in the tissue and also in the body circulation, until a threshold value is reached which inhibits mitosis, and hence further growth, in this particular tissue. (See Bullough and Laurence, 1960, 1964.)

is disturbed, and to restore it, the concentration of chalone escaping from the organ into the circulation falls. As a result, there will be not only less chalone in the remainder of the organ, but less chalone reaching any counterpart organ on the other side of the body. There will therefore be less inhibition of mitotic rate in this organ and its counterpart. These may both proliferate, therefore, until the concentration of chalone is restored to a value which inhibits mitosis and arrests growth once more.

Bullough (1971) has broadened the concept of chalones to include interactions between cells within each developing organ, controlling the phases of mitosis, differentiation and growth that occur in the

247

I

organ during embryonic and adult life. This concept is illustrated in Fig. 11.4. He suggests that when an optimal cell density is reached, the concentration of chalone is such as to stop further mitosis, and from then on the differentiating cells liberate chalone into the blood-stream. If chalones do indeed circulate in the blood, this is a point of comparison with hormones, though their effects are very different, being suppressive rather than stimulatory. A point of *contrast* with hormones is that chalones act only on the same type of cell that produces them, and not on any *other* target cell.

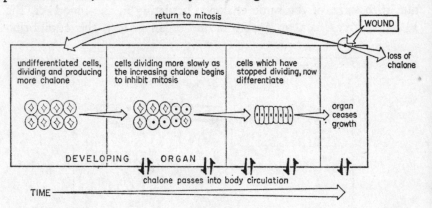

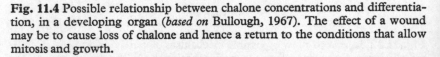

Fig. 11.4 Possible relationship between chalone concentrations and differentiation, in a developing organ (*based on* Bullough, 1967). The effect of a wound may be to cause loss of chalone and hence a return to the conditions that allow mitosis and growth.

We still need a good deal more experimental evidence before it can be assumed that chalones exist in all tissues. At first, crude aqueous extracts of several tissues were tried, and it is only in the last two or three years that the isolation and characterization of active components from these extracts has been achieved. Thus, Verly *et al.* (1971) have now extracted a polypeptide chalone from liver, and the chalone in epidermis has been found to be a glycoprotein (see O'Dell, 1972 for review). Chalones have been tested on cells in tissue culture as well as *in vivo*, and in amphibians as well as in mammals. Chopra and Simnett (1969, 1971) obtained an extract from the kidney of adult *Xenopus* frogs and found that this inhibited mitosis in cultures of larval pronephric kidney tissue. It also worked *in vivo*, inhibiting the compensatory hypertrophy that normally occurs in one pronephric kidney after the other one has been removed. The kidney extract had no effect on cultures of other organs of *Xenopus*, so appeared to be organ-specific, though it is interesting that its specificity was common to both pro- and mesonephric kidney tissue.

The possible existence of these mitotic inhibitors at all stages of development in animals could provide a means of growth control through cellular interactions within each organ or tissue, as Bullough envisaged. If the maturer cells of each tissue produce most chalone, they could block further mitosis in cells which are less mature, so that these then start to differentiate and the organ ceases its phase of growth. A further and more protective role has also been suggested for chalones, in limiting the growth of tumours. Rytomäa and Kiviniemi (1969) showed in rats that injections of aqueous extracts of epidermis into skin carcinomas reduced their size dramatically. Most tumours are due to over-active mitosis, the cells evidently having got out of control and started a burst of abnormal proliferative activity. We must now go on to consider some of the properties of these tumour cells, and how interactions with normal cells may be used to control them.

11.5 The properties of tumour cells, compared with those of normal embryonic cells

It is often said that tumour cells are adult cells that have reverted to an 'embryonic' condition. This is a somewhat misleading statement, however. Many tumours of embryonic tissues are known, which develop during embryonic life and are fully formed at birth. The cells in these tumours have never become normal 'adult' cells: moreover, they are clearly very unlike the normal cells of the embryo. So, while it remains true that some features of adult tumour cells are reminiscent of the embryonic condition, this comparison should not be taken too far.

One significant type of similarity observed between some tumours and embryonic cells is that they may contain the same tissue-specific proteins, such as antigens and isoenzymes. For instance, there are the 'onco-fetal antigens' which, as their name implies, are found in both fetal tissues and tumours. An example is α-foetoprotein, which is found in the liver of fetal and newborn mice, and in hepatomas, but is not present in the liver of older mice. The occurrences of these and many other antigenic similarities between fetal and cancerous cells in mammals have been reviewed by Alexander (1972). Recently, too, Artzt et al. (1973) have reported that immature teratocarcinoma cells carry surface antigens resembling those of

249

I*

cleavage stage mouse embryos. Resemblances in the antigenic patterns of human fetuses and adult carcinomas have also been demonstrated, by cross-reactions that show after antisera to them have been absorbed with extracts from normal organs (Klavins *et al.*, 1971). Many similarities between the isoenzymes of tumours and of embryonic tissue have also been observed: these are reviewed by Criss (1971). One example is 'Regan isoenzyme', a fetal form of alkaline phosphatase which is found in patients with bronchiocarcinoma.

Work on the antigens of mammal embryos has led to the recent discovery in mice of indigenous sources of potential cancerous growth, which are normally eliminated early in development, but may reappear in old age. Huebner *et al.* (1970) found that virus antigens of the *gs* type are present in early mouse embryos but disappear later in development. The antigens evidently remain latent in the cells in some precursor form, however, since they are detected again in mice of advanced age, when cancerous growths develop. Whether their apparent disappearance for the greater part of the mouse's life span is due to most of the antigenic cells being eliminated, or whether these cells remain but the part of their genome that responds to the virus is inactivated, is not yet known. A finding with similar implications is that of Pearson and Freeman (1968) who observed that some antigens of embryonic hamster cells resembled a transplantation antigen induced by treating the cells with polyoma virus. They suggested that the antigen which the virus induces is at least partly determined by the genome of the embryo. Since this antigen disappears from later embryos, like the *gs* antigen in mice, one must assume that the parts of the genome that are sensitive to the virus are suppressed later in embryonic life, but were active in the early embryo. One might then argue that the induction of cancer by viruses is due to a reactivation of these virus-sensitive embryonic genes.

A special type of tumour whose cells show remarkable resemblances to embryonic cells, in their ability to differentiate into a wide variety of tissues, is the teratoma (Plate 11.1). Teratomas form jumbled masses of several well-differentiated tissues such as bones, teeth and hair. They are sometimes found in the ovary or the testis, suggesting that a germ cell may have undergone parthenogenetic development in these cases. Testicular teratomas can also be obtained experimentally in certain strains of mice, by implanting either eggs or young embryos into the testis (Stevens, 1967, 1968, 1970). Whether the testis merely forms a favourable environment nutritionally, or whether its cells interact with the graft and form part of the teratoma, or even induce it to develop, is not clear. Evans (1972) was able to identify at least ten types of differentiated tissue in a particular strain of mouse testicular teratoma cells isolated in culture. He also obtained two types of clone from these, one of which showed tumour-like growth and multiple differentiation, while the other type of clone formed only epithelia. This suggests that in any group of teratoma cells, only some become pluripotent. One wonders if these are cells that derive from the inner cell mass region of the transplanted mouse embryo, while the epithelial-type clones may be derivatives of trophoblast. Probably the relative proportions of these two types of cell that develop determine how tumourous and how extensive is the growth of the teratoma. The cells within the tumour may interact and to some extent limit each other's development, just as do embryonic cells. It is possible that similar relationships exist between the cells of other types of tumour which show a variety of modes of differentiation. For instance, tumours derived from embryonic neural crest cells may show as melanomas (masses of pigment cells), as neuroblastomas (masses of neuroblasts), or as medulloblastomas (more mature nerve cells). All three of these types of tumour may on occasions 'convert' into differentiated tissue and cease their malignant growth. Possibly some internal 'chalone' has eventually prevailed in these cases, blocking further mitoses and promoting differentiation. Certain melanomas have been found to respond to chalone injections, as we saw above, and to reduce in size (Rytomäa and Kiviniemi, 1969).

11.6   Control of tumours by interaction with normal cells

Having seen some of the resemblances and differences between tumour cells and normal embryonic cells, it is tempting to try and suggest in what kinds of circumstances tumour cells could be converted to normal ones, and the tumour thus eliminated at an early stage in the life of the animal or human being. It is not necessarily sufficient just to have a means of killing the abnormal cells selectively, for exudates from the dead cells may still be capable of transmitting the tumour transformation to other cells, in some cases. This is true of tumours induced by viruses, of course. It is also true of certain lymphosarcomas in the frog *Xenopus*. Balls and Ruben and their colleagues have studied some of the interactions of these lymphosar-

251

coma cells with normal cells in *Xenopus* (see Balls and Ruben, 1968; Hadji-Azimi, 1970). Ruben and Balls (1964) thought it possible that cells undergoing rapid morphogenesis might exert an organizing effect on any cells introduced into their neighbourhood, and might restrain the development of a tumour and convert it instead into organized tissue. They induced lymphosarcoma cells to arise in the tissues of *Xenopus* larvae by implanting methyl-cholanthrene crystals at various sites. One of these sites was the regenerating limb bud, where they hoped that even if sarcoma was induced, it would be converted into normal limb tissue by the morphogenetic influence of the limb regenerate. Contrary to this hope, they found that these lymphosarcomas developed just as extensively in the regenerating limb as in other larval tissue. Ingram (1971) has been more successful since, however, and has been able to inhibit tumour growth by grafts to regenerating limbs in other amphibian species. So far, no one seems to have tried introducing these tumour cells into early embryos, to see if they can then be incorporated into normal developing tissues and their excessive proliferation controlled. This seems a crucial experiment that ought to be tried soon.

Ruben (1970) found that in *Xenopus*, tumour growth was sometimes enhanced at the time of metamorphosis: this was true even of tumours placed on the regressing tail. This suggests that the hormone thyroxine may have a stimulatory effect on tumour growth. If so, it should be possible to inhibit the growth of some tumours by anti-thyroid agents such as phenylthiourea and tri-iodothyronine. This seems another experiment worth trying if no one has as yet. Of course it is possible that the main reason for the good growth of lymphosarcomas during metamorphosis in *Xenopus* is the abundance of nutrients provided by the breakdown of larval tissues. It is a general feature of tumours that they live on breakdown products from the normal tissues round them, and often the outer cells of a tumour liberate proteolytic enzymes which digest the surrounding cells (Sylvèn and Malmgren, 1955). But there is other evidence also that animals about to undergo metamorphosis are specially favourable environments for tumour growth. Pflugfelder (1959) showed that if metamorphosis was delayed experimentally in *Xenopus* by the administration of potassium chlorate to the larvae, tumours formed spontaneously in some of the tissues. These tissues were 'poised' ready to proliferate rapidly, and so could perhaps more readily support the growth of tumours rather than offering any resistance to them.

One of the most precise ways of studying the properties of tumour cells and how their spread may be controlled is by tissue culture methods. Abercrombie, Heaysman and Karthauser (1957) showed that mouse sarcoma cells were able to move over monolayers of fibroblasts, when the two cell types were confronted in tissue culture. This was in contrast to the normal behaviour of fibroblasts, for these cells are inhibited from further movement when they make contact with other cells and thus stay as a monolayer in culture. Abercrombie and Ambrose (1958) obtained ciné films of the behaviour of sarcoma cells, showing that they progressed by a rolling movement, over the surfaces of the fibroblasts. The sarcoma cells appeared spherical under these conditions, with numerous tuft-like pseudopodia projecting from all over their surfaces. Recently Ambrose and Easty (1973) have extended their studies to the penetration of malignant cells into sheets of tissue in organ culture. They have shown that the cells make many attachments to the tissue they are about to invade. Studies with the electron microscope on cultured cells have shown up the ultrastructure of the projections from tumour cells and their attachments to the normal cells that they invade. Williams and Ratcliffe (1969) described in some detail the filamentous attachments of ascites tumour cells to rat diaphragm tissue. There was nothing in their findings to suggest that these attachments were different from the variety of attachments made by normal cells to each other, however.

It is possible that antigen-antibody type reactions may also govern the adhesions of tumour cells. One of the special features of these cells may be that they have more agglutinin-receptors exposed at their surfaces than normal. This could perhaps enable them to attach to a wider variety of tissues than do normal embryonic cells. Burger and Noonan (1970) found, for instance, that fibroblasts which have been treated with trypsin lose their contact inhibition properties. This appears to be because a number of agglutinin-receptor sites are exposed, since treating the cells with a split form of concanavalin A, containing monovalent agglutinins, causes the fibroblasts to lose their malignancy and invasiveness and again to show inhibition of movement on making contact with other fibroblasts.

The transformation of normal cells to tumourous ones, whether it occurs spontaneously during embryonic or adult life, or is induced by virus treatment, results from alterations in the activities of the genes in the cells. Some spontaneous tumours may arise as a direct result of gene mutations. It should therefore in theory be possible to convert a tumour cell back to normal by replacing its nucleus with a normal one. This is obviously an impracticable method for controlling the growth of whole tumours, since each nuclear transplant would be a very difficult operation. However, Harris and his co-workers (see Harris, 1971) have instituted a method of fusing tumour cells to normal ones in culture. The fusion is achieved by treating the cells with Sendai virus, which modifies the cell membranes and makes them break down on making contact with other cells, so that a 'heterokaryon' – i.e. a single cell with two different nuclei – results (Harris et al. 1966, 1969; Jacobson 1969). The nuclei in these cells have a mutual influence on the activities of the protoplasm. Thus, normal, fully-differentiated cells may be induced to resume synthesis of DNA, RNA and protein by fusing them to embryonic or tumour cells. More important clinically is the finding that the malignancy of tumour cells may be suppressed by fusion to normal cells. But in experiments so far, this suppression has not been permanent, and the cells have later reverted to the malignant condition. This seems to be because the nuclei of the normal cells suffer losses of chromosomes, and their influence over the activity of the tumour cell nucleus recedes when this happens. Nevertheless, Harris's approach holds out hopes of controlling some of the properties of tumour cells in this special mode of interaction with normal cells in heterokaryons. Another hope is that, since the heterokaryons carry some antigens characteristic of the tumour cells, patients might be able to be immunized against the tumours by injections of heterokaryons in small doses. This would be a very risky procedure in the present state of our knowledge, but is a possibility for the future.

11.7   Concluding remarks

A very broad spectrum of examples of protective interactions between cells has been covered in this chapter. The reader cannot fail to notice how often in the preceding paragraphs, and indeed throughout this book, cellular interactions are explained by postulating the existence of some surface 'marker' molecule on each type of tissue cell in the embryo. The language of immunology, which is only *known* to be appropriate for events in the immune system, seems so far to be the only terminology suitable for discussing cell affinities, cell movements, cell recognition and the differences in behaviour of malignant and normal cells on making contact with other cells in culture. The 'antigen' offers a wide variety of possible interactive mechanisms. As we saw in the last chapter, an even wider variety of interactions may be imputed to cyclic AMP. Another term which is fast finding place in discussions of control processes in development and cellular interactions is the 'chalone'. It is only fair to stress, now that we have come to the end of this array of examples of cellular interaction throughout animal development, that although there is experimental *support* for nearly all of the mechanisms that have been postulated, none of them has been proved beyond all doubt to be a general and normal phenomenon. Finally, one may question whether we can ever really know how cells interact when they are *not* being observed by developmental biologists.

References

ABERCROMBIE, M. & AMBROSE, E. J. (1958). 'Interference microscope studies of cell contacts in tissue culture', *Exptl. Cell Res.* **15**, 332–345.

ABERCROMBIE, M. & HEAYSMAN, J. E. M. (1954), 'Observations on the social behaviour of cells in tissue culture. II. "Monolayering" of fibroblasts', *Exptl. Cell Res.* **6**, 293–306.

ABERCROMBIE, M., HEAYSMAN, J. E. M. & KARTHAUSER, H. M. (1957), Social behaviour of cells in tissue culture. III. Mutual influence of sarcoma cells and fibroblasts', *Exptl. Cell Res.* **13**, 276–92.

ADELSON, J. W. (1971), 'Enterosecretory proteins', *Nature, Lond.* **229**, 321–5.

ADLER, R. (1970), 'Changes in reaggregation during neural tube differentiation', *Devl. Biol.* **21**, 403–423.

AJIRO, K. (1971), 'Critical studies on the mechanism of inductive differentiation of ectoderm cells of amphibian gastrula. I. Differentiation of ectoderm cells in small explants cultured in "conditioned" media and in RNA- and tissue extract-containing media', *Sci. Rept. Kyoik. Diag. (Tokyo) B,* **14**, 145–76.

AKRAM, H. & WENIGER, J. P. (1969), 'Activité hormonale des gonades d'embryons de poulet hypophysectomisés', *J. Embryol. exp. Morph.* **21**, 509–15.

ALESCIO, T. & DANI, A. M. (1971), 'The influence of mesenchyme on the epithelial glycogen and budding activity in mouse embryonic lung developing *in vitro*', *J. Embryol. exp. Morph.* **25**, 131–40.

ALESCIO, T. & DANI, A. M. (1972), 'Hydrocortisone-induced glycogen deposition and its dependence on tissue interaction in mouse embryonic lung developing *in vitro*', *J. Embryol. exp. Morph*, **27**, 155–62.

ALEXANDER, P. (1972), 'Foetal "antigens" in cancer', *Nature, Lond.* **235**, 137–40.

AMBROSE, E. J. & EASTY, D. M. (1973), 'Behaviour of malignant cells', *Differentiation* **1**, 277–84.

AMPRINO, R. (1965), 'Aspects of limb morphogenesis in the chicken'. In: *Organogenesis*, (eds. R. L. DeHaan & H. Ursprung) pp. 255–82. Holt, Rinehart & Winston, New York & London.

AMPRINO, R. & AMBROSI, A. (1973), 'Experimental analysis of chick embryo limb bud growth', *Arch. Biol.* **84**, 35–86.

REFERENCES

ANDERSON, J. T. (1973), *Embryology and Phylogeny in Annelids and Arthropods*. Pergamon Press, Oxford, New York & London.

ANDREW, A. (1970), 'The origin of intramural ganglia. III. The "vagal" source of enteric ganglion cells', *J. Anat*. **107**, 327–36.

ANDREW, A. (1971), 'The origin of intramural ganglia. IV. The origin of enteric ganglia: a critical review', *J. Anat*. **108**, 169–84.

ANGELICI, D. & POURTOIS, M. (1968), 'The role of acid phosphatase in the fusion of the secondary palate', *J. Embryol. exp. Morph*. **20**, 15–23.

AREY, L. B. (1954), *Developmental Anatomy*, (6th edition). W. B. Saunders Co., Philadelphia.

ARNOLD, J. M. (1968), 'The role of the egg cortex in Cephalopod development', *Devl. Biol*. **18**, 180–97.

ARNOLD, J. M. (1969), 'Cleavage furrow formation in a telolecithal egg (*Loligo pealli*)', *J. Cell Biol*. **41**, 894–904.

ARTZT, K., DUBOIS, P., BENNETT, D., CONDAMINE, H., BABINET, C. & JACOB, F. (1973), 'Surface antigens common to mouse cleavage embryos and primitive teratocarcinoma cells in culture', *Proc. Nat. Acad. Sci*. **70**, 2988–92.

ATTARDI, D. G. & SPERRY, R. W. (1963), 'Preferential selection of central pathways by regenerating optic fibers', *Exp. Neurol*. **7**, 46–64.

AUSTIN, C. R. (1968), *Ultrastructure of Fertilization*, Holt, Rinehart & Winston, New York.

AUSTIN, C. R. & SHORT, R. V. (1972), *Reproduction in Mammals* Book 3: *Hormones in Reproduction*. University Press, Cambridge.

AVE, K., KANAKAMI, I. & SHAMESHIMA, M. (1968), 'Studies on the heterogeneity of cell populations in amphibian presumptive epidermis, with reference to primary induction', *Devl. Biol*. **17**, 617–26.

BABCOCK, R. G. (1963), 'RNA and lens induction in *Xenopus laevis*', *Amer. Zoologist* **3**, 511–2.

BACHVAROVA, R., DAVIDSON, E. H., ALLFREY, V. G. & MIRSKY, A. E. (1966), 'Activation of RNA synthesis associated with gastrulation', *Proc. Nat. Acad. Sci. U.S.A*. **55**, 358–65.

BAKER, R. E. & JACOBSON, (1970), 'Development of reflexes from skin grafts in *Rana pipiens*: influence of size and position of grafts', *Devl. Biol*. **22**. 476–94.

BALINSKY, B. I. (1960), 'Ultrastructural mechanisms of gastrulation and neurulation'. In: *Symposium on Germ Cells and Early Stages of Development* (ed. S. Ranzi)., pp. 550–63. Fond. Baselli, Milan.

BALINSKY, B. I. (1966), 'Changes in ultrastructure of amphibian eggs following fertilization', *Acta Embryol. Morph. Exper*. **9**, 132–54.

BALINSKY, B. I. (1970), *An Introduction to Embryology*. (3rd edn). W. B. Saunders & Co., Philadelphia, London & Toronto.

BALINSKY, B. I. (1972), 'The fine structure of the amphibian limb bud', *Acta Embryol. Exper. Suppl*., p. 455–70.

BALLARD, K. T. & HOLT, S. J. (1968), 'Cytological and Cytochemical Studies on cell death and digestion in the foetal rat foot: the role of macrophages and hydrolytic enzymes', *J. Cell. Sci*. **3**, 245–53.

BALLS, M. & RUBEN, L. N. (1968), 'Lymphoid tumours in Amphibia: a Review', *Progr. exp. Tumor Res*. **10**, 238–60.

BARKLEY, D. S. (1969), 'Adenosine-3', 5'-phosphate: identification as acrasin in a species of cellular slime mold', *Science, N.Y*. **165**, 1133–4.

257

BARROS, C., & YANAGIMACHI, R. (1971), 'Induction of zona reaction in golden hamster eggs by cortical granule material', *Nature, Lond.* **233**, 268–9.

BAUTZMANN, H., HOLTFRETER, J., SPEMANN, H. & MANGOLD, O. (1932), 'Versuche zur Analyse der Induktionsmittel in der Embryonalentwicklung', *Naturwiss.* **20**, 971–4.

BEDFORD, J. M. (1972), 'An electron microscope study of sperm penetration into the rabbit egg after natural mating', *Am. J. Anat.* **133**, 213–53.

BEERMANN, W. (1973), 'Directed changes in the patternof Balbiani ring puffing in *Chironomus*: effects of a sugar treatment', *Chromosoma* **41**, 297–326.

BELL, E., GASSELING, M. T., SAUNDERS, J. W. & ZWILLING, E. (1962), 'On the role of ectoderm in limb development', *Devl. Biol.*. **4**, 177–96.

BELLAIRS, R. (1951), Development of early reptile embryos *in vitro*. *Nature, Lond.* **167**, 687.

BELLAIRS, R. (1953), 'Studies on the development of the foregut in the chick blastoderm. 2. The morphogenetic movements', *J. Embryol. exp. Morph.* **1**, 369–85.

BELLAIRS, R. (1963), 'The development of somites in the chick embryo', *J. Embryol. exp. Morph*, **11**, 697–714.

BELLAIRS, R. (1965), 'The relationship between oocyte and follicle in the hen's ovary as shown by electron microscopy', *J. Embryol. exp. Morph.* **13**, 215–33.

BERNFIELD, M. R. (1970), 'Collagen synthesis during epitheliomesenchymal interactions', *Devl. Biol.* **22**, 213–31.

BERNFIELD, M. R., COHN, R. H. & BANERJEE, S.D. (1973), 'Glycosaminoglycans and epithelial organ formation', *Amer. Zoologist* **13**, 1067–83.

BERNFIELD, M. R. & WESSELS, N. K. (1970), 'Intra- and extracellular control of epithelial morphogenesis', *Devl. Biol. Suppl.* **4**, 195–249.

BILLINGHAM, R. E., BRENT, L. & MEDAWAR, P. B. (1953), 'Actively acquired tolerance of foreign cells', *Nature, Lond.* **172**, 603–6.

BILLINGHAM, R. & SILVERS, W. K. (1967), 'Studies on the conservation of epidermal specificities of skin and certain mucosas in adult mammals', *J. exp. Med.* **125**, 429–446.

BITENSKY, M. W. & GORMAN, R. E. (1973), 'Cellular responses to cyclic AMP'. In: *Progress in Biophysics and Molecular Biology*, vol. **26**, 411–61.

BLACKLER, A. W. (1958), 'Contribution to the study of germ cells in the Anura', *J. Embryol. exp. Morph.* **6**, 491–503.

BLACKLER, A. W. (1960), 'Transfer of germ cells in *Xenopus laevis*', *Nature, Lond.* **185**, 859–60.

BLACKLER, A. W. (1962), 'Transfer of primordial germ cells between two subspecies of *Xenopus laevis*', *J. Embryol. exp. Morph.* **10**, 641–51.

BLUEMINK, J. G. (1971), 'Cytokinesis and cytochalasin-induced furrow regression in the first-cleavage zygote of *Xenopus laevis*', *Z. Zellforsch.* **121**, 102–26.

BODENSTEIN, D. (1946), 'Developmental relations between genital ducts and gonads in *Drosophila*', *Biol. Bull. mar. biol. lab. Woods Hole, Mass.* **91**, 288–94.

BODENSTEIN, D. (1955), 'Embryogenesis: progressive differentitaion – 4. Insects'. In: *Analysis of Development* (eds. B. H. Willier, P. A. Weiss & V. Hamburger), pp. 337–345. W. B. Saunders & Co., Philadelphia & London.

BONSOR, L. (1972), B.Sc. Dissertation, Dept. of Anatomy, Bristol University.

REFERENCES

BOUNOURE, L. (1934), 'Recherches sur la lignée germinale chez la grenouille rousse aux premiers stades du développement', *Annls. Sci. nat.* **10e**, sér. **17**, 67–248.

BOVERI, T. (1910), 'Ueber die Teilung centrifugierter Eier von *Ascaris megalocephala*', *W. Roux' Arch. EntwMech. Org.* **30**, 101–25.

BRADLEY, S. (1970), 'An analysis of self-differentiation of chick limb buds in chorio-allantoic grafts', *J. Anat.* **107**, 479–90.

BRAHMA, S. K. (1958), 'Experiments on the diffusibility of the amphibian evocator', *J. Embryol. exp. Morph.* **6**, 418–23.

BRAHMA, S. K. (1959), 'Studies on the process of lens induction in *Xenopus laevis*', *W. Roux' Arch. EntwMech. Org.* **151**, 181–7.

BRANT, J. W. A. & NALBODOV, A. V. (1956), 'Role of sex hormones in albumin secretion by the oviduct of chickens', *Poultry Sci.* **35**, 692–709.

BRAVERMAN, M., COHEN, C. & KATOH, A. (1969), Cytotoxicity of lens antisera to dissociated chick neural retina cells in tissue culture', *J. Embryol. exp. Morph.* **21**, 391–406.

BRETSCHER, P. A. & COHN, M. (1968), 'Minimal model for the mechanism of antibody induction and paralysis by antigen', *Nature, Lond.* **220**, 444–8.

BRETSCHER, P. & COHN, M. (1970), 'A theory of self-non-self discrimination', *Science, N.Y.* **169**, 1042–9.

BRYANT, P. J. (1970), 'Cell lineage relationships in the imaginal wing disc of *Drosophila melanogaster*', *Devl. Biol.* **22**, 389–411.

BRYANT, P. J. (1971), 'Regeneration and duplication following operations *in situ* on the imaginal discs of *Drosophila melanogaster*', *Devl. Biol.* **26**, 606–615.

BRYANT, P. J. & SCHNEIDERMAN, H. A. (1969), 'Cell lineage, growth, and determination in the imaginal leg discs of *Drosophila melanogaster*', *Devl. Biol.* **20**, 263–90.

BUEHR, M. L. & BLACKLER, A. W. (1970), 'Sterility and partial sterility in the South African clawed toad following the pricking of the egg', *J. Embryol. exp. Morph.* **23**, 375–84.

BULLOUGH, W. S. (1967), *The Evolution of Differentiation*, Academic Press, New York & London.

BULLOUGH, W. S. (1971), 'Ageing of mammals', *Nature, Lond.* **229**, 608–10.

BULLOUGH, W. S. & LAURENCE, E. B. (1960), 'The control of epidermal mitotic activity in the mouse', *Proc. Roy. Soc. (B)* **151**, 517–36.

BULLOUGH, W. S. & LAURENCE, E. B. (1964), 'Mitotic control by internal secretion: the role of the chalone-adrenalin complex', *Exptl. Cell Res.* **33**, 176–94.

BURGER, M. M. & NOONAN, K. D. (1970), 'Restoration of normal growth by covering of agglutinin sites on tumour cell surface', *Nature, Lond.* **228**, 512–5.

BURNET, F. M. (1968), 'Evolution of the immune process in vertebrates', *Nature, Lond.* **218**, 426–30.

BURNET, F. M. (1969), *Self and Not-Self*. University Press, Melbourne & Cambridge.

BURNS, R. K. (1950), 'Sex transformation in the opossum: some new results and a retrospect', *Arch. Anat. Micr. Morph. Exp.* **39**, 467–81.

BURNSIDE, B. (1971), 'Microtubules and microfilaments in newt neurulation', *Devl. Biol.* **26**, 416–441.

BURNSIDE, B. (1973), 'Microtubules and microfilaments in amphibian neurulation', *Amer. Zoologist* **13**, 989–1006.

CAMPBELL, J. C. (1965), 'An immuno-fluorescent study of lens regeneration in larval *Xenopus laevis*', *J. Embryol. exp. Morph.* **13**, 171–9.

CATHER, J. N. (1971), 'Cellular interactions in the regulation of development in annelids and molluscs', *Adv. Morphogenesis* **9**, 67–125.

CATHER, J. N. & VERDONK, N. H. (1974), 'The development of *Bithynia tentaculata* after removal of the polar lobe, *J. Embryol. exp. Morph.* **31**, 415–22.

CHANG, C. Y. & WITSCHI, E. (1956), 'Genic control and hormonal reversal of sex differentiation in *Xenopus*', *Proc. Soc. exp. Biol. Med.* **93**, 140–4.

CHARNIAUX-COTTON, H. (1965), 'Hormonal control of sex differentiation in invertebrates'. In: *Organogenesis* (eds. R. L. DeHaan & H. Ursprung), pp. 701–40. Holt, Rinehart, Winston, Philadelphia & London.

CHOPRA, D. P. & SIMNETT, J. D. (1969), 'Demonstration of an organ-specific mitotic inhibitor in amphibian kidney', *Exptl. Cell Res.* **58**, 319–22.

CHOPRA, D. P. & SIMNETT, J. D. (1971), 'Tissue-specific mitotic inhibition in the kidneys of embryonic grafts and partially nephrectomised host *Xenopus laevis*', *J. Embryol. exp. Morph.* **25**, 321–9.

CHULITSKAIA, E. V. (1970), 'Desynchronization of cell divisions in the course of egg cleavage and an attempt at experimental shift of its onset', *J. Embryol. exp. Morph.* **23**, 359–74.

CLAYTON, R. M. (1953), 'Antigens in the developing newt embryo', *J. Embryol. exp. Morph.* **1**, 25–42.

CLAYTON, R. M. & ROMANOVSKY, A. (1959), 'Passage of antigenic material between inductor and ectoderm', *Exptl. Cell Res.* **18**, 410–14.

CLEMENT, A. C. (1952), 'Experimental studies on germinal localization in *Ilyanassa*: I. The role of the polar lobe in determination of the cleavage pattern and its influence on later development', *J. exp. Zool.* **121**, 593–611.

COHEN, A. M. & HAY, E. D. (1971), 'Secretion of collagen by embryonic neuroepithelium at the time of spinal cord-somite interaction', *Devl. Biol.* **26**, 578–605.

COHEN, J. (1967), *Living Embryos. An Introduction to the Study of Animal Development*. 2nd. edition. Pergamon Press, Oxford & London.

COHEN, K. L. & BITENSKY, M. W. (1969), 'Inhibitory effects of alloxan on mammalian adenyl cyclase', *J. Pharmacol. exp. Ther.* **169**, 80–6.

COHEN, M. I. & MORRILL, G. A. (1969), 'Model for the electric field generated by unidirectional sodium transport in the amphibian embryo', *Nature, Lond.* **222**, 84–6.

COHEN, P. P. (1970), 'Biochemical differentiation during amphibian metamorphosis', *Science, NY.* **168**, 533–43.

COLE, R. J. & PAUL, J. (1966), 'The effects of erythropoietin on haem synthesis in mouse yolk sac and cultured foetal liver cells', *J. Embryol. exp. Morph.* **15**, 245–60.

COLWIN, L. H. & COLWIN, A. L. (1967), 'Membrane fusion in relation to sperm-egg association'. In: *Fertilization*, vol. I. (eds. C. B. Metz & A. Monroy), pp. 295–367. Academic Press, New York & London.

COONS, A. H. & KAPLAN, M. H. (1950), 'Localization of antigen in tissue cells', *J. exp. Med.* **91**, 1–14.

CORNER, M. A. & BOT, A. P. C. (1969), 'Electrical activity in the isolated forebrain of the chick embryo', *Brain Res.* **12**, 473–6.

COSTELLO, D. P. (1945), 'Experimental studies of germinal localization in Nereis. I. The development of isolated blastomeres', *J. exp. Zool.* **100**, 19–66.

COULOMBRE, A. J. (1965), 'The eye'. In: *Organogenesis*, (eds. R. L. DeHaan & H. Ursprung), pp. 219–251. Holt, Rinehart and Winston, New York, Toronto, London.

COULOMBRE, J. L. & COULOMBRE, A. J. (1970), 'Influence of mouse neural retina on regeneration of chick neural retina from chick embryonic pigmented epithelium', *Nature, Lond.* **228**, 559–60.

COULOMBRE, J. L. & COULOMBRE, A. J. (1971), 'Metaplastic induction of scales and feathers in the corneal anterior epithelium of the chick embryo', *Devl. Biol.* **25**, 464–78.

COUNCE, S. J. & WADDINGTON, C. H. (1972), *Developmental Systems.* I. *The Insects.* Academic Press, London & New York.

CRAIN, S. M. (1966), 'Development of "organotypic" bioelectric activities in central nervous tissues during maturation in culture', *Int. Rev. Neurobiol.* **9**, 1–43.

CRISS, W. E. (1971). 'A review of isozymes in cancer', *Cancer Res.* **31**, 1523–42.

CROFT, C. B. & TARIN, D. (1970), 'Ultrastructural studies of wound healing in mouse skin', *J. Anat.* **106**, 79–92.

CROISILLE, Y. & LE DOUARIN, N. M. (1965), 'Development and regeneration of the liver'. In *Organogenesis* (eds. R. L. DeHaan and H. Ursprung) pp. 421–466. Holt, Rinehart & Winston, New York.

CUNHA, G. R. (1972), 'Support of normal salivary gland morphogenesis by mesenchyme derived from accessory sexual glands of embryonic mice', *Anat. Rec.* **173**, 205–12.

CURTIS, A. S. G. (1957), 'The role of calcium in cell aggregation of *Xenopus* embryos', *Proc. Roy. Phys. Soc. Edin.* **26**, 25–32.

CURTIS, A. S. (1960), 'Cell Contacts: some physical considerations', *Amer. Nat.* **94**, 37–56.

CURTIS, A. S. G. (1961), 'Timing mechanisms in the specific adhesions of cells', *Exptl. Cell Res. Suppl.* **8**, 107–22.

CURTIS, A. S. G. (1962a), 'Morphogenetic interactions before gastrulation in the amphibian, *Xenopus laevis* – the cortical field', *J. Embryol. exp. Morph.* **10**, 410–22.

CURTIS, A. S. G. (1962b), 'Cell contact and adhesion', *Biol. Rev.* **37**, 82–129.

CZIHAK, G. & HÖRSTADIUS, S. (1970), 'Transplantation of RNA-labeled micromeres into animal halves of sea urchin embryos. A contribution to the problem of embryonic induction', *Devl. Biol.* **22**, 15–30.

DADAY, H. & CREASER, E. H. (1970), 'Isolation of a protein responsible for aggregation of avian embryonic cells', *Nature, Lond.* **226**, 970–1.

DALTON, H. C. (1950), 'Inhibition of chromatoblast migration as a factor in the development of genetic differences in pigmentation in white and black axolotls', *J. exp. Zool.* **115**, 151–74.

DAUGHADAY, W. H. & READER, C. (1966), 'Synchronous activation of DNA synthesis in hypophysectomised rat cartilage by growth hormone', *J. Lub. Clin. Med.* **68**, 357–68.

DAVEY, K. G. (1965), *Reproduction in the Insects.* Oliver & Boyd, Edinburgh & London.

DAVID, D. (1972), 'Les relations epithélio-mesenchymateuses au cours de l'organogénèse gastrique de foetus de lapin', *J. Embryol. exp. Morph.* **27**, 177–97.

DAWD, D. S. & HINCHLIFFE, J. R. (1971), 'Cell death in the "opaque patch" in the central mesenchyme of the developing chick limb: a cytological, cytochemical and electron microscopic analysis, *J. Embryol. exp. Morph.* **26**, 401-24.

DECK, J. D. (1971), The effects of infused material upon the regeneration of newt limbs: II. Extracts from newt brain and spinal cord, *Acta Anat.* **79**, 321-332.

DeHAAN, R. L. (1965), 'Morphogenesis of the vertebrate heart'. In: *Organogenesis* (eds. R. L. DeHaan & H. Ursprung), pp. 377-420. Holt, Rinehart & Winston, Philadelphia & London.

DeLONG, G. R. & SIDMAN, R. L. (1970), 'Alignment defect of reaggregating cells in cultures of developing brains of reeler mutant mice', *Devl. Biol.* **22**, 584-600.

DENNY, P. C. & REBACK, P. (1970), 'Active polysomes in sea urchin eggs and zygotes: evidence for an increase in translatable messenger RNA after fertilization', *J. exp. Zool.* **175**, 133-40.

DEOL, M. S. (1970), 'The origin of the acoustic ganglion and the effects of the gene dominant spotting (W^v) in the mouse', *J. Embryol. exp. Morph.* **23**. 773-84.

DETWILER, S. R. (1936), *Neuroembryology: an Experimental Study.* Haffner (Reprint) 1964, New York & London.

DEUCHAR, E. M. (1967), 'Inducing activity in extracts from *Xenopus* embryos and from isolated dorsal lips', *J. Embryol. exp. Morph.* **17**, 341-8.

DEUCHAR, E. M. (1970a), 'Effect of cell number on the type and stability of differentiation in amphibian ectoderm', *Exptl. Cell Res.* **59**, 341-3.

DEUCHAR, E. M. (1970b), 'Neural induction and differentiation with minimum numbers of cells', *Devl. Biol.* **22**, 189-99.

DEUCHAR, E. M. (1971), 'Transfer of the primary induction stimulus by small numbers of amphibian ectoderm cells', *Acta Embryol. exp.*, pp. 387-96.

DEUCHAR, E. M. & BURGESS, A. M. C. (1967), 'Somite segmentation in amphibian embryos: is there a transmitted control mechanism?' *J. Embryol. exp. Morph.* **17**, 349-57.

DEUCHAR, E. M. & HERRMANN, H. (1962), 'Uptake of amino acids into explanted chick embryos by epidermal and endodermal routes,' *Acta Embryol. Morph. exp.* **5**, 161-8.

DHOUAILLY, D. (1973), 'Dermo-epidermal interactions between birds and mammals: differentiation of cutaneous appendages', *J. Embryol. exp. Morph.* **30**, 587-603.

DIDIER, E. (1971), 'Le canal de Wolff induit la formation de l'ostium mullerien: démonstration expérimentale chez l'embryon de poulet', *J. Embryol. exp. Morph.* **25**, 115-29.

DIDIER, E. (1973), 'Researches on Mullerian duct morphogenesis in birds. II. Experimental study', *W. Roux' Arch. EntwMech. Org.* **172**, 287-302.

DIETERLEN-LIÈVRE, F. (1970), 'Tissus exocrine et endocrine du pancréas chez l'embryon de poulet: origine et interactions tissulaires dans la différenciation', *Devl. Biol.* **22**, 138-56.

DRIESCH, H. (1891), 'Entwicklungsmechanische Studien. I. Der Werth der beiden ersten Furchungszellen in der Echinodermenentwicklung', *Z. wiss. Zool.* **53**, 160-78.

DRIESCH, H. (1910), 'Neue Versuche über die Entwicklung verschmolzener Echinidenkeime', *W. Roux' Arch. EntwMech. Org.* **30**, 8-23.

REFERENCES

DUBOIS, R. (1969), 'Le mécanisme d'entrée des cellules germinales primordiales dans le réseau vasculaire, chez l'embryon de poulet', *J. Embryol. exp. Morph.* **21**, 255–70.

EDE, D. A. & AGERBAK, G. S. (1968), 'Cell adhesion and movement in relation to the developing limb pattern in normal and talpid mutant chick embryos', *J. Embryol. exp. Morph.* **20**, 81–100.

EDE, D. A. & FLINT, O. P. (1972), 'Patterns of cell division, cell death and chondrogenesis in cultured aggregates of normal and talpid mutant chick limb mesenchyme cells', *J. Embryol. exp. Morph.* **27**, 245–60.

EDE, D. A. & LAW, J. T., (1969), 'Computer simulation of vertebrate limb morphogenesis, *Nature, Lond.* **221**, 244–8.

EDWARDS, R. G. (1970), 'Immunology of conception and pregnancy', *Brit. Med. Bull,* **26**, 72–8.

EGUCHI, G. & OKADA, T. S. (1971), 'Ultrastructure of the differentiated cell colony derived from a singly isolated chondrocyte in *in vitro* culture' *Devl., Growth & Diffn.* **12**, 297–312.

ELLISON, M. L., AMBROSE, E. J. & EASTY, G. C. (1969), 'Differentiation in a transplantable rat tumour maintained in organ culture by spinal cord', *Exptl. Cell Res.* **55**, 198–204.

ELSDALE, T. & JONES, K. (1963), 'The independence and interdependence of cells in the amphibian embryo', *Brit. Soc. exp. Biol. Symposium* **17**, *Cell Differentiation.* Cambridge University Press.

ELSDALE, T. & BARD, J. (1972), 'Cellular interactions in mass cultures of human diploid fibroblasts', *Nature, Lond.* **236**, 152–5.

ENGLAND, M. (1969), 'Millipore filters studied in isolation and *in vitro* by transmission electron microscopy and stereoscan electron microscopy', *Exptl. Cell Res.* **54**, 222–30.

ERICSSON, R. J. (1969), 'Capacitation *in vitro* of rabbit sperm with mule eosinophils', *Nature, Lond.* **221**, 568–9.

ESTENSEN, R. D., HILL, H. R., QUIE, P. G., HOGAN, N. & GOLDBERG, N. D. (1973), 'Cyclic GMP and cell movement', *Nature, Lond.* **245**, 458–61.

ETTERIDGE, A. L. (1969), 'Determination of the mesonephric kidney', *J. exp. Zool.* **169**, 357–70.

EVANS, M. J. (1969), 'Studies on the ribonucleic acid of early amphibian embryos'. Ph.D. Thesis, University College London.

EVANS, M. J. (1972), 'The isolation and properties of a clonal tissue culture strain of pluripotent mouse teratoma cells', *J. Embryol. exp. Morph.* **28**, 163–76.

FALLON, J. F. & SAUNDERS, J. W. (1968), '*In vitro* analysis of the control of cell death in a zone of prospective necrosis from the chick wing bud', *Devl. Biol.* **18**, 553–70.

FARBMAN, A. I. (1968), 'Electron microscope study of palate fusion in mouse embryos', *Devl. Biol.* **18**, 93–116.

FARBMAN, A. I. (1969), 'Fine structure of degenerating taste buds after denervation', *J. Embryol. exp. Morph.* **22**, 55–68.

FAULHABER, I. & GEITHE, H. P. (1972), 'Nachweis deuterencephalspino-caudaler Induktionsfähigkeit in Gastrulaextrakten von *Xenopus laevis* nach Chromatographie am Hydroxyapatit oder Elektrofokussierung. *Revue Suisse Zool.* **79** *(suppl).* 103–17.

FELL, H. B. & CANTI, R. G. (1934), 'Experiments on the development *in vitro* of the avian knee joint', *Proc. Roy. Soc. (B),* **116**, 316–51.

263

FERRIER, V. (1967a), 'Étude cytologique des premiers stades du développement de quelques hybrides létaux d'Amphibiens Urodèles', *J. Embryol. exp. Morph.* **18**, 227–51.

FERRIER, V. (1967b), 'Données cytologiques sur la fécondation hybride d'oeufs de *Pleurodeles waltlii* par des spermatozoides d'*Ambystoma*', *C. R. Acad. Sci. Paris* **265**, 441–3.

FICQ, A. (1954), 'Analyse de l'induction neurale chez les Amphibiens au moyen d'organisateurs marqués', *J. Embryol. exp. Morph*, **2**, 194–203.

FLOOD, P. R. (1970), 'The connection between spinal cord and notochord in Amphioxus (*Branchiostoma lanceolata*)', *Z. Zellforsch. mikr. Anat.* **103**, 115–28.

FOL, H. (1879), 'Recherches sur la fécondation et la commencement de l'hénogénie chez divers animaux', *Mem. Soc. Phys. Hist. Nat. Genève* **26**, 89–397.

FORER, A., EMMERSON, J. & BEHNKE, O. (1972), 'Cytochalasin B: does it affect actin-like filaments?' *Science, N.Y.* **175**, 774–6.

FRIEDBERG, F. & EAKIN, R. M. (1949), 'Studies on protein metabolism of the amphibian embryo. I. Uptake of radioactive glycine', *J. exp. Zool.* **131**, 307–22.

FUJISAWA, H. (1971), 'A complete reconstruction of the neural retina of the chick embryo grafted onto the chorio-allantoic membrane', *Devl., Growth & Diffn.* **13**, 25–36.

FULLILOVE, S. L. (1970), 'Heart induction: distribution of active factors in newt endoderm', *J. exp. Zool.* **175**, 323–6.

FURSHPAN, E. J. & POTTER, D. D. (1968), 'Low resistance junctions between cells in embryos and in tissue culture', *Curr. Top. Devl. Biol.* **3**, 95–125.

GABIE, V. & ANDREW, A. (1969), 'Staging of pigment cells in cultures of *Xenopus laevis* neural crest', *Acta Embryol. Exper.*, 137–146.

GALLERA, J. (1965), 'Quelle est la durée nécessaire pour déclencher des inductions neurales chez le poulet?' *Experientia* **21**, 218–9.

GALLERA, J., NICOLET, G. & BALLMANN, M. (1968), 'Induction neurale chez les oiseaux à travers un filtre millipore: étude au microscope optique et électronique', *J. Embryol. exp. Morph.* **19**, 439–50.

GALLIEN, L. (1953), 'Inversion totale du sexe chez *Xenopus laevis* Daud. à la suite d'un traitement gynogène par le benzoate d'oestradiol administré pendant la vie larvaire', *C. R. Acad. Sci. Paris*, **237**, 1365–6.

GARBER, B., KOLLAR E. J. & MOSCONA, A. A. (1968), 'Aggregation *in vivo* of dissociated cells: III. Effect of state of differentiation of cells on feather development in hybrid aggregates of embryonic mouse and chick skin cells', *J. exp. Zool.* **168**, 455–72.

GASSELING, M. T. & SAUNDERS, J. W. (1964), 'Effect of the "posterior necrotic zone" of the early chick wing bud on the pattern and symmetry of limb outgrowth', *Amer. Zoologist.* **4**, 303–4.

GAZE, R. M. (1970), *The Formation of Nerve Connections: a Consideration of Neuronal Specificity, Modulation and Comparable Phenomena*. Academic Press, New York & London.

GAZE, R. M., CHUNG, S. H. & KEATING, M. J. (1972), 'Development of the retinotectal projection in *Xenopus*', *Nature (New Biol.)* **236**, 133–135.

GAZE, R. M. & JACOBSON, M. (1963), 'A study of the retinotectal projections during regeneration of the optic nerve in the frog', *Proc. Roy. Soc. Lond. B* **157**, 420–448.

GAZE, R. M., JACOBSON, M. & SZÉKELY, G. (1965), 'On the formation of connexions by compound eyes in *Xenopus*', *J. Physiol.* **176**, 409–17.

GAZE, R. M. & KEATING, M. J. (1972), 'The visual system and "neuronal specificity" ', *Nature, Lond.* **237**, 375–8.

GEIGY, R. (1931), 'Action de l'ultraviolet sur le pole germinale dans l'oeuf de *Drosophila melanogaster*', *Rev. Suisse Zool.* **38**, 187–288.

GEILENKIRCHEN, W. L. M., VERDONK, N. H. & TIMMERMANS, L. P. M. (1970), 'Experimental studies on morphogenetic factors localized in the first and second polar lobe of *Dentalium* eggs', *J. Embryol. exp. Morph.* **23**, 237–43.

GLASSTONE, S. (1936), 'Development of tooth germs *in vitro*,' *J. Anat.* **70**, 260–6.

GLÜCKSMANN, A. (1951), 'Cell deaths in normal vertebrate ontogeny', *Biol. Rev.* **26**, 59–86.

GLUECKSOHN-SCHOENHEIMER, S. (1945), 'The embryonic development of mutants of the Sd-strain in mice', *Genetics* **30**, 29–38.

GOETTERT, L. (1966), 'Differenzierungsleistungen von explantiertem Urodelenektoderm (*Ambystoma mexicanum* und *Triturus alpestris* Laur.) nach verschieden Langer Unterlagerungszeit', *W. Roux' Arch. EntwMech. Org.* **157**, 75–100.

GOLDSCHMIDT, R. B., HANNAH, A. & PITERNICK, L. K. (1951), 'The podoptera effect in *Drosophila melanogaster*', *Univ. Calif. Publ. Zool.* **55**, 67–294.

GOLOSOV, N. & GROBSTEIN, C. (1962), 'Epitheliomesenchymal interaction in pancreatic morphogenesis', *Devl. Biol.* **4**, 242–55.

GOMOT, L. (1970), 'Sex differentiation of mollusks in organ culture', In: *Invertebrate Organ Cultures* (ed. H. Lutz), pp. 105–133. Gordon & Breach, New York & London.

GONDOS, B. & BHIRALEUS, P. (1970), 'Pronuclear relationship and association of maternal and paternal chromosomes in flushed rabbit ova', *Z. Zellforsch. mikr. Anat.* **111**, 149–59.

GOODWIN, B. (1971), 'A model of early amphibian development', *Brit. Soc. exp. Biol. Symposium* **25** (eds. D. D. Davies & M. Balls), pp. 417–28. Cambridge University Press.

GOOS, H. J. TH. (1969), 'Hypothalamic neurosecretion and metamorphosis in *Xenopus laevis*', *Z. Zellforsch. mikr. Anat.* **97**, 449–58.

GOSS, R. J. (1964), *Adaptive Growth*. Logos/Academic Press, London & New York.

GRAHAM, C. F. (1971), 'The design of the mouse blastocyst', *Brit. Soc. exp. Biol. Symposium* **25** (eds. D. D. Davies & M. Balls), pp. 371–8. Cambridge University Press.

GREGG, K. W. (1969), 'Cortical response antigens released at fertilization from sea urchin eggs and their relation to antigens of the jelly coat', *Biol. Bull.* **137**, 146–54.

GROBSTEIN, C. (1953a), 'Morphogenetic interaction between embryonic mouse tissues separated by a membrane filter', *Nature, Lond.* **172**, 869.

GROBSTEIN, C. (1953b), 'Epithelio-mesenchymal specificity in the morphogenesis of mouse submandibular rudiments *in vitro*', *J. exp. Zool.* **124**, 383–414.

GROBSTEIN, C. (1957), 'Some transmission characteristics of the tubule-inducing influence on mouse metanephrogenic mesenchyme', *Exptl. Cell Res.* **13**, 575–87.

GROSS, J. (1964), 'Studies on the biology of connective tissues: the remodelling of collagen in metamorphosis', *Medicine* **43**, 291–304.

GUERRIER, P. (1970), 'Les caractères de la ségmentation et de la détermination de la polarité dorsoventrale dans le développement de quelques Spiralia. I. Les formes à premier clivage égal', *J. Embryol. exp. Morph.* **23**, 611–637.

GUSTAFSON, T. & KINNANDER, H. (1956), 'Microaquaria for time-lapse cinematographic studies of morphogenesis in swimming larvae and observations on sea urchin gastrulation', *Exptl. Cell Res.* **11**, 36–51.

GUSTAFSON, T. & WOLPERT, L. (1961), 'Cellular mechanisms in the morphogenesis of the sea urchin larva', *Exptl. Cell. Res.* **22**, 509–20.

HADJI-AZIMI, I. (1970), 'Transmission of the "lymphoid tumour" of *Xenopus laevis* by injection of cell-free extracts', *Experientia* **26**, 894–5.

HADORN, E. (1965), 'Problems of determination and transdetermination', *Brookhaven Symp. Biol.* **18**, 148–161.

HADORN, E. & GLOOR, H. (1946), 'Transplanatation zur Bestimmung des Anlagemusters in der weiblichen Genital-Imaginalscheib von *Drosophila melanogaster*', *Rev. Suisse Zool.* **53**, 495–510.

HADORN, E. BERTANI, G. & GALLERA, J. (1949), 'Regulationsfähigkeit und Feldorganisation der männlichen Genital-Imaginalscheibe von *Drosophila melanogaster*', *W. Roux' Arch. EntwMech. Org.* **144**, 31–70.

HAGEDORN, H. H. & FALLON, A. M. (1973), 'Ovarian control of vitellogenin synthesis by the fat body in *Aedes aegypti*', *Nature, Lond.* **244**, 103–5.

HAGET, A. (1953), 'Analyse expérimentale des facteurs de la morphogénèse embryonnaire chez la Coléoptère Leptinotarsa', *Bull. Biol.* **87**, 123.

HAMBURGER, V. (1962), Specificity in neurogenesis. *J. Cell Comp. Physiol.* **60**, suppl. 1, 81–92.

HAMBURGER, V. & LEVI-MONTALCINI, R. (1949), 'Proliferation, differentiation and degeneration in the spinal ganglion of the chick embryo under normal and experimental conditions', *J. exp. Zool.* **111**, 457–502.

HAMBURGER, V. & HAMILTON, H. L. (1951), 'A series of normal stages in the development of the chick embryo', *J. Morph.* **88**, 49–92.

HAMILTON, L. (1969), 'The formation of somites in *Xenopus*', *J. Embryol. exp. Morph.* **22**, 253–64.

HAMILTON, T. H. & TENG, C. (1965), 'Sexual stabilization of Mullerian ducts in the chick embryo'. In: *Organogenesis* (eds. R. L. DeHaan & H. Ursprung). Holt, Rinehart & Winston, Philadelphia & London.

HARA, K. (1971), 'Cinematographic observation of "surface contraction waves" (SCW) during the early cleavage of Axolotl embryos', *W. Roux' Arch. EntwMech. Org.* **167**, 183–6.

HARRIS, H. (1971), 'Cell fusion and the analysis of malignancy', *Proc. Roy. Soc.* (*B*) **179**, 1–20.

HARRIS, H., MILLER, O. J., KLEIN, G., WORST, P. & TACHIBANA, T. (1969), 'Suppression of malignancy by cell fusion', *Nature, Lond.* **223**, 363–8.

HARRIS, H., WATKINS, J. F., FORD, C. E. & SCHOEFL, G. I. (1966), 'Artificial heterokaryons of animal cells from different species', *J. Cell Sci.* **1**, 1–30.

HARRISON, R. G. (1918), 'Experiments on the development of the forelimb of *Amblystoma*, a self-differentiating, equipotential system', *J. exp. Zool.* **25**, 413–62.

REFERENCES

HAUSER, R. (1969), 'Abhängigkeit der normalen Schwanzenregeneration bei *Xenopus*larven von einer diencephaler Faktor im Zentralkanal', *W. Roux' Arch. Entw Mech. Org.* **163**, 221–47.

HAUSER, R. (1972), 'Morphogenetic action of the subcommisural organ on tail regeneration in *Xenopus* larvae', *W. Roux' Arch. Entw Mech. Org.* **169**, 170–84.

HAUSER, R. & LEHMANN, F. E. (1962), 'Regeneration in isolated tails of Xenopus larvae', *Experientia* **18**, 83–4.

HAY, E. D. (1973), 'Origin and role of collagen in the embryo', *Amer. Zoologist* **13**, 1085–1117.

HAYASHI, Y. (1965), 'Differentiation of the beak epithelium as studied by a xenoplastic induction system', *Jap. J. exp. Morph.* **19**, 116.

HAYES, B. P. & ROBERTS, A. (1973), 'Synaptic junction development in the spinal cord of an amphibian embryo: an electron microscope study', *Z. Zellforsch.* **137**, 251–69.

HESS, O. (1956), 'Die Entwicklung von Exogastrula-Keimen bei dem Süsswasser-Prosobranchier *Bithynia tentaculata*', *W. Roux' Arch. EntwMech. Org.* **148**, 474–88.

HILFER, S. R. (1973), 'Extracellular and intracellular correlates of organ initiation in the embryonic chick thyroid', *Amer. Zoologist* **13**, 1023–38.

HILFER, S. R., HILFER, E. K. & TISZARD, L. B. (1967), 'The relationship between cytoplasmic organization and the epithelio-mesodermal interaction in the embryonic chick thyroid', *J. Morph.* **123**, 199–212.

HILLMAN, N., SHERMAN, M. I. & GRAHAM, C. F. (1972), 'The effect of spatial arrangement on cell determination during mouse development', *J. Embryol. exp. Morph.* **28**, 263–78.

HINCHLIFFE, J. R. & EDE, D. A. (1968), 'Abnormalities in bone and cartilage development in the talpid mutant fowl', *J. Embryol. exp. Morph.* **19**, 327–39.

HINCHLIFFE, J. R. & EDE, D. A. (1973), 'Cell death and the development of limb form and skeletal pattern in wingless (ws) chick embryos', *J. Embryol. exp. Morph.* **30**, 753–772.

HOBSON, L. B. (1941), 'On the ultrastructure of the neural plate and tube of the early chick embryo', *J. exp. Zool.* **88**, 107–15.

HOLTFRETER, J. (1933), 'Die totale Exogastrulation, eine Selbstablösung des Ektoderms vom Entomesoderm', *W. Roux' Arch. EntwMech. Org.* **128**, 584–633.

HOLTFRETER, J. (1938), 'Differenzierungspotenzen isolierter Teile der Anurengastrula', *W. Roux' Arch. EntwMech. Org.* **138**, 657–738.

HOLTFRETER, J. (1944), 'A study of the mechanics of gastrulation', *J. exp. Zool.* **95**, 171–212.

HORNBRUCH, A. & WOLPERT, L. (1970), 'Cell division in the early growth and morphogenesis of the chick limb', *Nature, Lond.* **226**, 764–6.

HÖRSTADIUS, S. (1973), *Experimental Embryology of Echinoderms.* Clarendon Press, Oxford.

HÖRSTADIUS, S. & JOSEFSSON, L. (1972), 'Morphogenetic substances from sea urchin eggs. Isloation of animalizing substances from developing eggs of *Paracentrotus lividus*', *Acta Embryol. exp.*, p. 7–23.

HORTON, J. D. (1969), 'Ontogeny of the immune responses to skin allografts in relation to lymphoid organ development in the amphibian *Xenopus laevis* (Daudin)', *J. exp. Zool.* **170**, 449–66.

HUEBNER, R. J., KELLOFF, G. J., SARMA, P. S., LANE, W. T. & TURNER,

267

H. C. (1970), 'Group-specific antigen expression during embryogenesis of the genome of the C-type RNA virus: implications for ontogenesis and oncogenesis', *Proc. Nat. Acad. Sci. (U.S.A.)* **67**, 366–76.

HUGHES, A. F. W. & LEWIS, P. R. (1961), 'Effect of limb ablation on neurones in *Xenopus* larvae', *Nature, Lond.* **189**, 333–4.

HUGHES, A. F. W. & TSCHUMI, P. A. (1958), 'The factors controlling the development of the dorsal root ganglia and ventral horn in *Xenopus laevis* (Daudin)', *J. Anat.* **92**, 498–526.

HUGHES, A. F. W. & TSCHUMI, P. A. (1960), 'Heterotopic grafting of the spinal cord in *Xenopus laevis*' (Daudin)', *Jl. R. microsc. Soc.* **79**, 155–64.

HUNT, E. A. (1932), 'The differentiation of chick limb buds in chorioallantoic grafts with special reference to the muscles', *J. exp. Zool.* **62**, 57–92.

ILLMENSEE, K. (1972) 'Developmental potentialities of nuclei from cleavage, pre-blastoderm and syncytial blastoderm transplanted into unfertilized eggs of *Drosophilia melanogaster*', *W. Roux' Arch. EntwMech. Org.* **170**, 269-98.

ILLMENSEE, K. (1973), 'The potentialities of transplanted early gastrula nuclei of *Drosophila melanogaster*. Production of imago descendants by germ-line transplantation', *W. Roux' Arch. Entw Mech. Org.* **171**, 331–43.

INGRAM, A. J. (1971), 'The reactions to carcinogens in the axolotl (*Ambystoma mexicanum*) in relation to the "regeneration field control" hypothesis', *J. embryol. exp. Morph.* **26**, 425–41.

INOUE, K. (1961), 'Serologically active groups of amphibian embryos', *J. Embryol. exp. Morph.* **9**, 563–85.

INOUE, S., HARDY, J. P., COUSINEAU, G. H. & BAL, A. K. (1967), 'Fertilization membranes structure analysis with the surface replica method', *Exptl. Cell Res.* **48**, 248–51.

IWAYAMA, T. & NADA, O. (1969), 'Histochemical observations on phosphatase activities of degenerating and regenerating taste buds', *Anat. Rec.* **163**, 31–8.

JACOB, F. & MONOD, J. (1961), 'Genetic regulatory mechanisms in the synthesis of proteins', *J. molec. Biol.* **3**, 318–27.

JACOBSON, A. G. & DUNCAN, J. T. (1968), 'Heart induction in salamanders', *J. exp. Zool.* **167**, 79–104.

JACOBSON, C.-O. (1968), 'Selective affinity as a working force in neurulation movements', *J. exp. Zool.* **168**, 125–36.

JACOBSON, C.-O. (1969a), 'Production of artificial heterokaryons from mammalian neurons and various undifferentiated cells', *Zool. Bid. Uppsala* **38**, 241–8.

JACOBSON, C.-O. (1969b), 'Reactivation of DNA synthesis in mammalian neuron nuclei after fusion with cells of an undifferentiated fibroblast line', *Exp. Cell Res.* **53**, 316–8.

JACOBSON, C.-O. (1970), 'Experiments on β-mercaptoethanol as an inhibitor of neurulation movements in amphibian larvae', *J. Embryol. exp. Morph.* **23**, 463–71.

JACOBSON, C.-O. & LÖFBERG, J. (1969), 'Mesoderm movements in the amphibian neurula', *Zool. Bid. Uppsala* **38**, 233–9.

JACOBSON, M. (1965), 'Development of neuronal specificity in retinal ganglion cells of *Xenopus*', *Devl. Biol.* **17**, 202–18.

JACOBSON, M. & BAKER, R. E. (1968), 'Neuronal specification of cutaneous nerves through connections with skin grafts in the frog', *Science (N.Y.)*, **160**, 543.

JANNERS, M. Y. & SEARLS, R. L. (1970), 'Changes in rate of cellular proliferation during the differentiation of cartilage and muscle in the mesenchyme of the embryonic chick wing', *Devl. Biol.* **23**, 136–65.

JENSEN, E. V. & DESOMBRE, E. R. (1973), 'Oestrogen-receptor interaction', *Science (N.Y.)* **182**, 126–34.

JEON, K. W. & KENNEDY, J. R. (1973), 'The primordial germ cells in early mouse embryos: light and electron microscope studies', *Devl. Biol.* **31**, 275–84.

JOHANSSEN, O. A. & BUTT, F. H. (1941), *Embryology of Insects and Myriapods.* McGraw-Hill, New York.

JOHNEN, G. (1964), 'Experimentelle Untersuchungen über die Bedeutung des Zeitfaktors beim Vorgang der Neuralen Induktion', *W. Roux' Arch. Entw Mech. Org.* **155**, 302–14.

JOHNSON, G. S., FRIEDMAN, R. M. & PASTAN, I. (1971), 'Restoration of several morphological characteristics of normal fibroblasts in sarcoma cells treated with adenosine $3'5'$-cyclic monophosphate and its derivatives', *Proc. Nat. Acad. Sci. U.S.A.* **68**, 425–9.

JOHNSON, K. E. (1969), 'Altered contact behaviour of presumptive mesodermal cells from hybrid amphibian embryos arrested at gastrulation', *J. exp. Zool.* **170**, 325–32.

JOHNSON, K. E. (1972), 'The extent of cell contact and the relative frequency of small and large gaps between presumptive mesoderm cells in normal gastrulae of *Rana pipiens* and in the arrested gastrulae of the *Rana pipiens* ♀ x *Rana catesbeiana* ♂ hybrid', *J. exp. Zool.* **179**, 229–38.

JONES, K. W. & ELSDALE, T. R. (1963), 'The culture of small aggregates of amphibian embryonic cells *in vitro*', *J. Embryol. exp. Morph.* **11**, 135–54.

JONES, R. O. (1970), 'Ultrastructural analysis of hepatic haematopoiesis in the foetal mouse', *J. Anat.* **107**, 301–14.

JÖRGENSEN, M. (1913), 'Zellenstudien. I. Morphologische Beiträge zum Problem des Eiwachstums', *Arch. Zellforsch.* **10**, 1–126.

JOSEFSSON, L. & HÖRSTADIUS, S. (1969), 'Morphogenetic substances from sea urchin eggs. Isolation of animalizing and vegetalizing substances from unfertilized eggs of *Paracentrotus lividus*, *Devl. Biol.* **20**, 481–500.

JOSEPH, J. & DYSON, M. (1970), 'The effect of abdominal wounding on the rate of tissue regeneration', *Experientia* **26**, 66–7.

JOST, A. (1948), 'Activité androgène du testicule foetal de Rat greffé sur l'adulte castré', *Compt. Rend. Soc. Biol. Paris* **142**, 196–8.

KALT, M. R. (1971a, b), 'The relationship between cleavage and blastocoel formation in *Xenopus laevis*. I. Light microscopic observations. II. Electron microscopic observations', *J. Embryol. exp. Morph.* **26**, 37–49 and 51–66.

KARFUNKEL, P. (1971), 'The role of microtubules and microfilaments in neurulation in *Xenopus*', *Devl. Biol.* **25**, 30–56.

KEAN, T. & OVERTON, J. (1969), 'Staining of intercellular material in re-aggregating chick liver and cartilage cells', *J. exp. Zool.* **171**, 161–74.

KELLEY, R. O. (1969), 'An electron microscope study of choradamesoderm-neurectoderm association in gastrulae of a toad, *Xenopus laevis*', *J. exp. Zool.* **172**, 153–80.

KELLEY, R. O. (1970), 'An electron microscope study of mesenchyme during development of interdigital spaces in man', *Anat. Rec.* **168**, 43–54.

KIDDER, G. M., RUBEN, L. N. & STEVENS, J. M. (1973), 'Cytodynamics and ontogeny of the immune response of *Xenopus laevis* against sheep erythrocytes', *J. Embryol. exp. Morph.* **29**, 73–85.

KING, R. C. & DEVINE, R. L. (1958), 'Oogenesis in adult *Drosophila melanogaster*. VII. The submicroscopic morphology of the ovary', *Growth* **20**, 121–57.

KIRBY, D. R. S. (1962), 'The effect of the uterine environment on the development of mouse eggs', *J. Embryol. exp. Morph.* **10**, 496–509.

KIREMIDJIAN, L. & KOPAC, M. J. (1972), 'Changes in cell adhesiveness associated with the development of *Rana pipiens* pronephros', *Devl. Biol.* **27**, 116–30.

KLAVINS, J. V., MESA-TEJADA, R. & WEISS, M. (1971), 'Human carcinoma antigens reacting with anti-embryonic antibodies', *Nature* (*New Biol.*) **234**, 153–4.

KOCH, W. E. (1967), '*In vitro* differentiation of tooth rudiments of embryonic mice. I. Transfilter interactions of embryonic incisor tissues', *J. exp. Zool.* **165**, 155–70.

KOCH, W. E., KOCH, B. A. & LEDBURY, P. A. (1970), '*In vitro* differentiation of tooth rudiments of embryonic mice. II. Growth of isolated thirds of embryonic mouse incisors', *Anat. Rec.* **166**, 517–28.

KOCHER-BECKER, U., DREWS, U. & DREWS, U. (1970), 'Histochemischer Cholinesterase-Nachweis im axialen Mesoderm von *Triturus cristatus*', *W. Roux' Arch. Entw Mech. Org.* **165**, 163x73.

KOCHER-BECKER, U. & TIEDEMANN, H. (1971), 'Induction of mesodermal and endodermal structures and primordial germ cells in *Triturus* ectoderm by a vegetalizing factor from chick embryos', *Nature, Lond.* **233**, 65–6.

KOLLAR, E. J. & BAIRD, G. R. (1969), 'The influence of the dental papilla on the development of tooth shape in embryonic mouse tooth germs', *J. Embryol. exp. Morph.* **21**, 131–48.

KOLLAR, E. J. & BAIRD, G. R. (1970), 'Tissue interactions in embryonic mouse tooth germs. I. Reorganization of the dental epithelium during tooth germ reconstruction. II. The inductive role of the dental papilla', *J. Embryol. exp. Morph.* **24**, 159–71 & 173–86.

KONIJN, T. M., VAN DE MEENE, J. G. C., BONNER, T. J. & BARKLEY, D. S. (1967), 'The acrasin activity of adenosine-3′, 5′-cyclic phosphate', *Proc. Nat. Acad. Sci. U.S.A.* **58**, 1152–4.

KRATCHOWIL, K. (1971), '*In vitro* analysis of the hormonal basis for the sexual dimorphism in the embryonic development of the mouse mammary gland', *J. Embryol. exp. Morph.* **25**, 141–53.

KRATCHOWIL, K. (1972), 'Tissue interaction during embryonic development: general properties'. In *Tissue interactions in carcinogenesis* (ed. D. Tarin), pp. 1–48. Academic Press, New York & London.

KRAUSE, G. (1953), 'Die Aktionsfolge zur Gestaltung des Keimstreifs von *Tachycines* (Saltatoria)', *W. Roux' Arch. Entw Mech. Org.* **146**, 275–370.

LAMONT, M. D. & VERNON, C. A. (1967), 'The migration of neurones from chick embryonic dorsal root ganglia in tissue culture', *Exptl. Cell. Res.* **47**, 661–2.

LANDESMAN, R. & GROSS, P. R. (1968), 'Patterns of macromolecule synthesis during development of *Xenopus laevis*. 1. Incorporation of radioactive precursors into dissociated embryos', *Devl. Biol.* **18**, 571–89.

LANDSBOROUGH THOMSON, A. (1910), *Britain's Birds and their Nests*, W. & R. Chambers, Edinburgh.

LASH, J. W. (1968), 'Chondrogenesis: genotypic and phenotypic expression', *J. Cell Physiol.* **72**, Suppl. 1, 35–46

REFERENCES

LASH, J. W., HOLTZER, S. & HOLTZER, H. (1957), 'An experimental analysis of the development of the spinal column. VI. Aspects of cartilage induction', *Exp. Cell Res.* **13**, 292–303.

LAVIETES, B. B. (1970), 'Cellular interaction and chondrogenesis *in vitro*', *Devl. Biol.* **21**, 584–610.

LAWRENCE, I. E. & BURDEN, H. W. (1973), 'Catecholamines and morphogenesis of the chick neural tube and notochord', *Am. J. Anat.* **137**, 199–208.

LAWSON, K. A. (1972), 'The role of mesenchyme in morphogenesis and functional differentiation of the salivary epithelium', *J. Embryol. exp. Morph.* **27**, 497–513.

LAZAR, G. Y. (1973), 'The development of the optic tectum in *Xenopus laevis*: a Golgi study', *J. Anat.* **116**, 347–355.

LEBOWITZ, P. & SINGER, M. (1970), 'Neurotrophic control of protein synthesis in the regenerating limb of the newt, *Triturus*', *Nature, Lond.* **225**, 824–7.

LEDOUARIN, N. & HOUSSAINT, E. (1968), 'La synthèse du glycogène dans des hepatocytes de mammifère adulte associés expérimentalement au mésenchyme du métanephros d'embryon de poulet', *Compt. Rend. Soc. Biol. Paris*, **162**, 1196–9.

LEDOUARIN, N. & TEILLET, M. A. (1971), 'Localisation, par la méthode des greffes interspécifiques, du territoire neural dont dérivent les cellules adrénales surrénaliennes chez l'embryon d'oiseau', *Compt. Rend. Acad. Sci. Paris* **272**, 481–4.

LEDOUARIN, N. M. & TEILLET, M. A. (1973), 'The migration of neural crest cells to the wall of the digestive tract in avian embryos', *J. Embryol. exp. Morph.* **30**, 31–48.

LEFKOWITZ, R. J., ROTH, J. & PASTAN, I. (1970), 'Effects of calcium on ACTH stimulation of the adrenal: separation of hormone binding from adenyl cyclase activation', *Nature, Lond.* **228**, 864.

LEVI-MONTALCINI, R. (1958), 'Chemical stimulation of nerve growth'. In: *Chemical Basis of Development* (eds. W. D. McElroy & B. Glass) pp. 645–664. Johns Hopkins Press. Baltimore, Md.

LILIEN, J. E. (1968), 'Enhancement of cell aggregation *in vitro*', *Devl. Biol.* **17**, 657–91.

LILLIE, F. R. (1919), *Problems of Fertilization.* University Press, Chicago.

LIPTON, B. H. (1973), 'Mechanisms of somite morphogenesis', *Anat. Rec.* **175**, 372.

LIPTON, B. H. & JACOBON, A. G. (1974), Experimental analysis of the mechanisms of somite morphogensis. *Devl. Biol.* **38**, 91–103.

LJUNKVIST, H. I. (1967), 'Light and electron microscopical study of the effect of oestrogen on the chicken oviduct', *Acta Endocrinol. (Copenhagen)* **56**, 391–402.

LOCKE, M. (1967), 'The development of patterns in the integument of insects', *Adv. Morph.* **6**, 33–88.

LUTZ, H. (1970), *Invertebrate Organ Cultures.* Gordon & Breach, New York & London.

MADDRELL, S. H. P. (1967), 'Neurosecretion in insects'. In: *Insects and Physiology.* (eds. J. W. L. Beament & J. E. Treherne). Oliver & Boyd, London.

MANDARON, P. (1970), 'Développement in vitro des disques imaginaux de la

271

Drosophile. Aspects morphologiques et histologiques', *Devl. Biol.* **22**, 298–320.

MANNING, M. J. (1971), 'The effect of early thymectomy on histogenesis of the lymphoid organs in *Xenopus laevis*', *J. Embryol. exp. Morph.* **26**, 219–29.

MANTON, S. M. (1928), 'On the embryology of a mysid crustacean *Hemimysis lamornae*', *Phil. Trans. R. Soc. Lond.* (*B*), **216**, 363–463.

MARSLAND, D. & LANDAU, J. V. (1954) 'The mechanism of cytokinesis: temperature and pressure studies on the cortical gel system in various marine invertebrates', *J. exp. Zool.* **125**, 507–539.

MATO, M., AIKAWA, E. & KATAHIRA, M. (1969), 'Further studies on "cell reaction" at the lower surface of the nasal septum of human embryos during the fusion of the palate', *Acta Anat.* **71**, 154–60.

MAUGER, A. (1970), 'Le développement du plumage dorsal de l'embryon de poulet étudié à l'aide d'irradiations aux rayons X', *Devl. Biol.* **22**, 412–32.

MAYNARD-SMITH, J. (1960), 'Continuous, quantized and modal variation', *Proc. Roy. Soc.* (*B*), **152**, 397–409.

McCABE, J. A. & SAUNDERS, J. W. (1971), 'The induction of anteroposterior polarity in chick limbs developed from reaggregated limb bud mesoderm', *Anat. Rec.* **169**, 372.

McGARRY, M. P. & VANABLE, J. W. (1969), 'The role of cell division in *Xenopus laevis* skin gland development', *Devl. Biol.* **20**, 291–303.

McLAREN, A. (1969), 'Stimulus and response during early pregnancy in the mouse', *Nature, Lond.* **221**, 739–41.

McLAREN, A. (1972), 'Numerology of development', *Nature, Lond.* **239**, 274–6.

McLEAN, J. M. & SCOTHORNE, R. J. (1972), 'The fate of skin allografts in the rabbit uterus', *J. Anat.* **112**, 423–32.

McMURDO, H. L. & ZALIK, S. E. (1970), 'Embryonic cell surface: electrophoretic mobilities of blastula cells', *Experientia* **26**, 406–8.

MEDAWAR, P. R. (1954), 'The significance of the inductive relationships in the development of vertebrates', *J. Embryol. exp. Morph.* **2**, 172–4.

MESSIER, P. E. (1972), 'The occurrence of nuclear migration under thiol treatment effective in inhibiting neurulation', *J. Embryol. exp. Morph.* **27**, 577–84.

METZ, C. B. (1967), 'Gamete surface components and their role in fertilization'. In: *Fertilization*, vol. 1. pp. 163–236. (eds. C. B. Metz and A. Monroy), Academic Press, New York & London.

METZ, C. B. & TYLER, A. (1959), personal communication.

MICHAEL, M. J. & FABER, J. (1971), 'Morphogenesis of mesenchyme from regeneration blastemas in the absence of digit formation in *Ambystoma mexicanum*', *W. Roux' Arch. Entw Mech. Org.* **168**, 174–80.

MILAIRE, J. (1965), 'Aspects of limb morphogenesis in mammals'. In: *Organogenesis* (eds. R. L. DeHaan & H. Ursprung), pp. 283–300. Holt, Rinehart & Winston, Philadelphia & London.

MILAIRE, J. & MULNARD, J. (1968), 'Le rôle de l'epiblaste dans la chondrogénèse du bourgeon de membre chez la souris', *J. Embryol. exp. Morph.* **20**, 215–36.

MILLER, W., GORMAN, R. E. & BITENSKY, M. W. (1971), 'Cyclic adenosine monophosphate: function in photoreceptors', *Science, N.Y.* **174**, 295–7.

MINER, N. (1956), 'Integumental specification of sensory fibres in the development of cutaneous local sign', *J. Comp. Neurol.* **105**, 161-70.

MINTZ, B. (1957), 'Embryological development of primordial germ cells in the

REFERENCES

mouse: influence of a new mutation Wj', *J. Embryol. exp. Morph.* **5**, 396–403.

MINTZ, B. (1964), 'Formation of genetically mosaic mouse embryos, and early development of lethal (t^{12}/t^{12}) normal mosaics'. *J. exp. Zool.* **157**, 273–91.

MINTZ, B. (1967), 'Gene control of mammalian pigmentary differentiation. 1. Clonal origin of melanocytes', *Proc. Nat. Acad. Sci. U.S.A.* **58**, 344–51.

MINTZ, B. (1970), 'Gene expression in allophenic mice'. In: *Symposium of the International Society for Cell Biology*, 9 (ed. H. Padykula), on: *Control Mechanisms in the Expression of Cellular Phenotypes*. pp. 15–42. Academic Press, New York & London.

MITCHISON, J. M. & SWANN, M. M. (1952), 'Optical changes in the membranes of the sea-urchin egg at fertilization, mitosis and cleavage', *J. exp. Biol.* **29**, 357–62.

MIURA, Y. & WILT, F. H. (1969). Tissue interaction and the formation of the first erythroblasts of the chick embryo. *Devl. Biol.* **19**, 201–11.

MONROY, A. (1967), 'The activation of the egg.' In: *Fertilization*, vol. 1 (eds, C. B. Metz & A. Monroy.) Academic Press, New York.

MOORE, A. R. (1937), 'On the centering of the nuclei in centrifuged eggs as a result of fertilization and artificial membrane formation', *Protoplasma* **27**, 544–51.

MOORE, W. J. & MINTZ, B. (1972), 'Clonal model of vertebral column and skull development derived from genetically mosaic skeletons in allophenic mice', *Devl. Biol.* **27**, 55–70.

MORI, Y. (1932), 'Entwicklung isolierter Blastomeren und teileweise abgetöteter älterer Keime von *Clepsine sexoculata*', *Z. wiss Zool.* **141**, 399–431.

MOSCONA, A. A. (1962), 'Analysis of cell recombinations in experimental synthesis of tissues *in vitro*', *J. Cell Comp. Physiol.* **60**, suppl. 1. 65–80.

MOSCONA, A. A. & PIDDINGTON, R. (1966), 'Stimulation by hydrocortisone of premature changes in the developmental pattern of glutamine synthetase in embryonic retina', *Biochim. biophys. Acta* **121**, 409–11.

MOSCONA, A. A. & PIDDINGTON, R. (1967), 'Enzyme induction by corticosteroids in embryonic cells: steroid structure and inductive effect', *Science, N.Y.*, **158**, 496–7.

MUCHMORE, W. B. (1968), 'The influence of neural tissue on the early development of somitic muscle in ventrolateral implants in *Ambystoma*', *J. exp. Zool.* **169**, 251–8.

NAKANO, E. (1969), In *Fertilization*, vol. II. pp. 295–325 (eds. C. B. Metz & A. Monroy). Academic Press, New York & London.

NEEDHAM, A. E. (1952), *Regeneration and Wound Healing*. Methuen Monograph.

NEW, D. A. T. & BRENT, R. L. (1972), 'Effect of yolk-sac antibody on rat embryos grown in culture', *J. Embryol. exp. Morph.* **27**, 543–53.

NICOLET, G. (1970), 'Is the presumptive notochord responsible for somite genesis in the chick?' *J. Embryol. exp. Morph.* **24**, 467–78.

NIEUWKOOP, P. D. (1969), 'The formation of the mesoderm in urodelean amphibians. I. Induction by the endoderm', *W. Roux' Arch. Entw Mech. Org.* **162**, 341–73.

NIEUWKOOP, P. D. & FABER, J. (1956), *Normal Table of Xenopus laevis Daudin*. North-Holland Publ. Co., Amsterdam.

NIEUWKOOP, P. D. & UBBELS, G. A. (1972), 'The formation of the mesoderm in urodelean amphibians. IV. Qualitative evidence for the purely "ectodermal" origin of the entire mesoderm and of the pharyngeal endoderm', *W. Roux' Arch. Entw Mech. Org.* **169**, 185–99.

NIU, M. C. & TWITTY, V. (1953), 'The differentiation of gastrula ectoderm in medium conditioned by axial mesoderm', *Proc. Nat. Acad. Sci. U.S.A.* **39**, 985–9.

NOVAK, V. J. A. (1967), 'The juvenile hormone and the problem of animal morphogenesis'. In: *Insects and Physiology*. (eds. J. W. L. Beament & J. E. Treherne), pp. 105–53. Oliver & Boyd, London.

NOVIKOV, A. B. (1950), Morphogenetic substances or organizers in annelid development', *J. exp. Zool.* **85**, 127–51.

NYHOLM, M., SAXÈN, L., TOIVÖNEN, S. & VAINIO, T. (1962), 'Electron microscopy of transfilter neural induction', *Exptl. Cell Res.* **28**, 209–12.

O'DELL, D. S. (1972), Chalones: a report on the International Conference held at Brook Lodge, Augusta, Michigan during 5–7 June 1972. *Fed. Europ. Biol. Socs. Letters* **26**, 1–5.

O'HARE, M. J. (1972a), 'Differentiation of chick embryo somites in chorio-allantoic culture', *J. Embryol. exp. Morph.* **27**, 215–28.

O'HARE, M. J. (1972b), 'Chondrogenesis in chick embryo somites grafted with adjacent and heterologous tissue', *J. Embryol. exp. Morph.* **27**, 229–34.

O'HARE, M. J. (1972c), 'Aspects of spinal cord induction of chondrogenesis in chick embryo somites', *J. Embryol. exp. Morph.* **27**, 235–43.

OHTSU, K., NAITO, K. & WILT, F. H. (1964), 'Metabolic basis of visual pigment conversion in metamorphosing *R. catesbeiana*', *Devl. Biol.* **10**, 216–32.

OLSSON, R. (1955), 'Structure and development of Reissner's fibre in the caudal end of the *Amphioxus* and some lower vertebrates', *Acta zool. Stockh.* **36**, 167–98.

O'MALLEY, B. W., SPELLBERG, T. C., SCHRADER, W. T., CHYTIL, F. & STEGGLES, A. L. (1971), 'Mechanism of interaction of a hormone-receptor complex with the genome of a eukaryotic target cell', *Nature, Lond.* **235**, 141–4.

ORFILA, C. & DEPARIS, P. (1970), 'Evolution des homogreffes cutanées chez la larve de *Pleurodeles waltlii* Michah', *Compt. Rend. Soc. Biol. Paris*, **164**, 1124–5.

OUWENWEEL, W. J. (1971), 'Determination, regulation and positional information in insect development', *Acta Biochem.* **21**, 115–31.

PACKARD, D. S. (1973), 'Physical factors which influence the shape of chick somites', *Anat. Rec.* **175**, 405.

PALMER, J. F. & SLACK, C. (1969), 'Effect of "halothane" on electrical coupling in pregastrulation embryos of *Xenopus laevis*', *Nature, Lond.* **223**, 1286–7.

PAUL, J. & GILMOUR, R. S. (1968), 'Organ-specific restriction of transcription in mammalian chromatin', *J. molec. Biol.* **34**, 305–16.

PEARSE, A. G. E. (1969), 'The cytochemistry and ultrastructure of polypeptide hormone-producing cells of the APUD series and the embryologic, physiologic and pathologic implications of the concept', *J. Histochem. Cytochem.* **17**, 303–13.

PEARSE, A. G. E. (1971), 'The endocrine polypeptide cells of the APUD series', *Mem. Soc. Endocrinol.* **19**, 543–56.

PEARSON, G. & FREEMAN, G. (1968), 'Evidence suggesting a relationship

REFERENCES

between polyoma virus induced transplantation antigen and normal embryonic antigen', *Cancer Res.* **28**, 1665–73.

PENNERS, A. (1938), 'Abhängigkeit der Formbildung vom Mesoderm im Tubifex-Embryo', *Z. wiss. Zool.* **150**, 305–57.

PFLUGFELDER, O. (1959), 'Atypische Gewebsdifferenzierungen bei *Xenopus laevis* Daudin nach experimentelle Verhinderung der Metamorphose', *W. Roux' Arch. EntwMech. Org.* **151**, 229–41.

PIATT, J. (1942), 'Transplantation of aneurogenic forelimbs in *Amblystoma punctatum*', *J. exp. Zool.* **91**, 79–101.

PIKO, L. (1969). In: *Fertilization*, vol. II pp. 326–44 (eds. C. B. Metz & A. Monroy). Academic Press, New York & London.

PINOT, M. (1970), 'Le rôle du mésoderme somitique dans la morphogénèse précoce des membres de l'embryon de poulet', *J. Embryol. exp. Morph.* **23**, 109–51.

PRESTIGE, M. (1967), 'The control of cell number in the lumbar ventral horns during the development of *Xenopus laevis* tadpoles', *J. Embryol. exp. Morph.* **18**, 359–87.

PRICE, D. & ORTIZ, E. (1965), The role of foetal androgen in sex differentiation in mammals. In: *Organogenesis*, (eds. R. L. DeHaan & H. Ursprung). pp. 629–52. New York & London, Holt, Rinehart & Winston.

RAJAN, K. T. & HOPKINS, A. M. (1970), 'Human digits in organ culture', *Nature, Lond.* **227**, 621–2.

RAKIC, P. & SIDMAN, R. L. (1969), 'Telencephalic origin of pulvinar neurons in the fetal human brain', *Z. Anat. Entw-Gesch.* **129**, 53-82.

RAVEN, C. P. (1954), *An Outline of Developmental Physiology.* Pergamon Press, London.

RAVEN, C. P. (1958), 'Abnormal development of the foregut in *Limnaea stagnalis*', *J. exp. Zool.* **139**, 189–246.

RAVEN, C. P. (1967), 'The distribution of special cytoplasmic differentiations of the egg during early cleavage in *Limnaea stagnalis*', *Devl. Biol.* **16**, 407–37.

RAWLES, M. (1963), 'Tissue interactions in scale and feather development as studied in dermal-epidermal recombinations', *J. Embryol. exp. Morph.* **11**, 765–89.

RELEXANS, J. C. (1973), 'Transplantation de gonades indifférenciées chez l'hermaphrodite simultané *Eisenia foetida* (Oligochète, Lumbricidae). Mise en évidence de facteurs locaux (inducteurs ?) de la différenciation sexuelle', *J. Embryol. exp. Morph.* **30**, 143–61.

REVERBERI, G. (1961), 'The embryology of ascidians', *Adv. Morphogen.* **1**, 55–101.

REVERBERI, G. (1972). *Experimental Embryology of Marine and Fresh-water Invertebrates.* North-Holland Publ. Co., Amsterdam.

REYNAUD, G. (1969), 'Transfert de cellules germinales primordiales de dindon à l'embryon de poulet par injection intravasculaire', *J. Embryol. exp. Morph.* **21**, 485–507.

ROBERTS, A. (1969), 'Conducted impulses in the skin of young tadpoles', *Nature, Lond.* **222**, 1265–6.

ROBERTS, A. & STIRLING, C. A. (1971), 'The properties and propagation of a cardiac-like impulse in the skin of young tadpoles', *Z. vergl. Physiol.* **71**, 295–310.

ROGULSKA, T., OŻDŻEŚNKI, W. & KOMAR, A. (1971), 'Behaviour of mouse

275

primordial germ cells in the chick embryo', *J. Embryol. exp. Morph.* **25**, 155–64.

ROMANOFF, A. L. (1960), *The Avian Embryo.* MacMillan Co., New York.

ROSS, P. (1969), 'Wound Healing', *Scientific American* **220**(6), 40–62.

ROTH, S. (1968), 'Studies on intercellular adhesive selectivity', *Devl. Biol.* **18**, 602–31.

ROUX, W. (1888), 'Uber die künstliche Hervorbringung "halber" Embryonen durch Zerstörung einer der beiden ersten Furchungszellen, sowie über die Nachentwicklung der fehlenden Korperhälfte', *Virchows Arch.* **114**, 22–60.

RUBEN, L. N. (1970), 'Immunological maturation and lymphoreticular cancer transformation in larval *Xenopus laevis*, the South African clawed toad', *Devl. Biol.* **23**, 43–58.

RUBEN, L. N. & BALLS, M. (1964), 'The implantation of lymphosarcoma of *Xenopus laevis* into regenerating and non-regenerating forelimbs of that species', *J. Morph.* **115**, 225–38.

RUNNSTRÖM, J. (1952), 'The cell surface in relation to fertilization'. In *Symp. Soc. exp. Biol.* **6**, 39–88.

RYTOMÄA, T. & KIVINIEMI, K. (1969), 'Chloroma regression induced by the granulocyte chalone', *Nature, Lond.* **222**, 995–6.

SABATINI, D. D., BENSCH, K. & BARRNETT, R. J. (1963), 'Cytochemistry and electron microscopy. The preservation of cellular ultrastructure and enzymatic activity by aldehyde fixation', *J. Cell Biol.* **17**, 19–58.

SALVATORELLI, G. (1969), 'Nouvelles observations sur les facteurs qui contrôlent l'erythropoïèse *in vitro*', *J. Embryol. exp. Morph.* **22**, 15–25.

SAUNDERS, J. W. (1966), 'Death in embryonic systems', *Science, N.Y.* **154**, 604–12.

SAUNDERS, J. W. (1972), 'Developmental control of three-dimensional polarity in the avian limb', *Ann. N.Y. Acad. Sci.* **193**, 29–42.

SAXÈN, L. (1961), 'Transfilter neural induction of amphibian ectoderm', *Devl. Biol.* **3**, 140–52.

SAXÈN, L., SAXÈN, E., TOIVÖNEN, S. & SALIMAKI, K. (1957), 'Quantitative investigation on the anterior pituitary-thyroid mechanism during frog metamorphosis', *Endocrinology* **61**, 35–44.

SAXÈN, L. & TOIVÖNEN, S. (1962), *Primary Embryonic Induction.* Logos/ Academic Press, New York & London.

SCHMATOLLA, E. (1972), 'Dependence of tectal neuron differentiation on optic innervation in teleost fish', *J. Embryol. exp. Morph.* **27**, 555–76.

SCHOTTÉ, O. E. & HARLAND, M. (1943), 'Effects of denervation and amputation of hindlimbs in anuran tadpoles', *J. exp. Zool.* **93**, 453–93.

SCHROEDER, T. (1970), 'Neurulation in *Xenopus laevis.* An analysis and model based upon light and electron microscopy', *J. Embryol. exp. Morph.* **23**, 427–62.

SCHROEDER, T. E. (1973), 'Cell constriction: contractile role of microfilaments in division and development', *Amer. Zoologist* **13**, 949–960.

SCHUBIGER, G. (1971), 'Regeneration, duplication and transdetermination in fragments of the leg disc of *Drosophila melanogaster*', *Devl. Biol.* **26**, 277–95.

SCHWIND, J. (1933), 'Tissue specificity at the time of metamorphosis in frog larvae', *J. exp. Zool.* **66**, 1–14.

SEARLS, R. L. & JANNERS, M. Y. (1971), 'The initiation of limb bud outgrowth in the embryonic chick', *Devl. Biol.* **24**, 198–213.

REFERENCES

SEIDEL, F. (1960), 'Die Entwicklungsfähigkeiten isolierter Furchungszellen aus dem Ei des Kaninchens *Oryctolagus cuniculus*', *W. Roux' Arch. Entw Mech. Org.* **152**, 43–130.

SEIDEL, F., BOCK, E. & KRAUSE, G. (1940), 'Die Organisation des Insekteneies (Reaktionsablauf, Induktionsvorgänge, Eitypen)', *Naturwiss.* **28**, 433.

SELMAN, G. G. & PERRY, M. M. (1970), 'Ultrastructural changes in the surface layers of the newt egg in relation to the mechanism of its cleavage', *J. Cell Sci.* **6**, 207–27.

SENGEL, P. (1971), 'The organogenesis and arrangement of cutaneous appendages in birds', *Adv. Morphogen.* **9**, 181–230.

SENGEL, P., DHOUAILLY, D. & KIENY, M. (1969), 'Aptitude des constituants cutanés de l'aptérie médioventrale du poulet à former des plumes', *Devl. Biol.* **19**, 436–46.

SENGEL, P. & PATAU, M. P. (1969), 'Experimental conditions in which feather morphogenesis predominates over scale morphogenesis in bird embryos', *Nature, Lond.* **222**, 693–4.

SHAVER, J. R., BARCH, S. H. & UMPIERRE, C. C. (1970), 'Interspecific relationships of oviducal materials as related to fertilization in Amphibia', *J. Embryol. exp. Morph.* **24**, 209–25.

SHEFFIELD, J. B. (1970), 'Studies on the aggregation of embryonic cells: initial cell adhesions and the formation of intercellular junctions', *J. Morph.* **132**, 245–64.

SHEFFIELD, J. B. & FISCHMAN, D. A. (1970), 'Intercellular junctions in the developing neural retina of the chick embryo', *Z. Zellforsch. mikr. Anat.* **104**, 405–18.

SHEFFIELD, J. B. & MOSCONA, A. A. (1970), 'Electron microscope analysis of aggregation of embryonic cells: the structure and differentiation of aggregates of neural retina cells', *Devl. Biol.* **23**, 36–61.

SHERIDAN, J. D. (1973) 'Functional evaluation of low resistance junctions: influence of cell shape and size', *Amer. Zoologist* **13**, 1119–1128.

SHIMADA, Y., FISCHMAN, D. A. & MOSCONA, A. A. (1969), 'The development of nerve-muscle junctions in monolayer cultures of embryonic spinal cord and skeletal muscle cells', *J. Cell Biol.* **43**, 382–7.

SINGER, M. (1952), 'The influence of the nerve in regeneration of the amphibian extremity', *Quart. Rev. Biol.* **27**, 169–200.

SINGH, R. P. & CARR, D. H. (1968), 'Congenital anomalies in embryos with normal chromosomes', *Biol. Neonat.* **13**, 121–8.

SLACK, C. & PALMER, J. F. (1969), 'The permeability of intercellular junctions in the early embryo of *Xenopus laevis*, studied with a fluorescent tracer', *Exptl. Cell Res.* **55**, 416–9.

SLAVKIN, H. C. & BAVETTA, L. A. (1970), 'Epithelial and mesenchymal interactions with intercellular matrix RNA', *J. Cell Biol.* **47**, 195a.

SLAVKIN, H. C., BRINGAS, P., LeBARON, R., CAMERON, J. & BAVETTA, L. A. (1969a), 'The fine structure of the extracellular matrix during epithelio-mesenchymal interactions in the rabbit embryonic incisor', *Anat. Rec.* **165**, 237–56.

SLAVKIN, H. C., BRINGAS, P., CAMERON, J., LeBARON, R. & BAVETTA, L. A. (1969b), 'Epithelial-mesenchymal cell interactions with extracellular matrix material *in vitro*', *J. Embryol. exp. Morph.* **22**, 395–405.

SLAVKIN, H. C., FLORES, P., BRINGAS, P. & BAVETTA, L. A. (1970), 'Epithelial-mesenchymal interactions during odontogenesis I. Isolation of

277

several intercellular matrix low molecular weight methylated RNA's', *Devl. Biol.* **23**, 276–96.

SMILEY, G. R. & DIXON, A. D. (1968), 'Fine structure of midline epithelium in the developing palate of the mouse', *Anat. Rec.* **161**, 293–310.

SMILEY, G. R. & KOCH, W. E. (1972), 'An *in vitro* and *in vivo* study of single palatal processes,' *Anat. Rec.* **173**, 405–16.

SOLURSH, M., MEIER, S. & VAEREWYCK, S. (1973), 'Modulation of extra-cellular matrix production by conditioned medium', *Amer. Zoologist* **13**, 1051–1060.

SPEMANN, H. (1901), 'Entwicklungsphysiologische Studien am Tritonei', *W. Roux' Arch. EntwMech. Org.* **12**, 224–64.

SPEMANN, H. (1903), 'Uber die Linsenbildung bei defekter Augenblase', *Anat. Anz.* **23**, 457–64.

SPERRY, R. W. (1943), 'Visuomotor coordination in the newt (*Triturus viridescens*) after regeneration of the optic nerves', *J. Comp. Neurol.* **79**, 33–55.

SPERRY, R. W. (1944), 'Optic nerve regeneration with return of vision in anurans', *J. Neurophysiol.* **7**, 57–69.

SPERRY, R. W. (1965), 'Embryogenesis of behavioral nerve nets'. In: *Organogenesis* (eds. R. L. DeHaan & H. Ursprung), pp. 161–86. Holt, Rinehart & Winston, New York & London.

SPERRY, R. W. & MINER, M. (1949), 'Formation within sensory nucleus V of synaptic associations mediating cutaneous localization', *J. Comp. Neurol.* **90**, 403–23.

SPIEGEL, M. (1954), 'The role of specific surface antigens in cell adhesion. II. Studies on embryonic amphibian cells', *Biol. Bull. mar. lab. Woods Hole, Mass.* **107**, 149–55.

SPIRIN, A. S. (1966), 'On "masked" forms of messenger RNA in early embryogenesis and in other differentiating systems', *Curr. Top. Devel. Biol.*, **1**, 1–38.

SPOONER, B. S. (1973), 'Microfilaments, cell shape changes and morphogenesis of salivary epithelium', *Amer. Zoologist* **13**, 1007–1022.

SPOONER, B. S. & WESSELS, N. K. (1970), 'Mammalian lung development: interactions in primordium formation and bronchial morphogenesis', *J. exp. Zool.* **175**, 445–54.

SPOONER, B. S. & WESSELS, N. K. (1972), 'An analysis of salivary gland morphogenesis: role of cytoplasmic microfilaments and microtubules', *Devl. Biol.* **27**, 38–54.

SPRATT, N. T. (1957), 'Analysis of the organizer center in the early chick embryo. III. Regulative properties of the chorda and somite centers', *J. exp. Zool.* **135**, 319–54.

STALSBERG, H. (1969), 'The origin of heart asymmetry: right and left contributions to the early chick embryo heart', *Devl. Biol.* **19**, 109–27.

STALSBERG, H. & DEHAAN, R. L. (1968), 'Endodermal movements during foregut formation in the chick embryo', *Devl. Biol.* **18**, 198–215.

STALSBERG, H. & DEHAAN, R. L. (1969), 'The precardiac areas and formation of the tubular heart in the chick embryo', *Devl. Biol.* **19**, 128–59.

STAMBROOK, P. J. & FLICKINGER, R. A. (1970), 'Changes in chromosomal DNA replication patterns in developing frog embryos', *J. exp. Zool.* **174**, 101–14.

STANISSTREET, M. (1972), 'An immunochemical study of the proteins of the oocytes and early embryos of the South African clawed toad, *Xenopus laevis*', *Ph.D. Thesis*, Anatomy Dept., Bristol University.

REFERENCES

STANISSTREET, M. & DEUCHAR, E. M. (1972), 'Appearance of antigenic materials in gastrula ectoderm after neural induction', *Cell Differentiation* 1, 15–18.

STAVY, L. & GROSS, P. R. (1969), 'Protein synthesis *in vitro* with fractions of sea urchin eggs and embryos', *Biochim. Biophys. Acta* 182, 193–202.

STEVENS, L. C. (1967), 'The biology of teratomas', *Adv. Morphogen.* 6, 1–31.

STEVENS, L. C. (1968), 'The development of teratomas from intratesticular grafts of tubal mouse eggs', *J. Embryol. exp. Morph.* 20, 329–41.

STEVENS, L. C. (1970), 'The development of transplantable teratocarcinomas from intratesticular grafts of pre- and postimplantation mouse embryos', *Devl. Biol.* 21, 364–82.

STOCUM, D. L. (1968), 'The urodele limb regeneration blastema: a self-organizing system. 1. Differentiation *in vitro*. 2. Morphogenesis and differentiation of autografted whole and fractional blastemas', *Devl. Biol.* 18, 441–56 and 457–80.

STONE, L. S. (1954), 'Further experiments on lens regeneration in eyes of the adult newt, *Triturus viridescens*', *Anat. Rec.* 120, 599–624.

STRAZNICKY, K. (1973), 'The formation of the optic fibre projection after partial tectal removal in *Xenopus*', *J. Embryol. exp. Morph.* 29, 397–409.

STRAZNICKY, K. & GAZE, R. M. (1971), 'The growth of the retina in *Xenopus laevis*: an autoradiographic study', *J. Embryol. exp. Morph.* 26, 67–79.

STRAZNICKY, K. & GAZE, R. M. (1972), 'The development of the tectum in *Xenopus laevis*: an autoradiographic study', *J. Embryol. exp. Morph.* 28, 87–115.

STRAZNICKY, K., GAZE, R. M. & KEATING, M. J. (1971), 'The retinotectal projections after uncrossing the optic chiasma in *Xenopus* with one compound eye', *J. Embryol. exp. Morph.* 26, 523–42.

STREHLER, B. L. (1962), *Time, Cells and Aging.* Academic Press, New York.

SUDARAWATI, S. & NIEUWKOOP, P. D. (1971), 'Mesoderm formation in the anuran, *Xenopus laevis* (Daudin)', *W. Roux' Arch. Entw Mech. Org.* 166, 189–204.

SYLVÈN, B. & MALMGREN, H. (1955), 'Topical distribution of proteolytic activities in some transplanted mice tumors', *Exptl. Cell Res.* 8, 575–7.

SZÈKELY, G. (1966), 'Embryonic determination of neural connections', *Adv. Morphogen.* 5, 181–219.

SZÈKELY, G. (1971), personal communication.

SZOLOSSI, D. (1970), 'Cortical cytoplasmic filaments of cleaving eggs: a structural element corresponding to the contractile ring', *J. Cell. Biol* 44, 192–209.

TAKAYANAGI, N. (1958), 'Immunochemical analysis of cell differentiation', *J. Juzen Med. Soc.* (Japan), 60, 701–51.

TARIN, D. (1971), 'Histological features of neural induction in *Xenopus laevis*', *J. Embryol. exp. Morph.* 26, 543–70.

TARKOWSKI, A. K. (1964), 'Patterns of pigmentation in experimentally produced mouse chimaerae', *J. Embryol. exp. Morph.* 12, 575–85.

TARKOWSKI, A. K. (1965), 'Embryonic and postnatal development of mouse chimaeras'. In: *Preimplantation Stages of Pregnancy* (CIBA symposium: eds. G. E. W. Wolstenholme & M. O'Connor). pp. 183–93. Churchill Ltd., London.

TATA, J. A. (1971), 'Hormonal regulation of metamorphosis', *Brit. Soc. exp. Biol. Symposium* 25, on *Control Mechanisms of Growth and Differentiation*,

279

(eds. D. D. Davies & M. Balls), pp. 163–82. Cambridge University Press.

TEGNER, M. J. & EPEL, D. (1973), 'Scanning electron microscope pictures of sperm-egg interactions in the sea-urchin', *Science*, N.Y. **179**, 685–8.

TELFER, W. H. (1961), 'The route of entry and localisation of blood proteins in oocytes of Saturniid moths', *J. Biophys. Biochem. Cytol.* **9**, 747–59.

THIBAULT, C. (1969), '*In vitro* fertilization of the mammal egg', In: *Fertilization*, vol. II. (eds. C. B. Metz & A. Monroy), Chap. 9. Academic Press, New York.

THOMAS, N. & CORDALL, K. (1972), personal communication.

THOMAS, N. & DEUCHAR, E. M. (1971), 'Synthesis of high-molecular-weight RNA in *Xenopus* ectoderm after neural induction. *Acta. Embryol. exper.*, pp. 195–200.

THOMAS, P. J. (1973), 'Steroid hormones and their receptors', *J. Endocrinol.* **57**, 333–59.

THOMAS, R. J. (1968), 'Cytokinesis during early development of a teleost embryo: *Brachydanio rerio*', *J. Ultrastr. Res.* **24**, 232–8.

TIEDEMANN, H. (1967), 'Biochemical aspects of primary induction and determination'. In: *The Biochemistry of Animal Development*, vol. 2. (ed. R. Weber), pp. 4–56. Academic Press, New York.

TOWNES, P. L. & HOLTFRETER, J. (1955), 'Directed movements and selective adhesion of embryonic amphibian cells', *J. exp. Zool.* **128**, 53–150.

TSCHUMI, P. A. (1957), 'Growth of the hind limb bud of *Xenopus laevis* and its dependence upon the epidermis', *J. Anat.* **91**, 149–73.

TUNG, T. C., WU, S. C. & TUNG, Y. Y. F. (1962), 'Experimental studies on neural induction in Amphioxus', *Scientia sinica* **11**, 805–9.

TURING, A. M. (1952), 'The chemical basis of morphogenesis', *Phil. Trans. R. Soc.* (B), **237**, 37–72.

TURKINGTON, R. W. (1971), 'Hormonal regulation of cell proliferation and differentiation'. In: *Developmental Aspects of the Cell Cycle* (eds. I. L. Cameron, G. M. Padilla & A. M. Zimmerman), pp. 315–55. Academic Press, New York.

TWITTY, V. C. (1932), 'Influence of the eye on the growth of its associated structures, studied by means of heteroplastic transplantation', *J. exp. Zool.* **61**, 333–74.

TWITTY, V. C. & NIU, M. C. (1948), 'Causal analysis of chromatophore migration', *J. exp. Zool.*' **108**, 53–120.

TYLER, A. (1955), 'Gametogenesis, fertilization and parthenogenesis'. In: *Analysis of Development* (eds. B. H. Willier, P. A. Weiss & V. Hamburger), pp. 170–212. W. B. Saunders Co., Philadelphia & London.

ULLMANN, S. L. (1973), Oogenesis in *Tenebrio molitor*: Histological and auto-radiographical observations on pupal and adult ovaries', *J. Embryol. exp. Morph.* **30**, 179–217.

UNSWORTH, B. & GROBSTEIN, C. (1970), 'Induction of kidney tubules in mouse metanephrogenic mesenchyme by various embryonic mesenchymal tissues', *Devl. Biol.* **21**, 547–56.

VACQUIER, V. D., EPEL, D. & DOUGLAS, L. A. (1971), 'Sea urchin eggs release protease activity at fertilization', *Nature, Lond.* **237**, 34–6.

VAINIO, T. (1956), 'The antigenic properties of the blastoporal lip of the *Triturus* gastrula', *Annls. Med. exp. Biol. Fenn.* **34**, 71–80.

VAINIO, T., SAXÈN, L. & TOIVÖNEN, S. (1960), 'Transfer of the antigenicity of guinea pig bone marrow implants to the graft tissue in explantation experiments', *Experientia* **16**, 27–9.

VAN DEN BIGGELAAR, J. A. M. (1971), 'Development of division asynchrony and bilateral symmetry in the first quartet of micromeres in eggs of *Lymnaea*', *J. Embryol. exp. Morph.* **26**, 351–99.

VAN GANSEN, P. & SCHRAM, A. (1969), 'Etude des ribosomes et du glycogène des gastrules de *Xenopus laevis* par cytochimie ultrastructurale', *J. Embryol. exp. Morph.* **22**, 69–98.

VAN WYK, J. J., HALL, K. & WEAVER, R. P. (1969), 'Partial purification of sulphation factor and thymidine factor from plasma', *Biochim. Biophys. Acta* **192**, 560–2.

VENERONI, G. & MURRAY, M. R. (1969), 'Formation *de novo* and development of neuromuscular junctions *in vitro*', *J. Embryol. exp. Morph.* **21**, 369–82.

VERDONK, N. H. & CATHER, J. N. (1973), 'The development of isolated blastomeres in *Bithynia tentaculata*', *J. exp. Zool.* **186**, 47–62.

VERLY, W. G., DESCHAMPS, Y., PUSHPATHADAM, J. & DESROSIERS, M. (1971), 'The hepatic chalone. 1. Assay method for the hormone and purification of the rabbit liver chalone', *Canad. J. Biochem.* **49**, 1376–83.

VON WOELLEWARTH, C. (1969), 'Auslösung von Situs Inversus durch Materieldefekte im lateralen Ektoderm der Gastrula bei *Triturus alpestris*', *W. Roux' Arch. Entw Mech. Org.* **162**, 336–40.

WADDINGTON, C. H. (1938), 'The morphogenetic function of a vestigial organ in the chick', *J. exp. Biol.* **15**, 371–6.

WADDINGTON, C. H. (1956), *Principles of Embryology*. Allen & Unwin, London.

WADDINGTON, C. H. (1962), *New Patterns in Genetics and Development*. Columbia University Press, New York & London.

WADDINGTON, C. H. (1966), *Principles of Development and Differentiation*. MacMillan, New York; Collier-MacMillan Ltd., London.

WADDINGTON, C. H. & DEUCHAR, E. M. (1953), 'Studies on the mechanism of meristic segmentation. I. The dimensions of somites', *J. Embryol. exp. Morph.* **1**, 349–56.

WADDINGTON, C. H. & GOODHART, C. B. (1949), 'Localisation of adsorbed carcinogens within the amphibian cell', *Quart. J. Micr. Sci.* **90**, 209–19.

WADDINGTON, C. H. & PERKOWSKA, E. (1965), 'Synthesis of ribonucleic acid by different regions of the early amphibian embryo', *Nature, Lond.* **207**, 1244–6.

WADDINGTON, C. H. & SCHMIDT, G. A. (1933), 'Induction by heteroplastic grafts of the primitive streak in birds', *W. Roux' Arch. Entw Mech. Org.* **128**, 521–63.

WADDINGTON, C. H. & SIRLIN, J. L. (1964), 'The incorporation of labelled amino acids into amphibian embryos', *J. Embryol. exp. Morph.* **2**, 340–7.

WALLACE, R. A. & JARED, D. W. (1969), 'Studies on amphibian yolk. VIII. The oestrogen-induced hepatic synthesis of a serum lipophosphoprotein and its selective uptake by the ovary and transformation into yolk platelet protein in *Xenopus laevis*', *Devl. Biol.* **19**, 498–526.

WATANABE, K. (1970), 'Changes in the capacity for clonal growth and differentiation *in vitro* of the vertebral cartilage cells with embryonic development. 1. Culture in the standard medium', *Devel., Growth & Diffn.* **12**, 79–88.

WATANABE, K. (1971), 'Changes in the capacity for clonal growth and differentiation *in vitro* of the vertebral cartilage cells with embryonic development.

2. Vitalizing effect of conditioned medium on the cells of younger embryos'. *Devel., Growth. & Diffn.* **13**, 107–118.

WEBER, R. (1962), 'Induced metamorphosis in isolated tails of *Xenopus laevis*', *Experientia* **18**, 84–5.

WEBER, R. (1967), 'Biochemistry of amphibian metamorphosis'. In: *The Biochemistry of Animal Development*, vol. 2 (ed. R. Weber), pp. 227–302. Academic Press, New York.

WEISS, P. (1947), 'The problem of specificity in growth and development', *Yale J. Biol. and Med.* **19**, 235–78.

WEISS, P. & ROSETTI, F. (1957), 'Growth responses of the opposite sign among different neuron types exposed to thyroid hormone', *Proc. Nat. Acad. Sci. U.S.A.* **37**, 540–56.

WESSELS, N. K. (1964), 'DNA synthesis, mitosis and differentiation in pancreatic acinar cells *in vitro*', *J. Cell Biol.* **20**, 415–33.

WESSELS, N. K. (1970), 'Mammalian lung development: interactions in formation and morphogenesis of tracheal buds', *J. exp. Zool.* **175**, 455–66.

WESSELS, N. K. & COHEN, J. H. (1968), 'Effects of collagenase on developing epithelia *in vitro*: lung, ureteric bud and pancreas', *Devl. Biol.* **18**, 294–309.

WESSELS, N. K. & EVANS, J. (1968), 'The ultrastructure of oriented cells and extracellular materials between developing feathers', *Devl. Biol.* **18**, 42–61.

WESTON, J. A. (1963), 'A radioautographic analysis of the migration and localization of trunk neural crest cells in the chick', *Devl. Biol.* **6**, 279–310.

WHITTEN, J. M. (1969), 'Cell death during early morphogenesis: parallels between insect limb and vertebrate limb development', *Science, N.Y.* **163**. 1456–7.

WIGGLESWORTH, V. B. (1940), 'Local and general factors in the development of "pattern" in *Rhodnius prolixus* (Hemiptera)', *J. exp. Biol.* **17**, 180–200.

WIGGLESWORTH, V. B. (1954), *Physiology of Insect Metamorphosis*. Cambridge Monographs on Experimental Biology, Camb. Univ. Press.

WILLIAMS, C. (1967), 'The present status of the brain hormone'. In: *Insects and Physiology*. (eds. J. W. L. Beament & J. E. Treherne). pp. 133–9. Oliver & Boyd, London.

WILLIAMS, A. E. & RATCLIFFE, N. A. (1969), 'Attachment of ascites tumour cells to rat diaphragm as seen by scanning electron microscopy', *Nature, Lond.* **222**, 893–4.

WILSON, E. B. (1904), 'Experimental studies on germinal localisation. I. The germ regions in the egg of *Dentalium*. II. Experiments on the cleavage mosaic in *Patella* and *Dentalium*', *J. exp. Zool.* **1**, 1–268.

WILSON, E. B. (1925), *The Cell in Development and Heredity*. 3rd edn. MacMillan Co., New York.

WITSCHI, E. (1938), 'Studies on sex differentiation and sex determination in amphibians. V. Range of the cortex-medulla antagonism in parabiotic twins of Ranidae and Hylidae', *J. exp. Zool.* **78**, 113–45.

WOLFF, E. (1968), 'Specific interactions between tissues during organogenesis', *Curr. Top. Devel. Biol.*, **3** (eds. A. A. Moscona & A. Monroy) pp. 65–94. Academic Press, New York & London.

WOLFF, E. & WOLFF, E. (1951), 'The effects of castration on bird embryos', *J. exp. Zool.* **116**, 59–97.

WOLPERT, L. (1971), 'Positional information and pattern formation', *Curr. Top. Devel. Biol.* **6** (eds. A. A. Moscona & A. Monroy), pp. 183–224. Academic Press, New York & London.

REFERENCES

YANAGIMACHI, R. & NODA, Y. D. (1970), 'Electron microscope studies of sperm incorporation into the golden hamster egg', *Am. J. Anat.* **128**, 429–62.

ZALEWSKI, A. A. (1970), 'Continuous trophic influence of chromatolysed gustatory neurons on taste buds', *Anat. Rec.* **167**, 165–74.

ZWILLING, E. (1963), 'Formation of endoderm from ectoderm in *Cordylophora*'. *Biol. Bull. mar. biol. Lab. Woods Hole Mass.* **124**, 368–78.

ZWILLING, E. (1972), 'Limb morphogenesis', *Devl. Biol.* **28**, 1–11.

Index